BASIC CALCULUS

CONTEMPORARY UNDERGRADUATE MATHEMATICS SERIES
Robert J. Wisner, Editor

THE NATURE OF MODERN MATHEMATICS. Karl J. Smith

MATHEMATICS FOR THE LIBERAL ARTS STUDENT, SECOND EDITION. Fred Richman, Carol Walker, and Robert J. Wisner

X-RATED ALGEBRA: FOR MATURE BEGINNING STUDENTS. Patricia Fernandez and Richard R. Miller

INTERMEDIATE ALGEBRA. Edward D. Gaughan

COLLEGE ALGEBRA. Edward D. Gaughan

TRIGONOMETRY: CIRCULAR FUNCTIONS AND THEIR APPLICATIONS. James E. Hall

ANALYTIC GEOMETRY: A REFRESHER. Karl J. Smith

INTRODUCTION TO SYMBOLIC LOGIC. Karl J. Smith

METRIC MATH: THE MODERNIZED METRIC SYSTEM (SI). James R. Smart

A PROGRAMMED STUDY OF INTUITIVE GEOMETRY (WITH APPLICATIONS IN METRIC UNITS). Ruric E. Wheeler and Ed R. Wheeler

MODERN MATHEMATICS: AN ELEMENTARY APPROACH, THIRD EDITION. Ruric E. Wheeler

A PROGRAMMED STUDY OF NUMBER SYSTEMS. Ruric E. Wheeler and Ed R. Wheeler

FUNDAMENTAL COLLEGE MATHEMATICS: NUMBER SYSTEMS AND INTUITIVE GEOMETRY. Ruric E. Wheeler

MODERN MATHEMATICS WITH APPLICATIONS TO BUSINESS AND SOCIAL SCIENCE. Ruric E. Wheeler and W. D. Peeples, Jr.

FINITE MATHEMATICS: AN INTRODUCTION TO MATHEMATICAL MODELS. Ruric E. Wheeler and W. D. Peeples, Jr.

INTRODUCTORY GEOMETRY: AN INFORMAL APPROACH, SECOND EDITION. James R. Smart

MODERN GEOMETRIES. James R. Smart

LINEAR ALGEBRA. James E. Scroggs

ESSENTIALS OF ABSTRACT ALGEBRA. Charles M. Bundrick and John J. Leeson

AN INTRODUCTION TO ABSTRACT ALGEBRA. A. Richard Mitchell and Roger W. Mitchell

MODULES: A PRIMER OF STRUCTURE THEOREMS. Tom Head

BASIC CALCULUS. Darel W. Hardy

DIFFERENTIAL EQUATIONS AND RELATED TOPICS FOR SCIENCE AND ENGINEERING Robert W. Hunt

THE ANALYSIS AND SOLUTION OF PARTIAL DIFFERENTIAL EQUATIONS. Robert L. Street

CALCULUS OF SEVERAL VARIABLES. E. K. McLachlan

PROBABILITY. Donald R. Barr and Peter W. Zehna

THEORY AND EXAMPLES OF POINT-SET TOPOLOGY. John Greever

AN INTRODUCTION TO ALGEBRAIC TOPOLOGY. John W. Keesee

EXPLORATIONS IN NUMBER THEORY. Jeanne Agnew

NUMBER THEORY: AN INTRODUCTION TO ALGEBRA. Fred Richman

BASIC CALCULUS

DAREL W. HARDY

Colorado State University

BROOKS/COLE PUBLISHING COMPANY
Monterey, California
A Division of Wadsworth Publishing Company, Inc.

Illustrations by David A. Strassman

Gift; 1978-5-31

© 1975 by Wadsworth Publishing Company, Inc.,
Belmont, California 94002. All rights reserved.
No part of this book may be reproduced, stored
in a retrieval system, or transcribed, in any
form or by any means—electronic, mechanical,
photocopying, recording, or otherwise—without
the prior written permission of the publisher:
Brooks/Cole Publishing Company, Monterey, California
93940, a division of Wadsworth Publishing Company, Inc.

ISBN: 0-8185-0138-3
L.C. Catalog Card No.: 74-83957
Printed in the United States of America

10 9 8 7 6 5 4 3 2 1

PREFACE

In writing this book I had several goals in mind. The first goal was to write a text for student consumption, not for the instructor who already knows the material. The book is written in straightforward English. Students who have used preliminary versions have stressed the point that this is a book that is readable.

Everyone has an intuitive concept of area, so it is easy to motivate the concept of an integral. On the other hand, only some of the students have sufficient backgrounds in physics to use motion and velocity to motivate the derivative. Other interpretations of the derivative are also sophisticated and make an early treatment of differentiation difficult for many students. For these reasons, the integral is introduced before the derivative. Historically, the integral came long before the derivative, and it makes sense to follow the same order in a calculus class.

In this text, new ideas are introduced in a familiar setting as often as possible. For example, most of the basic concepts of calculus are introduced during the first five chapters in the familiar setting of polynomials, rational functions, and algebraic functions. The elementary transcendental functions are deferred until Chapter 6, when readers should be familiar with the basic concepts of calculus.

My second goal was to start with calculus in the first section of the book rather than force students to wade through $n!$ pages of preliminary material on "sets," "limits," "continuous functions," and so forth. I relented slightly, however, and Chapter 0 is the result. Chapter 0 sets the stage for later work by introducing some standard notation and definitions that are used throughout the text. The chapter is short and can be skimmed quickly and then referred to as needed. Concepts such as limits are introduced and discussed whenever they are needed in the text.

To write a book that could easily be adapted to a computer-oriented course was my third goal. Sequences are introduced early and many computational aspects of calculus are stressed. Integration before differentiation also facilitates the early use of computers. In recognition of the general availability of hand-held calculators, several optional sections that show how to take full advantage of these instruments are included.

Finally, my fourth goal was to include examples and exercises for students interested in biology, the social sciences, business, and economics, as well as physics and engineering. I hope that each reader will find at least some of the many examples and exercises both interesting and enlightening.

There are several ways in which this book can be used. By including all of the optional sections or by adding computer topics and applications, there is sufficient material for two semesters. Since the first six chapters make no mention of the trigonometric functions, they could be used for a terminal course for liberal arts or business students who have no trigonometry background. The text is also designed for use in a two-term sequence for science students, to be followed by a course in calculus of several variables.

I would like to thank a former student, Susan Rose, who originally inspired this approach because she didn't appreciate the traditional calculus course. Others who have been very helpful include Darrell Perkins, Duane Clow, and Ervin Deal of Colorado State University; Paul Fleming of Poudre High in Fort Collins; Michael H. Clapp of California State University, Fullerton; Garret J. Etgen of the University of Houston; Carl E. Hall of The University of Texas at El Paso; Conrad K. McKnight of the University of Southwestern Louisiana; Karl J. Smith of Santa Rosa Junior College; Fred N. Springsteel of the University of Montana; Dennis G. Zill of Loyola University of Los Angeles; series editor Robert J. Wisner of New Mexico State University; Jack N. Thornton, Konrad Kerst, and Vena Dyer of Brooks/Cole Publishing Company; and the many secretaries who helped type various versions of the manuscript.

Darel W. Hardy

CONTENTS

0 A QUICK REVIEW   1

    1. The Real Numbers   1
    2. Functions and Graphs   5

1 AREA AND INTEGRATION   12

    1. The Problem of Area   13
    2. Approximations to Area   20
    3. The Sigma Notation   24
    Mathematical Induction (Optional)   28
    4. Sequences   30
    5. The Existence of the Integral of a Continuous Function   36
    6. The Integral of a Sum   46
    Review Section   54

2 DIFFERENTIATION AND THE FUNDAMENTAL THEOREM 59

1. Rates of Change   60
2. Slope of the Tangent Line   62
3. Limits   68
The Binomial Theorem (Optional)   74
4. The Derivative   76
5. The Mean Value Theorem for Integrals   82
6. Finding the Function if Area Is Known   86
7. The Fundamental Theorem of Calculus   89
8. The Antiderivative   94
Review Section   96

3 ELEMENTARY TECHNIQUES OF DIFFERENTIATION AND INTEGRATION   99

1. The Derivative of a Sum   99
2. The Integral of a Sum   102
3. The Derivative of a Product   104
4. The Derivative of a Quotient   107
5. The Derivative of a Composition   110
6. Integration by Substitution   113
7. The Mean Value Theorem for Derivatives   119
8. Implicit Differentiation   122
Review Section   127

4 SKETCHING THE GRAPH OF A FUNCTION   129

1. Increasing and Decreasing Functions   130
2. The First Derivative Test for Extreme Values   136
3. The Second Derivative Test   141
4. Getting the Picture   146
Review Section   153

5 APPLICATIONS   156

1. Applied Maxima and Minima   156
2. Velocity and Acceleration   161
3. Related Rates   165
4. Newton's Method   168
Some Special Algorithms (Optional)   172
5. Volumes of Revolution   175
6. Differential Equations   180

6 LOGARITHMS AND EXPONENTIALS 185

1. The Natural Logarithm   185
2. The Exponential Function   193
3. Logarithms Base a 202
For Hand-Held Calculators Only (Optional)   212
Review Section   214

7 THE CIRCULAR FUNCTIONS 217

1. The Circular Functions   217
2. Derivatives of the Circular Functions   223
3. The Inverse Circular Functions   231
4. Applications of the Circular Functions   238
The Hyperbolic Functions (Optional)   245
For Hand-Held Calculators Only (Optional)   253
Review Section   255

8 METHODS OF INTEGRATION 258

1. Substitution   258
2. Integration by Parts   267
3. Completing the Square   271
4. Partial Fractions   275
5. Improper Integrals   283
6. The Trapezoidal Rule   288
7. Simpson's Rule   293

9 FURTHER APPLICATIONS 300

1. Arc Length   300
2. Center of Mass   304
3. Work   311
4. Miscellaneous Applications   314
Review Section   321

10 TOPICS IN ANALYTIC GEOMETRY 326

1. Parametric Equations   326
2. Polar Coordinates   331
3. The Conic Sections   338
4. Translation of Axes   346

5. Rotation of Axes   349
Review Section   354

11 POLYNOMIAL APPROXIMATIONS AND INFINITE SERIES   360

1. Taylor Polynomials   360
The Use of Taylor Polynomials for Computing (Optional)   369
2. Taylor Series   371
3. Tests for Convergence   376
4. Series Solutions to Differential Equations   384
Review Section   386

APPENDIX: MISCELLANEOUS FACTS AND FORMULAS   391

ANSWERS TO SELECTED PROBLEMS   396

INDEX   455

BASIC CALCULUS

CHAPTER 0

A QUICK REVIEW

It is expected that the reader has a basic background in algebra and geometry for Chapters 1 through 6, plus a background in trigonometry for Chapters 7 through 11. The purpose of Chapter 0 is to refresh the reader's memory on topics in algebra and geometry, as well as to establish basic notation and terminology for the remainder of the book. Perhaps the most appropriate way to treat Chapter 0 is to skim through it quickly and then refer back to it as needed in the remainder of the book.

1. THE REAL NUMBERS

The *system of real numbers* includes the *natural numbers* (1, 2, 3, 4, 5, and so forth), the *integers* (the natural numbers, their negatives, and zero), the *rational numbers* (numbers of the form p/q, where p and q are integers and q is not zero), and the *irrational numbers* (numbers such as $\sqrt{3}$ that cannot be written in the form p/q for any integers p and q).

It is possible to obtain a correspondence between real numbers and points on a line. Let 0 correspond to any point; then mark off equal distances in both directions and assign ..., $-3, -2, -1, 0, 1, 2, 3, \ldots$ as in Figure 0.1. (Three dots indicate that the list continues.) Let 5/2 correspond to the point halfway between 2 and 3, $-4/3$ to the point two-thirds the distance from -2 to -1, and so forth. Doing so will take care of all the rational numbers. This correspondence can be extended to include the irrational numbers. For example, $\sqrt{3}(=1.73205\ldots)$ corresponds to a point between the points corresponding to the rational numbers 1.73 and 1.74, and π (the ratio of the circumference of a circle to its diameter) corresponds to a point between the points labeled 3.14159 and 3.14160.

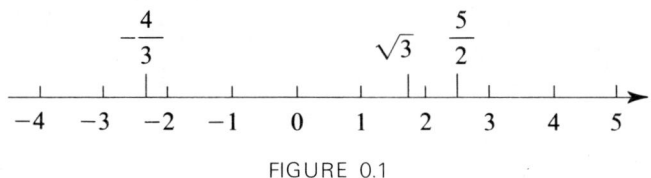

FIGURE 0.1

Lengths corresponding to some irrational numbers can actually be constructed. These constructions are based on the Pythagorean Theorem, which states that if a and b are legs of a right triangle and c is the hypotenuse (side opposite the right angle), then $a^2 + b^2 = c^2$ (see Figure 0.2). To construct a length corresponding to $\sqrt{2} = 1.41421\ldots$, we construct a right triangle with legs both equal to 1, as in Figure 0.3. The hypotenuse then has length $\sqrt{2}$. We can now use this length to construct $\sqrt{3} = 1.73205\ldots$ also by noting that $\sqrt{3}$ is the length of the hypotenuse of a right triangle with legs 1 and $\sqrt{2}$, since $(\sqrt{3})^2 = 1^2 + (\sqrt{2})^2$ (see Figure 0.4).

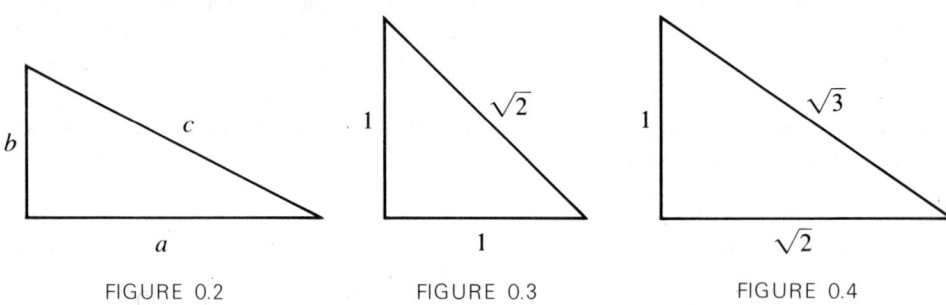

FIGURE 0.2   FIGURE 0.3   FIGURE 0.4

We say that $a < b$ (a is **less than** b) if the point labeled b lies to the right of the point labeled a. Also, $a \le b$ means that $a < b$ or $a = b$, and $b > a$ is equivalent to $a < b$. From algebra,

(1) if $a < b$, then $a + c < b + c$;

(2) if $a < b$ and $c > 0$, then $ac < bc$;

and

(3) if $a < b$ and $c < 0$, then $ac > bc$

for all real numbers a, b, and c.

EXAMPLE 1. Find x if $2x + 4 < 3$.

Solution. Using (1) with $c = -4$, we add -4 to both sides to get
$$2x < 3 - 4 = -1.$$
Now use (2) with $c = \frac{1}{2}$ to get
$$x < -\frac{1}{2}.$$
Hence x satisfies the inequality $2x + 4 < 3$ only if $x < -\frac{1}{2}$.

Given a real number a, the absolute value of a is the distance from the point labeled a to the point labeled 0. Thus the absolute value $|a|$ of a is given by

(4) $$|a| = \begin{cases} a & \text{if } a \geq 0 \\ -a & \text{if } a < 0 \end{cases}$$

so that $|a| \geq 0$ for every real number a. Equivalently,

(5) $$|a| = \sqrt{a^2}.$$

Notice that

(6) $$|ab| = |a| \cdot |b|$$

for all real numbers a and b.

EXAMPLE 2. $|5| = 5$, $|-3| = 3$, $|0| = 0$, and $|x - 3| = x - 3$ if $x \geq 3$ and $|x - 3| = -(x - 3) = 3 - x$ if $x < 3$.

It is sometimes convenient to replace an inequality of the form $|a| < b$ with the equivalent inequality $-b < a < b$, which means $-b < a$ and $a < b$. An inequality $|a| > b$ is equivalent to $a > b$ if $a \geq 0$ and $a < -b$ if $a < 0$.

EXAMPLE 3. Solve the inequality $|x - 4| < 3$.

$$-3 < x - 4 < 3$$
$$1 < x < 7$$

Solution. The given inequality is equivalent to

$$-3 < x - 4 < 3.$$

Adding 4 to each expression now yields

$$1 < x < 7.$$

Thus the real number x satisfies the inequality $|x - 4| < 3$ only if $1 < x < 7$—that is, only if x is some number between 1 and 7.

The expression $|a - b|$ can be thought of as the distance between the point labeled a and the point labeled b. Thus in Example 3 the distance between x and 4 is less than 3 if and only if x is some number between 1 and 7.

If $a < b$, then $[a, b]$ is the collection of all real numbers x such that $a \leq x \leq b$, and it is called the **closed interval** from a to b or simply the *interval* from a to b. (There are other types of intervals, but a consideration of closed intervals will be sufficient for our needs.)

If a and b are real numbers, then the symbol (a, b) is called an **ordered pair** of real numbers. By convention, $(a, b) \neq (c, d)$ unless $a = c$ and $b = d$; that is, ordered pairs are considered distinct unless they are identical. Thus $(2, \sqrt{3})$ and $(\sqrt{3}, 2)$ are different ordered pairs.

Two number lines can be placed perpendicular to each other in order to obtain a correspondence between ordered pairs of real numbers and points in a plane, as in Figure 0.5. The horizontal number line is called the **x-axis**, and the vertical number line is called the **y-axis**.

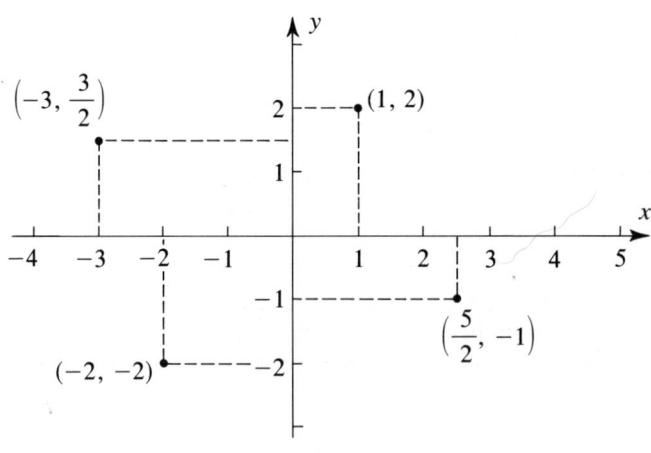

FIGURE 0.5

PROBLEMS

1. What is the absolute value of -5? Of $[(-3)^2 - 7]$?
2. By evaluating each side, verify that $|a + b| \leq |a| + |b|$ if
   a. $a = -4$ and $b = 5$;
   b. $a = 3$ and $b = 2$;
   c. $a = -1$ and $b = -1$.
3. Solve for x:
   a. $x^2 + 2 < 5$
   b. $|x + 3| < 1$
   c. $|3x + 4| < 7$
   d. $|2x + 5| < |3x - 2|$ [*Hint:* Consider (i) $3x - 2 > 0$ and (ii) $3x - 2 < 0$.]
4. Sketch two perpendicular number lines as in Figure 0.5. Locate and label the point corresponding to each ordered pair of real numbers: $(0,0)$, $(1,2)$, $(3,-4)$, $(-2,-3)$, $(2,-5)$, $(2,-1)$, $(-2,1)$.

2. FUNCTIONS AND GRAPHS

If A and B are sets of real numbers, then a ***function*** f from A to B is a rule that assigns to each number in A exactly one number in B. We write $x \to f(x)$ to indicate that the number x in A corresponds to the number $f(x)$ in B. The set A is called the ***domain*** of f, and the set B is called the ***range*** of f. The number $f(x)$ is read "f of x," "the ***value*** of f at x," or "the ***image*** of x under f."

Let f be the function, or rule, that assigns to each positive real number 5 times its square root. Thus $x \to 5\sqrt{x}$. In particular, $1 \to 5$, $2 \to 5\sqrt{2}$, $16 \to 20$, $5\sqrt{16}$ $a \to 5\sqrt{a}$, $(2c + d) \to 5\sqrt{2c + d}$, and $(\ ) \to 5\sqrt{(\ )}$, where the blanks can be filled in with any positive real number. Alternatively, we may write $f(1) = 5$, $f(2) = 5\sqrt{2}$, $f(16) = 20$, $f(a) = 5\sqrt{a}$, $f(2c + d) = 5\sqrt{2c + d}$, and $f(\ ) = 5\sqrt{(\ )}$, respectively. Note that f is a function from the positive real numbers to the positive real numbers.

The ***graph*** of a function f is the set of all points (x, y) in the plane such that $y = f(x)$. If f is a function, then each vertical line intersects (or touches) the graph of f once at most.

Suppose that L is a (nonvertical) straight line that passes through the points (x_1, y_1) and (x_2, y_2), where $x_1 \neq x_2$. If (x, y) is any other point on L, then by looking at similar triangles in Figure 0.6 we see that

$$\frac{y_2 - y_1}{x_2 - x_1} = \frac{y - y_1}{x - x_1}.$$

We call

(7) $$m = \frac{y_2 - y_1}{x_2 - x_1}$$

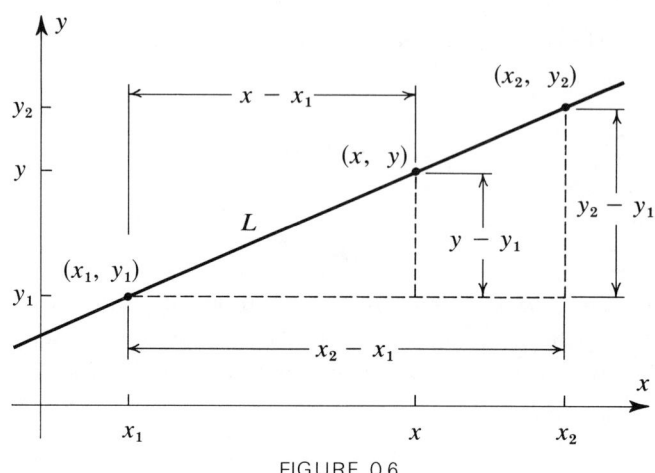

FIGURE 0.6

the *slope* of the line L. Thus

$$y - y_1 = \frac{y_2 - y_1}{x_2 - x_1}(x - x_1) = m(x - x_1).$$

Equivalently,

$$y = m(x - x_1) + y_1$$
$$= mx + (y_1 - mx_1)$$
$$= mx + b,$$

where $b = y_1 - mx_1$. Notice that L crosses the y-axis at the point $(0, b)$. The number b is called the **y-intercept**. The equation

(8) $$y - y_1 = m(x - x_1)$$

is called the **point-slope** form, and the equation

(9) $$y = mx + b$$

is called the **slope-intercept** form for the equation of a line. Thus the graph of the function $x \to mx + b$ is a line. If m is positive, then the graph slopes upward from left to right; and if m is negative, the graph slopes downward from left to right. The equation of a **horizontal line** is $y = b$ ($y = mx + b$ with $m = 0$), and the equation of a **vertical line** is $x = c$.

EXAMPLE 4. Find the equation of the line that passes through the points $(1, 2)$ and $(5, -8)$.

Solution. The slope is given by
$$m = \frac{2-(-8)}{1-5} = \frac{10}{-4} = -\frac{5}{2}.$$

By the point-slope form,
$$y - 2 = -\frac{5}{2}(x-1),$$

which simplifies to
$$y = -\frac{5}{2}x + \frac{9}{2}.$$

The following are additional examples of functions and their graphs.

EXAMPLE 5. Let w be the weight of a first-class letter and let $c(w)$ be the cost of mailing the letter in the United States in the year 1940, the author's favorite year. Then $c(w)$ equals 3 cents per ounce or fraction thereof. The graph is given in Figure 0.7. This postage function is from the positive real numbers to the set $\{3, 6, 9, 12, \ldots\}$.

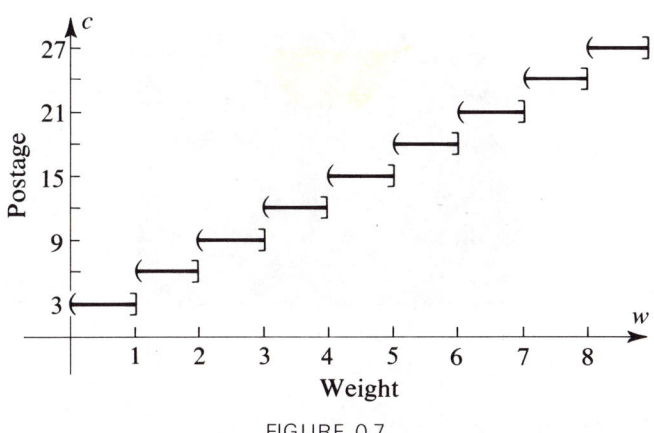

FIGURE 0.7

EXAMPLE 6. Sketch a graph of the function $x \to 5\sqrt{x}$.

Solution. Since $0 \to 0$, $1 \to 5$, $2 \to 5\sqrt{2}$, and $4 \to 10$, the points $(0, 0)$,

(1, 5), (2, 5√2), and (4, 10) all lie on the graph. A rough sketch can be made by connecting these points with a "smooth curve," as in Figure 0.8.

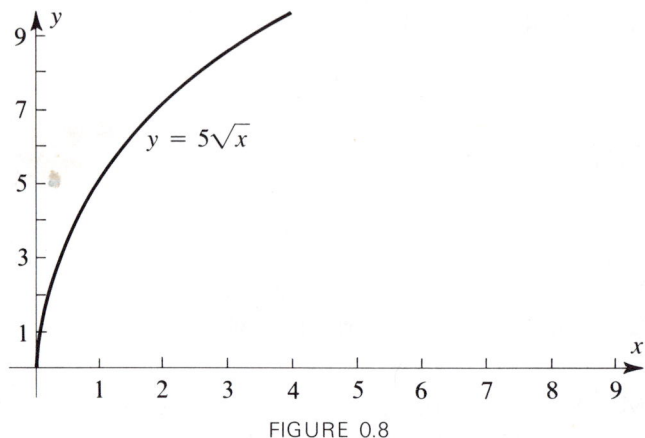

FIGURE 0.8

EXAMPLE 7. A certain pair of scales has a circular dial, as in Figure 0.9. If someone who weighs 150 pounds stands on the scales, the dial turns around one and one-half times and the arrow points to 50. If w is the actual weight and $s(w)$ is the number to which the arrow points, then s is a function whose graph is given in Figure 0.10.

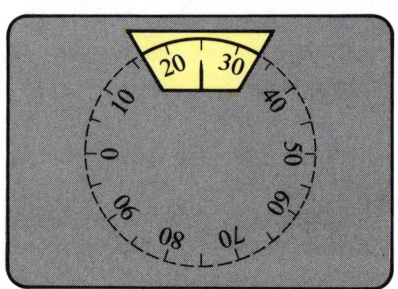

FIGURE 0.9

Notice that in the preceding example $s(w + 100) = s(w)$ for every w, and hence the graph forms a repeating pattern. Such a function is called *periodic* with *period* 100. Other periodic functions are studied in Chapter 7.

It is often convenient to construct new functions from old functions. If f and g are both functions from A to B, then the **sum** $f + g$, the **difference** $f - g$, the **product** fg, the **reciprocal** $1/f$, and the **quotient** f/g are defined, respectively, by

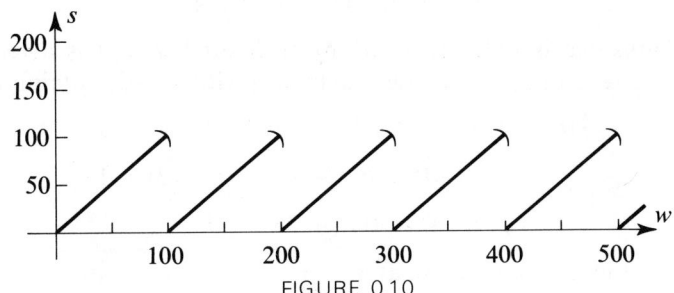

FIGURE 0.10

(10) $[f+g](x) = f(x) + g(x)$ for all x in A,

(11) $[f-g](x) = f(x) - g(x)$ for all x in A,

(12) $[fg](x) = f(x) \cdot g(x)$ for all x in A,

(13) $\left[\dfrac{1}{f}\right](x) = \dfrac{1}{f(x)}$ for all x in A such that $f(x) \neq 0$,

and

(14) $\left[\dfrac{f}{g}\right](x) = \dfrac{f(x)}{g(x)}$ for all x in A such that $g(x) \neq 0$.

EXAMPLE 8. Assume that $f(x) = 5\sqrt{x}$ and $g(x) = 1/x$ for every $x > 0$. Then $f+g$ is the function $x \to 5\sqrt{x} + 1/x$, or, equivalently, $(f+g)(x) = 5\sqrt{x} + 1/x$ for every $x > 0$. In words, $f+g$ is the function that associates with each positive real number x the sum of 5 times the square root of x with the reciprocal of x. Similarly, $f-g$ is the function $x \to 5\sqrt{x} - 1/x$, fg is the function $x \to 5\sqrt{x} \cdot (1/x) = 5/\sqrt{x}$, $1/f$ is the function $x \to 1/(5\sqrt{x})$, and f/g is the function $x \to 5\sqrt{x}/(1/x) = 5x^{3/2}$.

Suppose that $f(x) = 3x^2 + 2x + 4$. Then
$$f(0) = 3(0)^2 + 2(0) + 4 = 4,$$
$$f(1) = 3(1)^2 + 2(1) + 4 = 9,$$
$$f(-2) = 3(-2)^2 + 2(-2) + 4 = 12,$$
and
$$f(a) = 3(a)^2 + 2(a) + 4$$

for any real number a. In fact, we may write
$$f(\ ) = 3(\ )^2 + 2(\ ) + 4$$
where the blanks can be filled in with any real number. If t is a real number, then so is $t^2 - 3$. Hence we may fill in the blanks with $t^2 - 3$, obtaining
$$\begin{aligned} f(t^2 - 3) &= 3(t^2 - 3)^2 + 2(t^2 - 3) + 4 \\ &= 3(t^4 - 6t^2 + 9) + 2(t^2 - 3) + 4 \\ &= 3t^4 - 16t^2 + 25. \end{aligned}$$

If $g(t) = t^2 - 3$, then g is a function and
$$f(g(t)) = f(t^2 - 3) = 3t^4 - 16t^2 + 25.$$

Given any functions f and g, we can obtain a new function h defined by $x \to f(g(x))$, read "f evaluated at $g(x)$." Such a function h is called the *composition* of g with f and should not be confused with a product of functions. The composition of g with f is defined for all x such that $f(g(x))$ makes sense. [If $f(u) = \sqrt{u}$ and $g(x) = x + 1$, then $f(g(x)) = f(x + 1) = \sqrt{x + 1}$ makes sense only if $x \geq -1$.] The composition of functions is another useful means of creating new functions from old.

EXAMPLE 9. Let $f(x) = 1/x$ and $g(z) = z^2 + 1$. Then
$$f(g(z)) = f(z^2 + 1) = \frac{1}{z^2 + 1}$$

and
$$g(f(x)) = g\left(\frac{1}{x}\right) = \left(\frac{1}{x}\right)^2 + 1 = \frac{1}{x^2} + 1,$$

while
$$f(f(x)) = f\left(\frac{1}{x}\right) = \frac{1}{1/x} = x$$

and
$$g(g(z)) = g(z^2 + 1) = (z^2 + 1)^2 + 1 = z^4 + 2z^2 + 2.$$

EXAMPLE 10. Let $w(z) = \sqrt{z}$ and $u(x) = x^2 + 2$. Then
$$w(u(x)) = w(x^2 + 2) = \sqrt{x^2 + 2}$$
and
$$u(w(z)) = u(\sqrt{z}) = (\sqrt{z})^2 + 2 = z + 2.$$

PROBLEMS

1. Let g be the function that assigns to each real number the square root of one more than the square of the number. Write down a formula for the function g.

2. Describe in words what the function $x \to x + 1/x^2$ does to a nonzero real number. [Expressions such as $a + b/c$ are interpreted as $a + \dfrac{b}{c}$ and not $(a + b)/c$.]

In Problems 3 through 10, compute $(f + g)(x)$, $(fg)(x)$, $(f - g)(x)$, $(f/g)(x)$, $f(g(x))$, $g(f(x))$, $f(f(x))$, and $g(g(x))$.

3. $f(x) = x^2$, $g(x) = \sqrt{x + 1}$
4. $f(x) = x^3$, $g(x) = 1/x$
5. $f(x) = 3x^2 + 4$, $g(x) = 5 + 1/x$
6. $f(x) = \sqrt{x + 1}$, $g(x) = x^2 - 1$
7. $f(x) = x$, $g(x) = 1/x$
8. $f(x) = 1/x^2$, $g(x) = 1/x$
9. $f(x) = x + 1/x$, $g(x) = x - 1/x$
10. $f(x) = \sqrt{x^2 - 1}$, $g(x) = x^2 + 1$

In Problems 11 through 18, for each choice of f, find functions g and h such that $f(x) = h(g(x))$. [No fair using $g(x) = x$ and $h(x) = f(x)$, or $g(x) = f(x)$ and $h(x) = x$.]

11. $f(x) = x^2 + (1/x^2)$
12. $f(x) = (x^2 + 1)^{10}$
13. $f(x) = \sqrt{3x^2 + 4}$
14. $f(x) = (x + 1)^3 + \sqrt{x + 1}$
15. $f(x) = \sqrt{x} + 1/\sqrt{x}$
16. $f(x) = \sqrt{x} + \sqrt{1 + \sqrt{x}}$
17. $f(x) = 1/x + 1/x^2 + 1/x^3$
18. $f(x) = (x + 1)^{10} + x + 1$

19. Sketch the graph of the function $x \to x^2$ by plotting about six points and connecting them with a smooth curve.

20. Repeat Problem 19 for the function $s \to 1/s$, $s > 0$.

21. Repeat Problem 19 for the function $u \to 3u + 5$.

22. Repeat Problem 19 for the function $x \to \sqrt{x^2 + 1}$.

CHAPTER 1

AREA AND INTEGRATION

One of the basic problems in calculus is finding the exact area of an arbitrary region. Over 2000 years ago, the Greeks were able to use methods of calculus to compute the area of a circle, the area of an ellipse, and the area of a sector of a spiral. In fact, Archimedes' "Method of Equilibrium" is almost identical to the present-day method of computing areas by slicing a region into thin strips and estimating the areas of the strips. Using additional methods developed by Leibnitz, Newton, and others, we will be able to compute areas of many types of regions in the plane.

By reinterpreting the methods used to obtain areas, we will be able to find a variety of other applications for the calculus. In particular, these methods can be used to solve the following types of problems:

1. If $100 is put into a savings account earning 6% annual interest compounded daily, how much will the account be worth after 10 years if no additional money is deposited or withdrawn?
2. If a rock dropped into a deep well takes 5 seconds to hit bottom, how deep is the well?
3. How many cubic feet of helium are required to fill a spherical balloon of diameter 2 feet?

4. How much blood will flow through a section of a vein of known length if the diameter and pressure differential are known?

5. How much force due to water pressure is there on a triangular-shaped dam if the dam is 200 feet across at water level and the water is 50 feet deep?

1. THE PROBLEM OF AREA

Before we look at the preceding types of problems, we investigate one very basic problem: computing areas.

The area of a rectangle of width x and length y is equal to the product xy. (See Problem 15 at the end of this section.) A right triangle with legs x and y has area $\frac{1}{2}xy$, because two such triangles can be put together to form a rectangle of dimensions x by y, as in Figure 1.1.

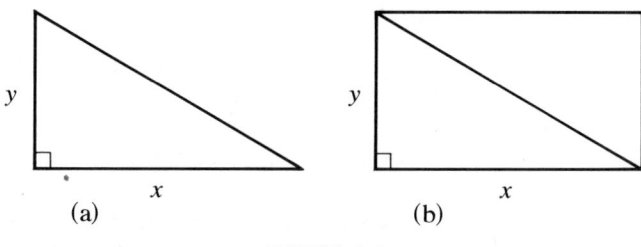

FIGURE 1.1

More generally, a triangle with base b and height h has area $\frac{1}{2}bh$. To see this formula, we divide an arbitrary triangle into two right triangles as in Figure 1.2. Then we conclude that

$$\begin{aligned}\text{Total area} &= \tfrac{1}{2}xh + \tfrac{1}{2}(b-x)h \\ &= \tfrac{1}{2}xh + \tfrac{1}{2}bh - \tfrac{1}{2}xh \\ &= \tfrac{1}{2}bh.\end{aligned}$$

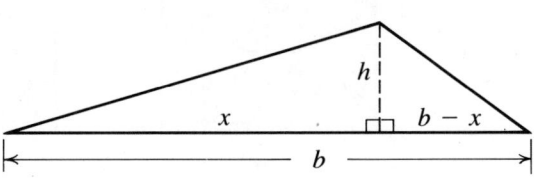

FIGURE 1.2

14 CHAPTER 1: AREA AND INTEGRATION

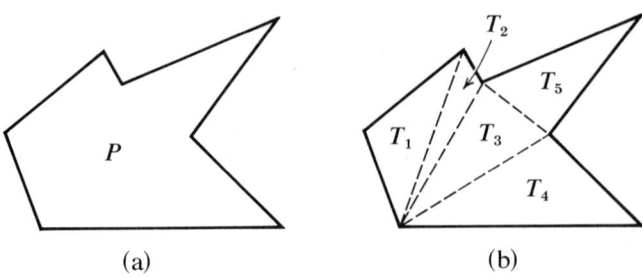

(a)   (b)

FIGURE 1.3

The problem of finding the area of a polygon can be solved by first subdividing the polygon into triangles, as in Figure 1.3, and then adding the areas of the triangles. Thus

$$\text{Area } P = \text{Area } T_1 + \text{Area } T_2 + \text{Area } T_3 + \text{Area } T_4 + \text{Area } T_5.$$

(Of course, it may not be easy to compute the base and height for all these triangles, but it is at least theoretically possible.)

The problem of finding the area of a region bounded by a curve that is not a polygon is more difficult. Our method of computing such areas involves approximations by rectangles.

We first simplify the problem slightly. Given the Region R in Figure 1.4, the area can be found if we know the areas of regions R_1 and R_2, because our intuitive concept of area tells us that Area R_1 equals Area R + Area R_2. Hence we will usually consider regions bounded on three sides by straight lines and bounded above by the graph of some function, as in Figure 1.5 (see Chapter 0 for a discussion of functions and graphs).

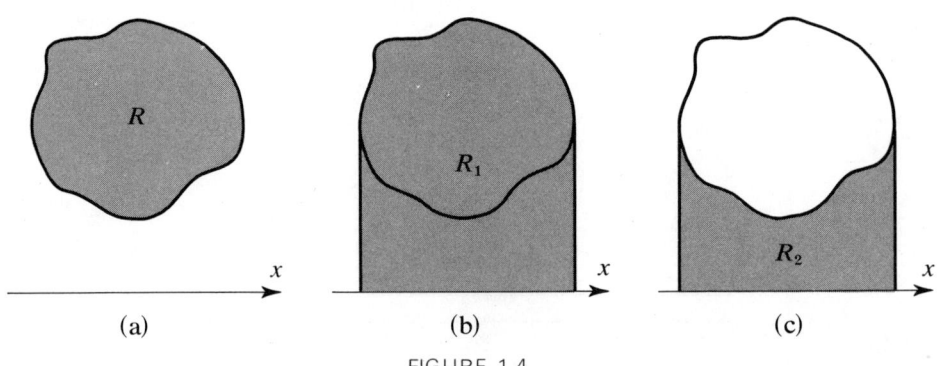

(a)   (b)   (c)

FIGURE 1.4

SECTION 1: THE PROBLEM OF AREA   15

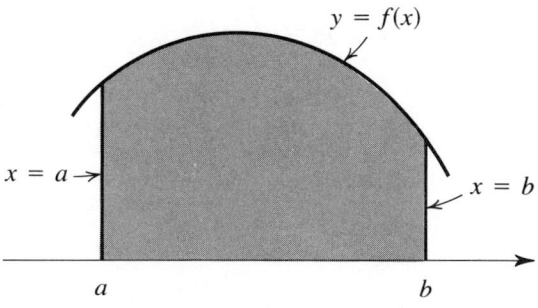

FIGURE 1.5

To simplify notation in what follows, recall that $[a, b]$ denotes the collection of all real numbers x such that $a \leq x \leq b$. (It is tacitly assumed that $a < b$.)

The following example shows how the area of a region can be approximated by the use of rectangles.

EXAMPLE 1. Estimate the area of the region bounded by $y = x^2$ and $y = 0$ between $x = 0$ and $x = 1$ as in Figure 1.6a.

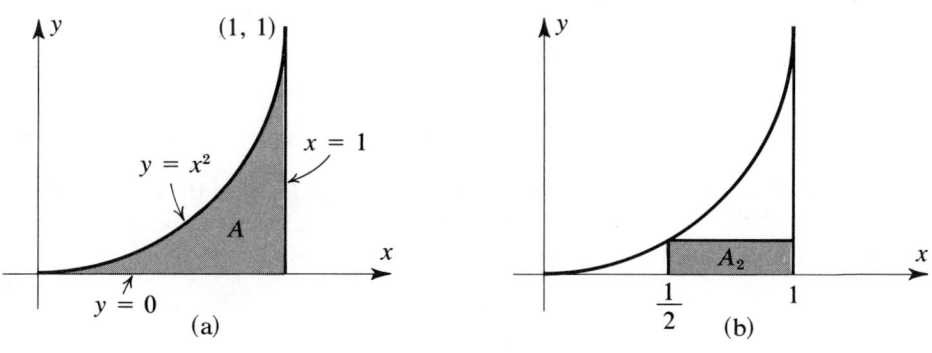

FIGURE 1.6

Solution. To approximate the area of this region, we first subdivide the interval $[0, 1]$ into the subintervals $[0, \frac{1}{2}]$ and $[\frac{1}{2}, 1]$. Then a rectangle A_2 of area $\frac{1}{2} \cdot \frac{1}{4} = \frac{1}{8}$ can be inscribed in Region A as in Figure 1.6b ($\frac{1}{2}$ is the base of the rectangle and $\frac{1}{4}$ is the height), so the area of Region A is greater than $\frac{1}{8}$.

If the interval $[0, 1]$ is subdivided into the three subintervals $[0, \frac{1}{3}]$, $[\frac{1}{3}, \frac{2}{3}]$, and $[\frac{2}{3}, 1]$, then from Figure 1.7a we see that the area of Region A is greater than

$$\text{Area } A_3 = \frac{1}{3} \cdot \frac{1}{9} + \frac{1}{3} \cdot \frac{4}{9}$$

$$= \frac{1}{3^3}(1^2 + 2^2),$$

since $\frac{1}{3}$ is the base of each rectangle and the heights of the rectangles are $\frac{1}{9}$ and $\frac{4}{9}$, respectively.

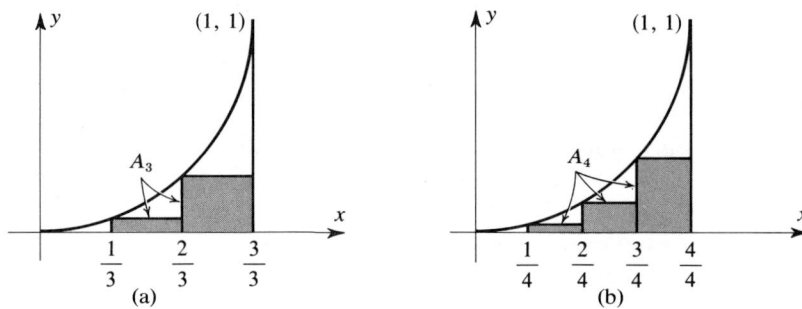

FIGURE 1.7

The interval $[0, 1]$ can be subdivided into the four subintervals $[0, \frac{1}{4}]$, $[\frac{1}{4}, \frac{2}{4}]$, $[\frac{2}{4}, \frac{3}{4}]$, and $[\frac{3}{4}, \frac{4}{4}]$ as in Figure 1.7b. The area of Region A is greater than

$$\text{Area } A_4 = \left(\frac{1}{4}\right)^2 \cdot \frac{1}{4} + \left(\frac{2}{4}\right)^2 \cdot \frac{1}{4} + \left(\frac{3}{4}\right)^2 \cdot \frac{1}{4}$$

$$= \frac{1}{4^3}(1^2 + 2^2 + 3^2)$$

$$= 0.21875.$$

Suppose now that the interval $[0, 1]$ is subdivided into the n subintervals $[0, 1/n]$, $[1/n, 2/n]$, ..., $[(n-1)/n, n/n]$, where n is a large positive integer. The rectangle with base on the interval $[k/n, (k+1)/n]$ has dimensions

$$\frac{k+1}{n} - \frac{k}{n} = \frac{1}{n}$$

by $(k/n)^2$ and hence has area equal to $(k/n)^2 \cdot (1/n)$. Thus the area of Region A is at least

$$\text{Area } A_n = \left(\frac{1}{n}\right)^2 \cdot \frac{1}{n} + \left(\frac{2}{n}\right)^2 \cdot \frac{1}{n} + \cdots + \left(\frac{n-1}{n}\right)^2 \cdot \frac{1}{n}$$

$$= [1^2 + 2^2 + \cdots + (n-1)^2]\frac{1}{n^3}.$$

We will show later in this chapter that

(1) $$1^2 + 2^2 + 3^2 + \cdots + m^2 = \frac{m(m+1)(2m+1)}{6}$$

for each positive integer m. Setting $m = n - 1$, we get

$$\text{Area } A_n = \frac{(n-1)n[2(n-1)+1]}{6} \cdot \frac{1}{n^3}$$

$$= \frac{n-1}{n} \cdot \frac{n}{n} \cdot \frac{2n-1}{6n}$$

$$= \frac{n-1}{n} \cdot \frac{2n-1}{2n} \cdot \frac{1}{3}$$

$$= \left(1 - \frac{1}{n}\right)\left(1 - \frac{1}{2n}\right) \cdot \frac{1}{3}.$$

If n is very large, then both $1 - 1/n$ and $1 - 1/2n$ are close to 1, so that Area A_n is close to $\frac{1}{3}$. For example, if $n = 100$, then

$$\text{Area } A_{100} = \left(1 - \frac{1}{100}\right)\left(1 - \frac{1}{200}\right) \cdot \frac{1}{3}$$

$$\approx 0.32835;$$

and, similarly,

$$\text{Area } A_{1000} \approx 0.3328335$$

and

$$\text{Area } A_{10000} \approx 0.3332833.$$

(The symbol $\approx$ means "approximately equal to.")

Since Area A_n tends to $\frac{1}{3}$ as n gets large, we conclude that the area of Region A is exactly $\frac{1}{3}$.

Notice that the foregoing computations involved replacing $(n-1)/n$ by $1 - 1/n$ and $(2n-1)/2n$ by $1 - 1/2n$. It is (or should be) obvious that $1 - 1/n$ is very close to 1 if n is a large integer. It is perhaps not quite so obvious that $(n-1)/n$ and $(2n-1)/2n$ are both close to 1 for large n.

If our method of approximating areas is any good, then it must give consistent answers when we use it to compute areas of polygons. In the following example we look at the area of a triangle.

EXAMPLE 2. Estimate the area of a right triangle with legs a and b by the method given in Example 1.

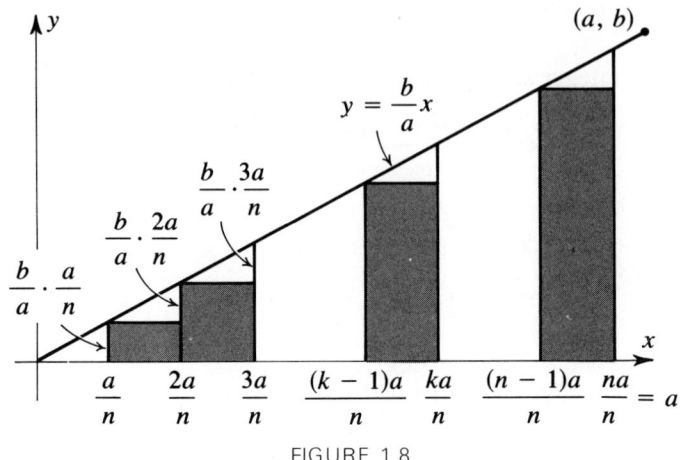

FIGURE 1.8

Solution. We first subdivide the interval $[0, a]$ into the n subintervals

$$\left[0, \frac{a}{n}\right], \quad \left[\frac{a}{n}, \frac{2a}{n}\right], \quad \left[\frac{2a}{n}, \frac{3a}{n}\right], \quad \ldots, \quad \left[\frac{(n-1)a}{n}, a\right].$$

The width of each rectangle in Figure 1.8 is equal to a/n. The height of each rectangle is determined by the left-hand endpoint of that subinterval. In particular, the rectangle sitting on the subinterval $[0, a/n]$ has height zero, and the rectangle sitting on the subinterval $[a/n, 2a/n]$ has height determined by $y = (b/a)x$ with $x = a/n$, so that

$$y = \frac{b}{a}x$$
$$= \frac{b}{a} \cdot \frac{a}{n}$$
$$= \frac{b}{n}.$$

Similarly, the other rectangles have heights $2b/n$, $3b/n$, and so forth. The area of the triangle can be approximated by

$$\text{Area } A_n = \frac{b}{n} \cdot \frac{a}{n} + \frac{2b}{n} \cdot \frac{a}{n} + \frac{3b}{n} \cdot \frac{a}{n} + \cdots + \frac{(n-1)b}{n} \cdot \frac{a}{n}$$

$$= \frac{ab}{n^2}[1 + 2 + 3 + \cdots + (n-1)].$$

Later in this chapter we will show that

(2) $$1 + 2 + 3 + \cdots + m = \frac{m(m+1)}{2}.$$

Using this fact (with $m = n - 1$), we see that

$$\text{Area } A_n = \frac{ab}{n^2} \cdot \frac{(n-1)n}{2}$$

$$= \frac{1}{2} ab \left(1 - \frac{1}{n}\right).$$

If we take n very large, then $1/n$ is very close to zero, and so the area of A_n is close to $\frac{1}{2}ab$. This result agrees with the area of a right triangle obtained by earlier methods.

PROBLEMS

In Problems 1 through 8, approximate the area of the region bounded by $y = f(x)$, $x = a$, $x = b$, and the x-axis by subdividing the interval $[a, b]$ into four equal parts. Draw an appropriate sketch in each case. (Plot the five points and connect them in a smooth fashion. You should become an expert at sketching graphs by the end of Chapter 4, but for now just do the best you can.)

1. $y = x + 1$, $a = 0$, $b = 4$
2. $y = 3x - 1$, $a = 1$, $b = 3$
3. $y = 2x^2$, $a = 0$, $b = 2$
4. $y = x^2 + 2$, $a = 1$, $b = 5$
5. $y = x - x^2$, $a = 0$, $b = 1$
6. $y = 1/x$, $a = 1$, $b = 5$
7. $y = \sqrt{x}$, $a = 1$, $b = 5$
8. $y = 1/(x^2 + 1)$, $a = 0$, $b = 4$

Find the exact area of the region bounded by $y = f(x)$, $x = a$, $x = b$, and the x-axis in Problems 9 through 14. Use the method outlined in Examples 1 and 2.

9. $y = x + 1$, $a = 0$, $b = 4$
10. $y = 3x - 1$, $a = 1$, $b = 3$
11. $y = 2x^2$, $a = 0$, $b = 2$
12. $y = x^2 + 2$, $a = 1$, $b = 5$
13. $y = x - x^2$, $a = 0$, $b = 1$
14. $y = 2x^2 + x + 1$, $a = 1$, $b = 4$

15. It is possible to start with squares rather than rectangles as building blocks for area. Assume that the area of a square with edge x is defined to be x^2. Compute the area of the region in Figure 1.9, first as a square with edge $x + y$ and then as a sum of areas of four smaller regions. Conclude that a rectangle with dimensions x by y has area xy.

16. Assume that a circle with radius r has circumference $2\pi r$. Approximate the area of the circle by looking at a large number of triangles, each with one vertex at the center of the circle. Conclude that the circle has area equal to πr^2. (This is essentially the "Method of Exhaustion" used by the early Greeks to obtain a formula for the area of a circle.)

17. It has been tacitly assumed that if Regions A and B overlap in at most a line, then the area of A and B taken together equals the sum of Area A plus Area B. Can you think

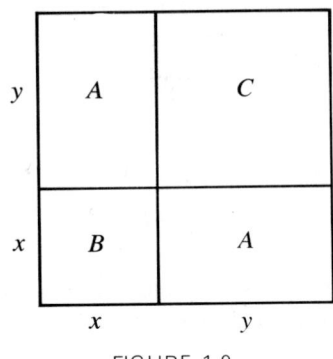

FIGURE 1.9

of any other tacit assumptions that might be made about area? [*Answers:* (i) For any region A, Area $A \geq 0$. (ii) If Region A is entirely contained inside Region B, then Area $A \leq$ Area B. (iii) Congruent regions have equal areas. (iv) The area of a rectangle of dimensions x by y is the product xy.]

2. APPROXIMATIONS TO AREA

It is difficult to compute some areas exactly. For such regions, it is usually satisfactory to estimate the area.

EXAMPLE 3. Estimate the area of the region bounded by $y = \sqrt{x}$, $x = 1$, $x = 5$, and the x-axis.

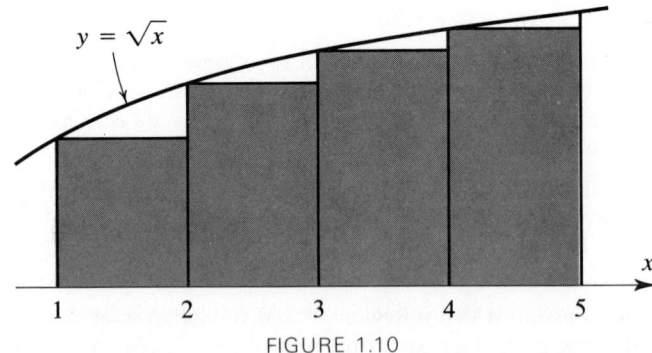

FIGURE 1.10

Solution. We subdivide the interval $[1, 5]$ into four equal parts and construct rectangles as in Figure 1.10. The base of each rectangle is 1, and the height of each rectangle is determined by the left-hand endpoint of each subinterval. An approximation to the area is given by

$$1 + \sqrt{2} + \sqrt{3} + 2 \approx 3 + 1.414 + 1.732$$
$$= 6.146.$$

By looking at Figure 1.10, we can tell that the approximation 6.146 is too small.

So far the height of each approximating rectangle has been determined by the value of the function at the left-hand endpoint. It is also possible to use right-hand endpoints, midpoints, or any other convenient points.

EXAMPLE 4. Estimate the area of the region bounded by $y = \sqrt{x}$, $x = 1$, $x = 5$, and the x-axis, using right-hand endpoints of the subintervals to determine the heights of the approximating rectangles.

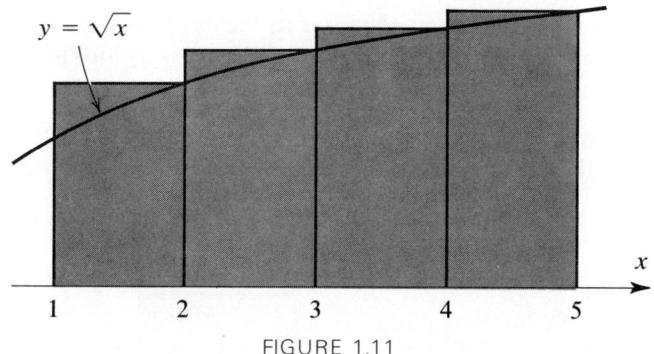

FIGURE 1.11

Solution. The height of each approximating rectangle is determined as in Figure 1.11. The sum of the areas of the rectangles is equal to

$$1 \cdot \sqrt{2} + 1 \cdot \sqrt{3} + 1 \cdot \sqrt{4} + 1 \cdot \sqrt{5} \approx 1.414 + 1.732 + 2 + 2.236$$
$$= 7.382.$$

Notice from Figure 1.11 that this approximation exceeds the actual area.

EXAMPLE 5. Estimate the area of the region bounded by $y = \sqrt{x}$, $x = 1$, $x = 5$, and the x-axis, using midpoints.

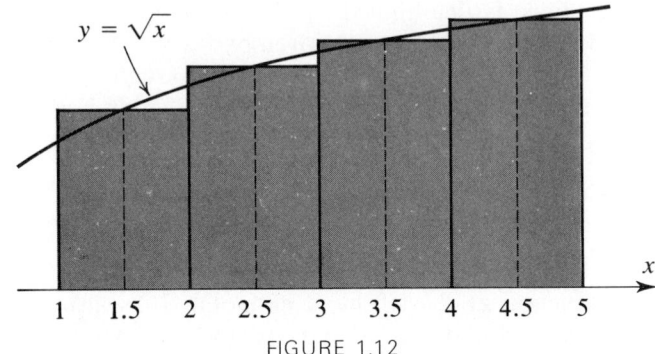

FIGURE 1.12

Solution. By looking at Figure 1.12, we see that the sum of the areas of the rectangles is given by

$$1 \cdot \sqrt{1.5} + 1 \cdot \sqrt{2.5} + 1 \cdot \sqrt{3.5} + 1 \cdot \sqrt{4.5}$$
$$\approx 1.225 + 1.581 + 1.871 + 2.121$$
$$= 6.798.$$

This approximation is much closer to the exact area than those given by the left-hand or right-hand methods, as one would guess by looking at Figure 1.12.

Occasionally it is convenient to pick points that are neither endpoints nor midpoints of the subintervals, as the following example illustrates.

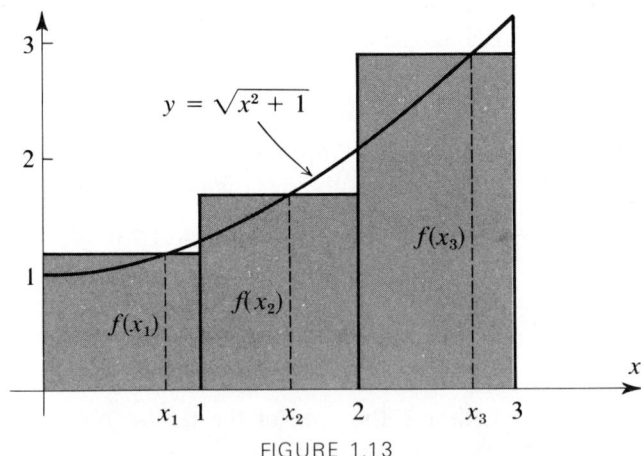

FIGURE 1.13

EXAMPLE 6. Approximate the area of the region bounded by the graph of $f(x) = \sqrt{x^2 + 1}$, $x = 0$, $x = 3$, and the x-axis.

Solution. Subdivide the interval $[0, 3]$ into the subintervals $[0, 1]$, $[1, 2]$, and $[2, 3]$. We wish to approximate the area of the region in Figure 1.13 by using three rectangles. Notice that if $x_1 = \sqrt{7/3}$, then $0 < x_1^2 < 1$, and hence $0 < x_1 < 1$. Also, $f(x_1) = 4/3$, a relatively simple number with which to compute. Now let $x_2 = \sqrt{3}$ and $x_3 = 2\sqrt{2}$. Notice that $1 < x_2 < 2$ and $2 < x_3 < 3$. Furthermore, $f(x_2) = 2$ and $f(x_3) = 3$. Each rectangle in Figure 1.13 has base 1. Thus the sum of the areas of the rectangles is given by

$$1 \cdot f(x_1) + 1 \cdot f(x_2) + 1 \cdot f(x_3) = \frac{4}{3} + 2 + 3$$

$$= \frac{19}{3}$$

$$\approx 6.333.$$

More refined methods can be used to show that 5.3 is a better approximation.

Example 6 is intended to illustrate that *any* point in an interval *can* be used to give the height of an approximating rectangle. Usually an endpoint or a midpoint is used for this purpose.

PROBLEMS

Estimate the area of the region bounded by $y = f(x)$, $x = a$, $x = b$, and the x-axis by subdividing the interval into n equal parts and using the left-hand method. Draw an appropriate sketch for each problem.

1. $y = x^3 - 1$, $a = 1$, $b = 4$, $n = 3$
2. $y = 1/x$, $a = 1$, $b = 3$, $n = 4$
3. $y = \sqrt{x}$, $a = 0$, $b = 5$, $n = 5$
4. $y = (x + 1)/(x - 1)$, $a = 2$, $b = 3$, $n = 4$
5. $y = 1/x^2$, $a = 1$, $b = 3$, $n = 4$
6 to 10. Estimate the areas of the regions described in Problems 1 through 5, using the right-hand method.
11 to 15. Repeat Problems 1 through 5, using the midpoint method.

3. THE SIGMA NOTATION

In order to simplify such expressions as

$$a_1 + a_2 + a_3 + a_4 + \cdots + a_{25}$$

and

$$f(x_1)(x_1 - x_0) + f(x_2)(x_2 - x_1) + \cdots + f(x_n)(x_n - x_{n-1}),$$

we introduce what is called *sigma notation*, first introduced by Leonhard Euler (1707–1783). Euler (pronounced "Oiler") was one of the most prolific mathematicians in history and is responsible for much of the standard mathematical notation in use today. The first sum is written in sigma notation as

$$\sum_{i=1}^{25} a_i$$

and the second as

$$\sum_{k=1}^{n} f(x_k)(x_k - x_{k-1}).$$

The symbolism $\sum_{i=1}^{25} a_i$ is read "The sum as i goes from 1 to 25 of a-sub-i." It is understood that the subscript i takes on successive integer values from 1 to 25. Other examples include

$$\sum_{k=1}^{n} k^2 = 1^2 + 2^2 + 3^2 + \cdots + n^2$$

$$\sum_{i=-1}^{2} i = -1 + 0 + 1 + 2$$
$$= 2$$

$$\sum_{k=1}^{6} 5 = 5 + 5 + 5 + 5 + 5 + 5$$
$$= 30$$

$$\sum_{i=3}^{5} i(i-1) = 3(2) + 4(3) + 5(4)$$
$$= 38$$

$$\sum_{k=2}^{2} k^2 = 2^2$$
$$= 4.$$

We will have need for a few computational rules for sigma notation. First, we see that

$$\sum_{i=1}^{n} (a_i + b_i) = (a_1 + b_1) + (a_2 + b_2) + \cdots + (a_n + b_n)$$
$$= (a_1 + a_2 + \cdots + a_n) + (b_1 + b_2 + \cdots + b_n)$$
$$= \sum_{i=1}^{n} a_i + \sum_{i=1}^{n} b_i.$$

In words, "The sum from 1 to n of a_i plus b_i equals the sum from 1 to n of a_i plus the sum from 1 to n of b_i."

Also, if c is any real number, then

$$\sum_{i=1}^{n} ca_i = ca_1 + ca_2 + \cdots + ca_n$$
$$= c(a_1 + a_2 + \cdots + a_n)$$
$$= c \sum_{i=1}^{n} a_i.$$

Thus the sum from 1 to n of a constant c times a_i equals the constant c times the sum from 1 to n of a_i.

The subscript i is usually called a "dummy" subscript and can be replaced by any other convenient symbol:

$$\sum_{i=1}^{n} a_i = \sum_{j=1}^{n} a_j = \sum_{\alpha=1}^{n} a_\alpha = a_1 + a_2 + \cdots + a_n.$$

If m is an integer less than or equal to n, then

$$\sum_{i=m}^{n} a_i = a_m + a_{m+1} + \cdots + a_n,$$

and

$$\sum_{i=m+1}^{n+1} a_{i-1} = a_{(m+1)-1} + a_{(m+2)-1} + \cdots + a_{(n+1)-1}$$
$$= a_m + a_{m+1} + \cdots + a_n.$$

Since both sums on the left are equal to the same thing,

$$\sum_{i=m}^{n} a_i = \sum_{i=m+1}^{n+1} a_{i-1}.$$

To summarize:

(3) $$\sum_{i=m}^{n} a_i = a_m + a_{m+1} + a_{m+2} + \cdots + a_{n-1} + a_n$$

(4) $$\sum_{i=m}^{n} [a_i + b_i] = \sum_{i=m}^{n} a_i + \sum_{i=m}^{n} b_i$$

(5) $$\sum_{i=m}^{n} [a_i - b_i] = \sum_{i=m}^{n} a_i - \sum_{i=m}^{n} b_i$$

(6) $$\sum_{i=m}^{n} [ca_i] = c \left[\sum_{i=m}^{n} a_i \right]$$

(7) $$\sum_{i=m}^{n} a_i = \sum_{k=m}^{n} a_k$$

(8) $$\sum_{i=m}^{n} a_i = \sum_{i=m+1}^{n+1} a_{i-1}$$

(9) $$\sum_{i=m}^{n} a_i = \sum_{i=m}^{k-1} a_i + \sum_{i=k}^{n} a_i \quad \text{(if } m < k \leq n)$$

(10) $$\sum_{i=1}^{n} a = na$$

In Section 1 we assumed two formulas, (2) and (1) respectively, which can now be written as follows:

$$\sum_{i=1}^{n} i = \frac{n(n+1)}{2}$$

$$\sum_{i=1}^{n} i^2 = \frac{n(n+1)(2n+1)}{6}$$

We will now derive these formulas. Notice that

$$\sum_{i=1}^{n} [i^2 - (i-1)^2] = (1^2 - 0^2) + (2^2 - 1^2) + (3^2 - 2^2) + \cdots$$

$$+ [(n-1)^2 + (n-2)^2] + [n^2 - (n-1)^2]$$

$$= n^2.$$

Such sums are called either **collapsing sums** or **telescoping sums**, depending on one's point of view. These sums appear more often than might be suspected and provide useful means of computing many expressions.

The preceding sum can be computed in another manner. Looking at the ith term,

$$i^2 - (i-1)^2 = i^2 - i^2 + 2i - 1$$
$$= 2i - 1,$$

so
$$\sum_{i=1}^{n} [i^2 - (i-1)^2] = \sum_{i=1}^{n} [2i - 1]$$
$$= 2\left(\sum_{i=1}^{n} i\right) - \left(\sum_{i=1}^{n} 1\right)$$
$$= 2\left(\sum_{i=1}^{n} i\right) - n.$$

Thus,
$$2\left(\sum_{i=1}^{n} i\right) - n = \sum_{i=1}^{n} [i^2 - (i-1)^2] = n^2.$$

Solving for $\sum_{i=1}^{n} i$, we get

$$\sum_{i=1}^{n} i = \frac{1}{2}(n^2 + n) = \frac{n(n+1)}{2},$$

which verifies the first formula.

Using the same technique, we see that

$$\sum_{i=1}^{n} [i^3 - (i-1)^3] = n^3,$$

and
$$\sum_{i=1}^{n} [i^3 - (i-1)^3] = \sum_{i=1}^{n} (i^3 - i^3 + 3i^2 - 3i + 1)$$
$$= 3\left(\sum_{i=1}^{n} i^2\right) - 3\left(\sum_{i=1}^{n} i\right) + \left(\sum_{i=1}^{n} 1\right)$$
$$= 3\left(\sum_{i=1}^{n} i^2\right) - \frac{3n(n+1)}{2} + n.$$

Hence
$$3\left(\sum_{i=1}^{n} i^2\right) - \frac{3n(n+1)}{2} + n = n^3.$$

Solving for $\sum_{i=1}^{n} i^2$, we get

$$\sum_{i=1}^{n} i^2 = \frac{1}{3}\left[n^3 + \frac{3n(n+1)}{2} - n\right]$$
$$= \frac{2n^3 + 3n^2 + 3n - 2n}{6}$$
$$= \frac{n(n+1)(2n+1)}{6}.$$

PROBLEMS

Evaluate the sums in Problems 1 through 16.

1. $\sum_{i=1}^{5} i$

2. $\sum_{i=-2}^{1} i^2$

3. $\sum_{i=1}^{20} (i^2 + 3i + 5)$

4. $\sum_{k=3}^{10} (k - 2)^2$

5. $\sum_{i=2}^{13} c(i - 1)^2$

6. $\sum_{k=2}^{4} \frac{1}{k}$

7. $\sum_{n=1}^{100} n$

8. $\sum_{i=5}^{10} (3i^2 + 4i)$

9. $\sum_{i=8}^{9} \frac{i+1}{i-1}$

10. $\sum_{i=-3}^{3} i$

11. $\sum_{k=1}^{10} (2k + 1)^2$

12. $\sum_{i=1}^{n} (ni + i^2)$

13. $\sum_{i=n}^{m} i^2, \ n \leq m$

14. $\sum_{k=11}^{10+n} (k - 10)^2, \ n \geq 1$

15. $\sum_{i=1}^{n^2} (i^2 + i + 1)$

16. $\sum_{j=-n}^{n} (j^2 + j + 1), \ n \geq 0$

17. Derive the formula $\sum_{k=1}^{n} k^3 = n^2(n + 1)^2/4$ by evaluating the sum $\sum_{k=1}^{n} [k^4 - (k - 1)^4]$ in two ways. [Notice that, combining this with the formula $\sum_{i=1}^{n} i = n(n + 1)/2$, we get $1^3 + 2^3 + 3^3 + \cdots + n^3 = (1 + 2 + 3 + \cdots + n)^2.$]

18. Evaluate $\sum_{i=1}^{n} i(i + 2)(i + 3)$.

19. Evaluate $\sum_{k=10}^{20} (k^3 + 3k + 5)$.

20. Derive a formula for $\sum_{k=1}^{n} k^4$.

21. Evaluate $\sum_{i=1}^{10} i^4$.

22. Evaluate $\sum_{j=1}^{5} (j^2 + 1)^2$.

23. Derive the formula $\sum_{i=1}^{n} i = n(n + 1)/2$ by adding $\sum_{i=1}^{n} i$ term by term to $\sum_{i=1}^{n} (n + 1 - i)$ and noting that $\sum_{i=1}^{n} i = \sum_{i=1}^{n} (n + 1 - i)$.

MATHEMATICAL INDUCTION (OPTIONAL)

There is an alternative technique for verifying such formulas as

$$\sum_{i=1}^{n} i = \frac{n(n + 1)}{2} \quad \text{and} \quad \sum_{i=1}^{n} i^2 = \frac{n(n + 1)(2n + 1)}{6}.$$

Suppose that for each positive integer n, $P(n)$ is a statement. The **Principle of Mathematical Induction** states that if

MATHEMATICAL INDUCTION

and
    (i) $P(1)$ is true

    (ii) the truth of $P(n)$ implies the truth of $P(n+1)$ for each positive integer n,

then $P(n)$ is true for all n.

This principle can be compared with a ladder with an infinite number of rungs. If a person

    (i) can stand on the first rung

and

    (ii) whenever he reaches the nth rung can also reach the $(n+1)$st rung,

then he can climb as high as he wishes.

Induction can also be compared with a string of dominoes where knocking down the first domino results in each successive domino falling in turn.

EXAMPLE 7. Verify by Mathematical Induction that

$$\sum_{i=1}^{n} i = \frac{n(n+1)}{2}.$$

Solution. Let $P(n)$ be the statement

$$\sum_{i=1}^{n} i = \frac{n(n+1)}{2}.$$

(i) $P(1)$ is true because

$$\sum_{i=1}^{1} i = 1 = \frac{1(1+1)}{2}.$$

(ii) Assume that $P(n)$ is true; that is, assume

$$\sum_{i=1}^{n} i = \frac{n(n+1)}{2}.$$

Then
$$\sum_{i=1}^{n+1} i = \sum_{i=1}^{n} i + (n+1)$$

$$= \frac{n(n+1)}{2} + (n+1)$$

$$= (n+1)\left(\frac{n}{2} + 1\right)$$

$$= \frac{(n+1)(n+2)}{2}.$$

CHAPTER 1: AREA AND INTEGRATION

But the equation

$$\sum_{i=1}^{n+1} i = \frac{(n+1)(n+2)}{2}$$

is exactly the statement $P(n+1)$. This shows that if $P(n)$ is true, then $P(n+1)$ is also true. By the Principle of Mathematical Induction, it follows that $P(n)$ must be true for every n; that is, the formula

$$\sum_{i=1}^{n} i = \frac{n(n+1)}{2}$$

must be valid for every positive integer n.

Mathematical Induction is a technique for verifying known formulas, but it does not help to discover new formulas. Techniques such as those given in Section 3 can be used to discover formulas.

PROBLEMS

Use the Principle of Mathematical Induction to verify each of the following formulas.

1. $\sum_{i=0}^{n} r^i = \dfrac{1 - r^{n+1}}{1 - r}$, $r \neq 1$, where $r^0 = 1$

2. $\sum_{i=1}^{n} (2i - 1) = n^2$

3. $\sum_{i=1}^{n} \dfrac{1}{i(i+1)} = \dfrac{n}{n+1}$

4. $\sum_{i=1}^{n} i^2 = \dfrac{n(n+1)(2n+1)}{6}$

5. $\sum_{i=1}^{n} i^3 = \dfrac{n^2(n+1)^2}{4}$

6. $\sum_{i=1}^{n} (2i - 1)2i = \dfrac{n(n+1)(4n-1)}{3}$

7. $\sum_{i=1}^{n} i^4 = \dfrac{n(n+1)(2n+1)(3n^2 + 3n - 1)}{30}$

8. $\dfrac{1}{2} + \dfrac{1}{4} + \dfrac{1}{8} + \cdots + \dfrac{1}{2^n} < 1$

4. SEQUENCES

A **sequence** is a function from the positive integers to the real numbers.

It is customary to describe a sequence by listing its values

$$a(1), \quad a(2), \quad a(3), \quad \ldots$$

or more commonly in the subscript notation

$$a_1, \quad a_2, \quad a_3, \quad \ldots.$$

We will often write $\{a_k\}_{k=1}^{\infty}$ or simply $\{a_k\}$ instead of $a_1, a_2, a_3, \ldots$. The notation $\{a_k\}_{k=1}^{\infty}$ is read "the sequence a-sub-k as k goes from one to infinity." (There is nothing special about the number one, but it is usually a convenient place to start. Occasionally it will be more convenient to start, say, at 5 and consider the sequence $\{a_k\}_{k=5}^{\infty}$.) The phrase "k goes from one to infinity" has no philosophical significance; it simply means that k takes on all possible positive integer values. The symbol a_k is called the kth **term** of the sequence.

EXAMPLE 8. $1, \frac{1}{2}, \frac{1}{3}, \frac{1}{4}, \ldots$ is the sequence whose kth term is $1/k$.

EXAMPLE 9. Let $a_k = 1 - 1/k$ be the kth term of a sequence. Then $a_1 = 0$, $a_2 = \frac{1}{2}$, $a_5 = \frac{4}{5}$, and $a_{13} = \frac{12}{13}$.

EXAMPLE 10. The positive integers $1, 2, 3, 4, \ldots$ form a sequence $\{k\}$ whose kth term is k.

EXAMPLE 11. In the sequence $-1, 1, -1, 1, -1, 1, \ldots$ the kth term is $(-1)^k$.

EXAMPLE 12. The sequence $5, 5, 5, 5, 5, \ldots$ is called a **constant sequence**. The kth term is 5.

EXAMPLE 13. The 10th term of the sequence $2, 3, 5, 7, 11, 13, 17, \ldots$ of primes is 29. What is the 15th term? There is no simple formula for the kth term of this sequence. (A prime is an integer greater than 1 whose only factors are itself and 1.)

We notice that in Example 8 the kth term of the sequence tends to get close to 0 as k becomes large. In Example 9 the kth term tends to 1. The sequences in Examples 10 and 11 do not seem to "settle down" and stay close to any fixed number. We say that the sequence $\{1/k\}$ converges to 0 and the sequence $\{1 - 1/k\}$ converges to 1, but the sequences $\{(-1)^k\}$ and $\{k\}$ do not converge.

In general, the sequence $\{a_n\}$ is said to **converge** to L if a_n is as close to L as desired whenever n is a sufficiently large integer. If there is no such number

L, then the sequence $\{a_n\}$ does not converge and the sequence is said to **diverge**. If the sequence $\{a_n\}$ converges to the number L, we write $\lim_{n \to \infty} a_n = L$. In this case, we also say that a_n tends to the limit L as n tends to infinity.

Notice that the constant sequence in Example 12 converges to 5; that is, $\lim_{n \to \infty} 5 = 5$. (5 is *always* close to 5.)

We leave a precise definition of limit to a later course. A great deal of useful calculus was developed long before Cauchy (1789–1857), Weierstrass (1815–1897), and others gave precise definitions and rigorous proofs that placed calculus on a solid foundation.

Suppose that

$$a_n = 3 - \frac{2}{n} + \frac{5}{n^3} \quad \text{and} \quad b_n = 2 + \frac{1}{n^3}.$$

Then

$$\lim_{n \to \infty} a_n = 3 \quad \text{and} \quad \lim_{n \to \infty} b_n = 2,$$

while

$$\lim_{n \to \infty} (a_n + b_n) = \lim_{n \to \infty} \left[\left(3 - \frac{2}{n} + \frac{5}{n^3} \right) + \left(2 + \frac{1}{n^3} \right) \right]$$

$$= \lim_{n \to \infty} \left(5 - \frac{2}{n} + \frac{6}{n^3} \right)$$

$$= 5$$

$$= 3 + 2$$

$$= \lim_{n \to \infty} a_n + \lim_{n \to \infty} b_n.$$

Thus in this example, at least, the limit of the sum equals the sum of the limits. In fact, we have the following theorem, which is stated without proof.

THEOREM 1. Suppose that the sequences $\{a_n\}$ and $\{b_n\}$ both converge. Then

(i) $\{a_n + b_n\}$ converges and $\lim_{n \to \infty} (a_n + b_n) = \lim_{n \to \infty} a_n + \lim_{n \to \infty} b_n$; that is, the limit of a sum is the sum of the limits.

(ii) $\{c \cdot a_n\}$ converges for each constant real number c, and $\lim_{n \to \infty} (c \cdot a_n) = c \lim_{n \to \infty} a_n$; that is, the limit of a constant multiple of a sequence is the constant multiplied by the limit of the sequence.

(iii) $\{a_n \cdot b_n\}$ converges and $\lim_{n \to \infty} (a_n \cdot b_n) = (\lim_{n \to \infty} a_n)(\lim_{n \to \infty} b_n)$; that is, the limit of a product is the product of the limits.

(iv) If $\lim_{n \to \infty} b_n \neq 0$ and $b_n \neq 0$ $(n = 1, 2, 3, \ldots)$, then $\{a_n/b_n\}$ converges and

$\lim_{n \to \infty} (a_n/b_n) = (\lim_{n \to \infty} a_n)/(\lim_{n \to \infty} b_n)$; that is, the limit of a quotient is the quotient of the limits.

Parts (i) and (ii) of Theorem 1 can be combined to get:

(v) $\lim_{n \to \infty} (c \cdot a_n + d \cdot b_n) = c \lim_{n \to \infty} a_n + d \lim_{n \to \infty} b_n$ for any constant real numbers c and d.

EXAMPLE 14. Test the sequence $\{5 + 1/n^2\}$ for convergence and compute the limit if it exists.

Solution. By part (i) of Theorem 1,

$$\lim_{n \to \infty} \left(5 + \frac{1}{n^2}\right) = \lim_{n \to \infty} 5 + \lim_{n \to \infty} \frac{1}{n^2}$$

if the limits on the right both exist. But $\lim_{n \to \infty} 5 = 5$ and $\lim_{n \to \infty} (1/n^2) = 0$, and hence the sequence does converge and $\lim_{n \to \infty} (5 + 1/n^2) = 5$.

EXAMPLE 15. Test the sequence $\{n + 1/n^2\}$ for convergence.

Solution. Since $n + 1/n^2 > n$, it follows that the sequence $\{n + 1/n^2\}$ increases without bound; hence it cannot settle down and stay close to some fixed number L. Thus the sequence $\{n + 1/n^2\}$ diverges.

EXAMPLE 16. Test the sequence

$$\left\{\frac{3n^3 + 5n^2 - 1}{2n^3 + 7n + 3}\right\}$$

for convergence and compute the limit if it exists.

Solution. Part (iv) cannot be used directly because then we would have something of the form ∞/∞. However, by dividing both numerator and denominator by n^3, we get

$$\lim_{n \to \infty} \frac{3n^3 + 5n^2 - 1}{2n^3 + 7n + 3} = \lim_{n \to \infty} \frac{3 + (5/n) - (1/n^3)}{2 + (7/n^2) + (3/n^3)},$$

and part (iv) of Theorem 1 can now be applied to get

$$\lim_{n \to \infty} \frac{3 + (5/n) - (1/n^3)}{2 + (7/n^2) + (3/n^3)} = \frac{\lim_{n \to \infty} [3 + (5/n) - (1/n^3)]}{\lim_{n \to \infty} [2 + (7/n^2) + (3/n^3)]},$$

provided that all these limits exist. By part (i),

$$\frac{\lim_{n\to\infty}[3+(5/n)-(1/n^3)]}{\lim_{n\to\infty}[2+(7/n^2)+(3/n^3)]} = \frac{\lim_{n\to\infty}3+\lim_{n\to\infty}(5/n)-\lim_{n\to\infty}(1/n^3)}{\lim_{n\to\infty}2+\lim_{n\to\infty}(7/n^2)+\lim_{n\to\infty}(3/n^3)},$$

which equals 3/2, since

$$\lim_{n\to\infty}(5/n) = \lim_{n\to\infty}(1/n^3) = \lim_{n\to\infty}(7/n^2) = \lim_{n\to\infty}(3/n^3) = 0.$$

Thus the sequence converges and

$$\lim_{n\to\infty}\frac{3n^2+5n^2-1}{2n^3+7n+3} = \frac{3}{2}.$$

EXAMPLE 17. Test the sequence

$$\left\{\frac{3n^2+5}{2n} - \frac{3n+1}{2}\right\}$$

for convergence.

Solution. Part (i) states that

$$\lim_{n\to\infty}\left(\frac{3n^2+5}{2n} - \frac{3n+1}{2}\right) = \lim_{n\to\infty}\frac{3n^2+5}{2n} - \lim_{n\to\infty}\frac{3n+1}{2}$$

if both these limits exist. However, $(3n+1)/2 > n$ and hence the sequence $\{(3n+1)/2\}$ diverges, so part (i) cannot be used directly. It is necessary first to simplify the expression

$$\frac{3n^2+5}{2n} - \frac{3n+1}{2}.$$

The common denominator is $2n$, and hence

$$\frac{3n^2+5}{2n} - \frac{3n+1}{2} = \frac{3n^2+5-3n^2-n}{2n}$$

$$= \frac{5-n}{2n}$$

$$= \frac{(5/n)-1}{2}.$$

It follows that

$$\lim_{n\to\infty}\left(\frac{3n^2+5}{2n} - \frac{3n+1}{2}\right) = \lim_{n\to\infty}\frac{(5/n)-1}{2}$$

$$= \frac{\lim_{n \to \infty} (5/n) - 1}{2}$$

$$= -\frac{1}{2}.$$

Thus the sequence converges to $-\frac{1}{2}$.

EXAMPLE 18. Test the sequence

$$\left\{ \frac{1 - \sqrt{n}}{1 + \sqrt{n}} \right\}$$

for convergence.

Solution. Again, part (iv) cannot be used directly, so we first divide numerator and denominator by $\sqrt{n}$ to get

$$\lim_{n \to \infty} \left(\frac{1 - \sqrt{n}}{1 + \sqrt{n}} \right) = \lim_{n \to \infty} \left[\frac{(1/\sqrt{n}) - 1}{(1/\sqrt{n}) + 1} \right].$$

By part (iv) of Theorem 1,

$$\lim_{n \to \infty} \frac{(1/\sqrt{n}) - 1}{(1/\sqrt{n}) + 1} = \frac{\lim_{n \to \infty} [(1/\sqrt{n}) - 1]}{\lim_{n \to \infty} [(1/\sqrt{n}) + 1]}$$

if these limits exist. Using part (i), we have

$$\frac{\lim_{n \to \infty} [(1/\sqrt{n}) - 1]}{\lim_{n \to \infty} [(1/\sqrt{n}) + 1]} = \frac{\lim_{n \to \infty} (1/\sqrt{n}) - \lim_{n \to \infty} 1}{\lim_{n \to \infty} (1/\sqrt{n}) + \lim_{n \to \infty} 1}$$

if all these limits exist. But $\lim_{n \to \infty} 1 = 1$ and $\lim_{n \to \infty} (1/\sqrt{n}) = 0$, and therefore

$$\lim_{n \to \infty} \left(\frac{1 - \sqrt{n}}{1 + \sqrt{n}} \right) = \frac{0 - 1}{0 + 1} = -1.$$

Thus the sequence does converge, and the limit of the sequence is -1.

PROBLEMS

Test the following sequences for convergence.

1. $\left\{ (-1)^n \cdot \frac{1}{n} \right\}_{n=1}^{\infty}$

2. $1 + \frac{1}{1}, 2 + \frac{1}{2}, 3 + \frac{1}{3}, \ldots, n + \frac{1}{n}, \ldots$

3. $\{a_n\}_{n=1}^\infty$, where $a_n = 3 + \dfrac{1}{n} + \dfrac{2}{n^2}$

4. $\left\{\dfrac{n^2+1}{n^2-1}\right\}_{n=2}^\infty$

5. $\left\{n^2 - \left(n + \dfrac{1}{n}\right)^2\right\}_{n=1}^\infty$

6. $\{5 + (-1)^n\}_{n=1}^\infty$

7. $\left\{\dfrac{\sqrt{n^2+1}}{n}\right\}$

8. $\left\{\dfrac{5n^3 + 4n + 2}{8n^3 + 6n^2 + 2n}\right\}$

[Hint: Note that

$$\dfrac{\sqrt{n^2+1}}{n} = \sqrt{\dfrac{n^2+1}{n^2}}.$$

]

[Hint: Divide numerator and denominator by n^3.]

9. $\left\{\dfrac{2n^2 + 3n}{n^3 + n + 1}\right\}$

10. $\left\{4 + \dfrac{n^3+1}{n^2}\right\}$

11. $\{\sqrt{n^2+1} - n\}$

12. $\{\sqrt{n^2+n} - n^2\}$

[Hint: Multiply and divide by $\sqrt{n^2+1} + n$.]

13. $\left\{\dfrac{(2n^2+1)(n+2)}{n^3}\right\}$

14. $\left\{\dfrac{n^4}{(3n+1)(n+2)(n-3)}\right\}_{n=4}^\infty$

15. $\left\{\dfrac{n^3+2}{3n^3} + \dfrac{n^2}{(n+1)(n-1.5)}\right\}$

16. $\{n^4 + 2n^2 - (n^2+1)^2\}$

5. THE EXISTENCE OF THE INTEGRAL OF A CONTINUOUS FUNCTION

Let us again consider the problem of finding the area under the graph of the function $y = f(x)$ from $x = a$ to $x = b$. [We will assume that $f(x) \geq 0$ for all x between a and b, since we want area.] As before, subdivide the interval $[a, b]$ into n equal subintervals $[x_{k-1}, x_k]$ ($k = 1, 2, 3, \ldots, n$), where

$$x_k = a + \dfrac{k(b-a)}{n}.$$

Let $\bar{x}_k$ be *any* point in the kth subinterval, as in Figure 1.14. The sum

$$\sum_{k=1}^n f(\bar{x}_k) \dfrac{b-a}{n}$$

is called a **Riemann sum**, in honor of the German mathematician Georg Bernard Riemann (1826–1866). Such a sum, it seems, should tend to a limit as n tends to infinity for any choice of $\bar{x}_k$ in the interval $[x_{k-1}, x_k]$. In this section we see that this fact is indeed true for many functions f, the so-called continuous functions.

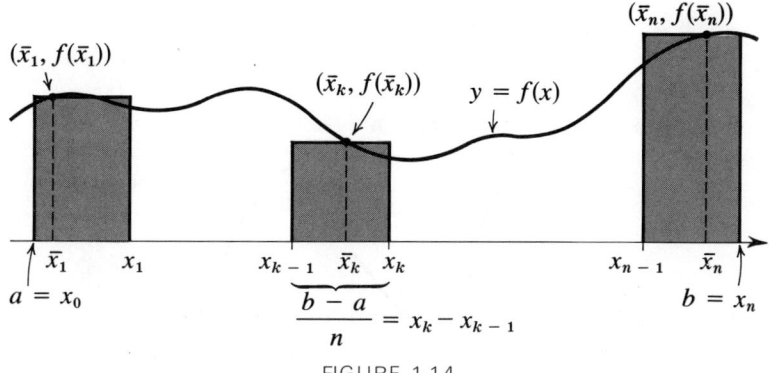

FIGURE 1.14

A function f from the interval $[a, b]$ into the real numbers is said to be **continuous** on $[a, b]$ if $|f(x_2) - f(x_1)|$ is as small as desired whenever $|x_2 - x_1|$ is sufficiently small, where x_1 and x_2 are in the interval $[a, b]$. (In mathematical literature, such functions are also called **uniformly continuous** on $[a, b]$.) Since $|c - d|$ is the distance between c and d, the preceding definition states that $f(x_2)$ and $f(x_1)$ must be close together whenever x_1 and x_2 are close together.

Roughly speaking, a function f is continuous on $[a, b]$ if the graph of f can be sketched on a blackboard without lifting the chalk from the board. Virtually every elementary function is continuous on *some* interval, so it is natural to look at such functions and see how they behave. The elementary functions include all the familiar functions, such as polynomial functions, rational functions, algebraic functions, logarithmic functions, exponential functions, and trigonometric functions.

The function f in Figure 1.15 is continuous on $[a, b]$ because its graph can be drawn in one "continuous" motion. The function g in Figure 1.16 is not

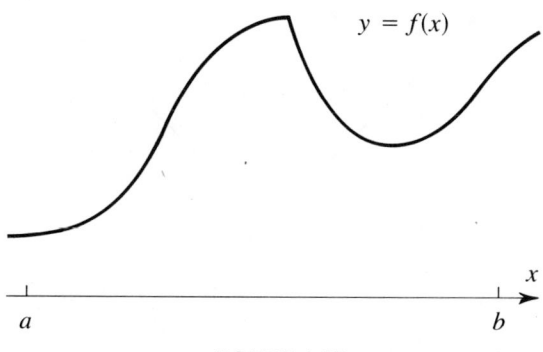

FIGURE 1.15

38   CHAPTER 1: AREA AND INTEGRATION

continuous on $[a, b]$ because of the "jumps" at c and d. To sketch this graph, it would be necessary to lift the chalk off the blackboard above c and again above d.

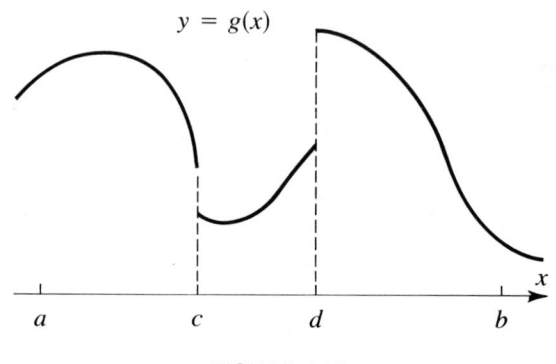

FIGURE 1.16

EXAMPLE 19.   Is the function f defined by

$$f(x) = \begin{cases} x & \text{if } x > 1 \\ 1 - x & \text{if } x \leq 1 \end{cases}$$

continuous on the interval $[-1, 5]$?

Solution.   The graph of f is given in Figure 1.17. Notice that $f(1) = 0$ but $f(x) = x$ for $x > 1$; that is, $f(x) > 1$ for $x > 1$. Hence, for x close to 1, it does not follow that $f(x)$ is close to $f(1) = 0$, so f is not continuous on $[-1, 5]$. There is a "jump" at $x = 1$ and the graph cannot be sketched completely without lifting the chalk from the blackboard.

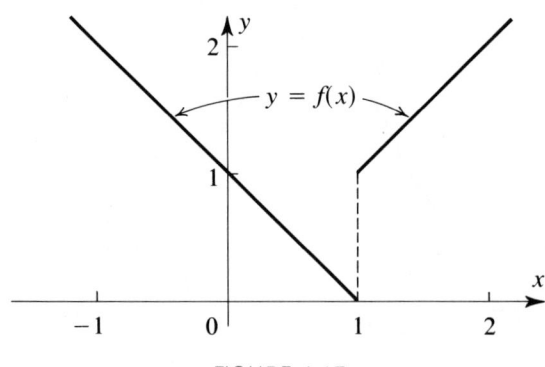

FIGURE 1.17

SECTION 5: THE INTEGRAL OF A CONTINUOUS FUNCTION

EXAMPLE 20. A number of functions cannot be expressed in terms of a simple formula. For example, Federal Income Tax in the United States for a married couple with no children and who itemize deductions for the calendar year 1973 is given in Table 1.1. Assume that Taxable Income I can be any number between 0 and 1,000,000 (not rounded to nearest dollars or to nearest pennies), and let $T(I)$ denote tax due under Table 1.1. Then T is a continuous function. This means that if a married couple has a small amount of additional income, the couple must pay a small amount of additional tax. The graph of T is given in Figure 1.18 for I between 0 and 20,000.

Table 1.1. Married Taxpayers Filing Joint Returns
(all amounts in dollars)

Over	But Not Over	$T(I)$	Of Excess Over
0	1,000	$0 + 14\%$	0
1,000	2,000	$140 + 15\%$	1,000
2,000	3,000	$290 + 16\%$	2,000
3,000	4,000	$450 + 17\%$	3,000
4,000	8,000	$620 + 19\%$	4,000
8,000	12,000	$1,380 + 22\%$	8,000
12,000	16,000	$2,260 + 25\%$	12,000
16,000	20,000	$3,260 + 28\%$	16,000
20,000	24,000	$4,380 + 32\%$	20,000
24,000	28,000	$5,660 + 36\%$	24,000
28,000	32,000	$7,100 + 39\%$	28,000
32,000	36,000	$8,660 + 42\%$	32,000
36,000	40,000	$10,340 + 45\%$	36,000
40,000	44,000	$12,140 + 48\%$	40,000
44,000	52,000	$14,060 + 50\%$	44,000
52,000	64,000	$18,060 + 53\%$	52,000
64,000	76,000	$24,420 + 55\%$	64,000
76,000	88,000	$31,020 + 58\%$	76,000
88,000	100,000	$37,980 + 60\%$	88,000
100,000	120,000	$45,180 + 62\%$	100,000
120,000	140,000	$57,580 + 64\%$	120,000
140,000	160,000	$70,380 + 66\%$	140,000
160,000	180,000	$83,580 + 68\%$	160,000
180,000	200,000	$97,180 + 69\%$	180,000
200,000	∞	$110,980 + 70\%$	200,000

We now give some concrete examples to show how the definition can be used to prove that a given function is or is not continuous.

EXAMPLE 21. Let $f(x) = x^2$, $a = 0$, and $b = 2$. Then for $0 \le x_1 \le 2$ and $0 \le x_2 \le 2$,

$$|f(x_2) - f(x_1)| = |x_2^2 - x_1^2|$$

$$
\begin{aligned}
&= |(x_2 - x_1)(x_2 + x_1)| \\
&= |x_2 - x_1| \cdot |x_2 + x_1| \\
&\leq |x_2 - x_1| \cdot (|x_2| + |x_1|) \\
&\leq |x_2 - x_1| \cdot (2 + 2) \\
&= 4|x_2 - x_1|,
\end{aligned}
$$

which can be made as small as desired by requiring $|x_2 - x_1|$ to be small. For example, $|f(x_2) - f(x_1)|$ can be made less than $1/100$ by requiring $|x_2 - x_1|$ to be less than $1/400$. Thus f is continuous on $[0, 2]$.

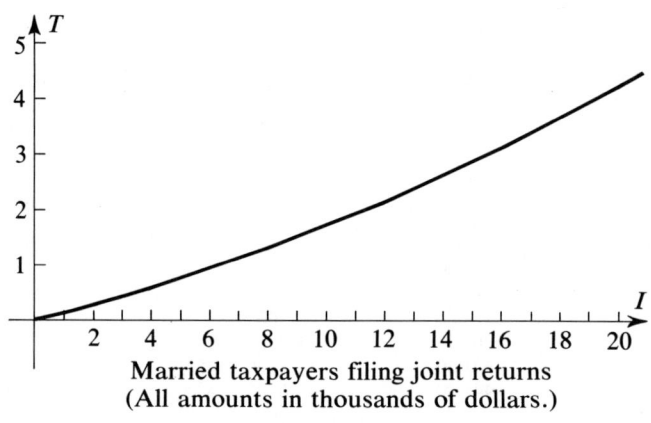

Married taxpayers filing joint returns
(All amounts in thousands of dollars.)

FIGURE 1.18

EXAMPLE 22. Let $f(x) = \sqrt{x}$, $a = 1$, and $b = 4$. Then for $1 \leq x_1 \leq 4$ and $1 \leq x_2 \leq 4$,

$$
\begin{aligned}
|f(x_2) - f(x_1)| &= |\sqrt{x_2} - \sqrt{x_1}| \\
&= |\sqrt{x_2} - \sqrt{x_1}| \cdot \frac{\sqrt{x_2} + \sqrt{x_1}}{\sqrt{x_2} + \sqrt{x_1}} \\
&= \frac{|x_2 - x_1|}{\sqrt{x_2} + \sqrt{x_1}} \\
&\leq \frac{|x_2 - x_1|}{1 + 1} \\
&= \frac{1}{2}|x_2 - x_1|.
\end{aligned}
$$

Hence $|f(x_2) - f(x_1)|$ can be made as small as desired by requiring $|x_2 - x_1|$ to be sufficiently small. In order to ensure that $|f(x_2) - f(x_1)| < 10^{-3}$, for example, it is sufficient to require that $|x_2 - x_1| < 2(10^{-3})$. The function $f(x) = \sqrt{x}$ is thus continuous on $[1, 4]$.

We have already stated that every elementary function is continuous on *some* interval $[a, b]$. Following, however, is an example of a function that is not continuous on $[-1, 1]$.

EXAMPLE 23. Let $f(x) = 1/x$, $a = -1$, and $b = 1$, as in Figure 1.19. Then f is not defined at zero, so f is not even a function from $[-1, 1]$ into the real numbers. The other part of the definition also fails. For if $x_1 < 0 < x_2$, then $|f(x_2) - f(x_1)| \geq 2$, no matter how small $|x_2 - x_1|$ is chosen.

The following basic existence theorem is stated without proof.

THEOREM 2. Let f be continuous on the interval $[a, b]$. Subdivide the interval $[a, b]$ into n subintervals, each of length $(b - a)/n$, and let $\bar{x}_k$ be any point in the kth subinterval. Then

(11) $$\lim_{n \to \infty} \sum_{k=1}^{n} f(\bar{x}_k) \frac{b - a}{n}$$

exists independently of how the points $\bar{x}_k$ are chosen.

The foregoing limit is usually denoted $\int_a^b f(x)\,dx$, or simply $\int_a^b f$, and is called the **Riemann integral** or the **definite integral** of f from a to b. The integral sign $\int$ was first used by Gottfried Wilhelm Leibnitz (1646–1716) and stands for *summa*, the Latin word for "sum." Note, however, that an integral is *not* a sum; it is the limit of a sum. The meaning of the symbol dx will be discussed in a later section. Roughly speaking, the symbols $\lim_{n \to \infty} \sum_{k=1}^{n}$ are replaced by $\int_a^b$, $f(\bar{x}_k)$ is replaced by $f(x)$, which can be thought of as the height of a rectangle, and $(b - a)/n$ is replaced by dx, the "change in x," which can be thought of as the width of a rectangle. If $f(x) \geq 0$ whenever $a \leq x \leq b$, then

$$\int_a^b f(x)\,dx = \lim_{n \to \infty} \sum_{k=1}^{n} f(\bar{x}_k) \frac{b - a}{n}$$

can be interpreted as the *area* of the region bounded by $x = a$, $x = b$, $y = 0$, and $y = f(x)$.

EXAMPLE 24. Compute $\int_0^1 x^2\,dx$.

42   CHAPTER 1: AREA AND INTEGRATION

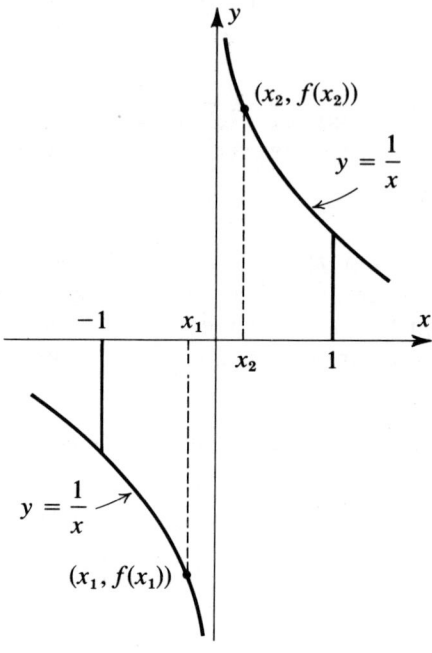

FIGURE 1.19

Solution. Since $\int_0^1 x^2\, dx$ can be interpreted as the area of the region bounded by $y = x^2$, $y = 0$, $x = 0$, and $x = 1$, we know by Example 1 of this chapter that this integral equals $\frac{1}{3}$.

EXAMPLE 25. Approximate $\int_1^2 1/x\, dx$.

Solution. Subdivide the interval $[1, 2]$ into four equal subintervals and let $\bar{x}_k$ be the midpoint of its subinterval, as in Figure 1.20. Then with $f(x) = 1/x$, we get

$$\sum_{k=1}^{4} f(\bar{x}_k) \frac{b-a}{4} = \left[\frac{1}{9/8} + \frac{1}{11/8} + \frac{1}{13/8} + \frac{1}{15/8}\right]\frac{1}{4}$$

$$= \frac{2}{9} + \frac{2}{11} + \frac{2}{13} + \frac{2}{15}$$

$$\approx 0.222 + 0.182 + 0.154 + 0.133$$

$$= 0.691.$$

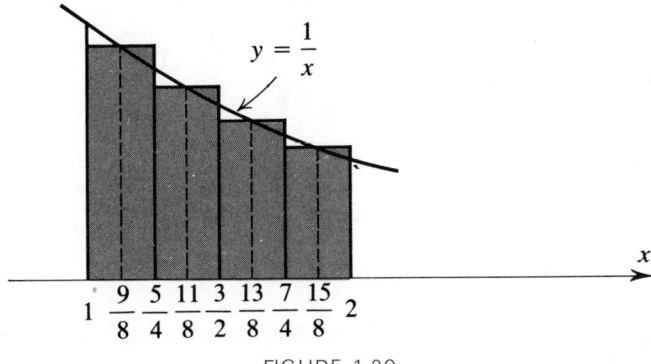

FIGURE 1.20

When we use ten equal subintervals, this method yields the approximation 0.6928. By subdividing the interval $[0, 1]$ into 100 equal subdivisions, we would get the approximation

$$\int_1^2 \frac{1}{x} dx \approx 0.69315.$$

Recall that if $a < b$ and $f(x) \geq 0$ for each x between a and b, then $\int_a^b f$ can be interpreted as the area of the region bounded by the lines $x = a$, $x = b$, and $y = 0$, and the graph of $y = f(x)$. However, the integral $\int_a^b f$ makes sense even when $f(x) < 0$ for some values of x. In Figure 1.21

$$\int_a^b f = \lim_{n \to \infty} \sum_{k=1}^n f(\bar{x}_k) \frac{b-a}{n}.$$

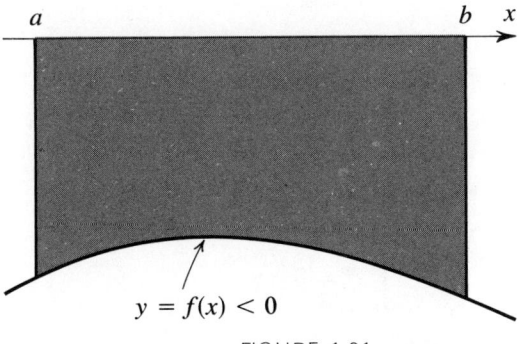

FIGURE 1.21

Since $f(x) < 0$ for all x between a and b, the area of a typical rectangle is equal to $-f(\bar{x}_k)(b-a)/n$. The total area is thus equal to

$$\lim_{n\to\infty} \sum_{k=1}^{n} [-f(\bar{x}_k)] \frac{b-a}{n} = -\lim_{n\to\infty} \sum_{k=1}^{n} f(\bar{x}_k) \frac{b-a}{n}$$

$$= -\int_a^b f(x)\,dx.$$

It follows that the integral $\int_a^b f$ is equal to the *negative* of the area of the shaded region.

In Figure 1.22 $\int_{-1}^{1} x^3\,dx = 0$. (The integral is the negative of the area from -1 to 0, plus the area from 0 to 1. These areas are equal, and hence the integral is zero.) In general, if $f(x)$ is positive for some x and negative for some x, then $\int_a^b f$ represents the difference between the area above the x-axis and the area below the x-axis.

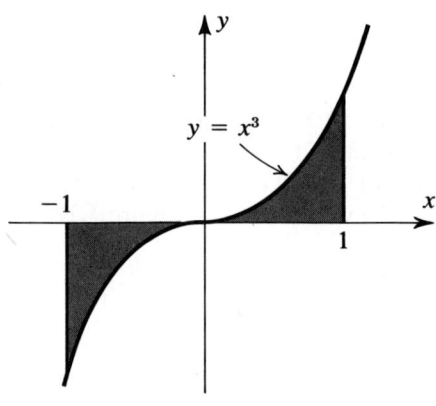

FIGURE 1.22

If $a > b$, define

(12) $$\int_a^b f = -\int_b^a f.$$

Also, define

(13) $$\int_a^a f = 0$$

if f is any function that is defined at a.

SECTION 5: THE INTEGRAL OF A CONTINUOUS FUNCTION

EXAMPLE 26. Compute $\int_2^1 x^2\, dx$.

Solution.

$$\int_2^1 x^2\, dx = -\int_1^2 x^2\, dx$$

$$= -\lim_{n\to\infty} \sum_{k=1}^{n} \left(1 + \frac{k}{n}\right)^2 \frac{1}{n}$$

$$= -\lim_{n\to\infty} \frac{1}{n} \sum_{k=1}^{n} \left(1 + \frac{2k}{n} + \frac{k^2}{n^2}\right)$$

$$= -\lim_{n\to\infty} \frac{1}{n} \left(n + \frac{2}{n}\sum_{k=1}^{n} k + \frac{1}{n^2}\sum_{k=1}^{n} k^2\right)$$

$$= -\lim_{n\to\infty} \frac{1}{n}\left[n + \frac{2}{n}\frac{n(n+1)}{2} + \frac{1}{n^2}\frac{n(n+1)(2n+1)}{6}\right]$$

$$= -\lim_{n\to\infty}\left(1 + \frac{n+1}{n} + \frac{n+1}{n}\cdot\frac{2n+1}{2n}\cdot\frac{1}{3}\right)$$

$$= -\left(1 + 1 + \frac{1}{3}\right)$$

$$= -\frac{7}{3}.$$

The x in $\int_a^b f(x)\, dx$ is really a "dummy variable," since it can be replaced by any other convenient symbol:

$$\int_a^b f(x)\, dx = \int_a^b f(t)\, dt = \int_a^b f(z)\, dz = \int_a^b f.$$

PROBLEMS

Compute the following integrals.

1. $\int_1^3 2x^2\, dx$

2. $\int_{-2}^{-1} 3x\, dx$

3. $\int_3^2 (x+2)\, dx$

4. $\int_4^7 x^3\, dx$

5. $\int_5^5 x^5\, dx$

6. $\int_{-1}^{0} (x+x^2)\, dx$

7. $\int_{-1}^{1} (x+x^3)\, dx$

8. $\int_a^b 1\, dx$

9. $\int_a^b x\,dx$

10. $\int_a^b x^2\,dx$

11. $\int_a^b x^3\,dx$

12. $\int_a^b (3x^2 + 2x)\,dx$

Rewrite the following limits as definite integrals but *do not evaluate*.

13. $\lim\limits_{n\to\infty} \sum\limits_{k=1}^{n} 4\left(1+k\dfrac{3}{n}\right)^3 \dfrac{3}{n}$ [*Hint:* Let $a = 1$ and $b = 4$.]

14. $\lim\limits_{n\to\infty} \sum\limits_{k=1}^{n} \dfrac{5/n}{4+k(5/n)}$ [*Hint:* Let $a = 4$ and $b = 9$.]

15. $\lim\limits_{n\to\infty} \sum\limits_{k=1}^{n} 5\sqrt{2+k\dfrac{2}{n}}\dfrac{2}{n}$

16. $\lim\limits_{n\to\infty} \sum\limits_{k=1}^{n} \left[\left(a+k\dfrac{1}{n}\right) + \sqrt{a+k\dfrac{1}{n}}\right]\dfrac{1}{n}$

6. THE INTEGRAL OF A SUM

In this section we consider two related problems. The first problem is to find the area under the graph of $y = x - x^2$, and the second is to find the area between the graphs of $y = x$ and $y = x^2$. The regions are sketched in Figure 1.23. These regions are not congruent (one region cannot be placed on the other in such a way that no edges overlap). However, it will be shown that these regions do have equal areas. This fact will have a useful interpretation in terms of integrals.

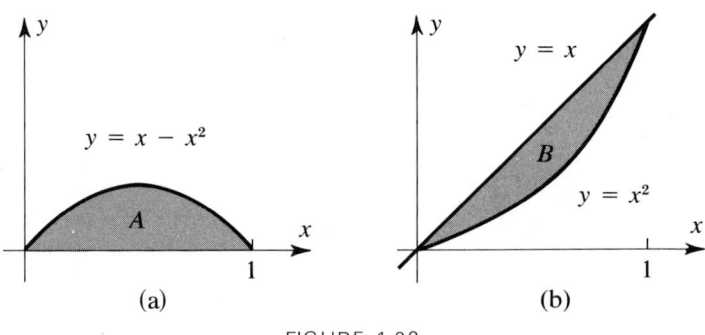

FIGURE 1.23

First, we subdivide the interval $[0, 1]$ into the n subintervals $[0, 1/n]$, $[1/n, 2/n]$, $[2/n, 3/n]$, ..., $[(n-1)/n, 1]$. We approximate the area of Region A by rectangles as in Figure 1.24. Notice that this time each rectangle is determined

SECTION 6: THE INTEGRAL OF A SUM

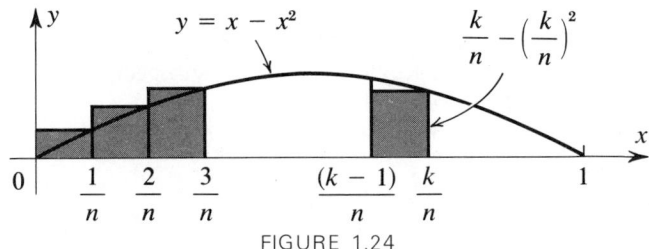

FIGURE 1.24

by the value of the function at the right-hand endpoint of the subinterval. The area of the rectangle with base on the interval $[(k-1)/n, k/n]$ is equal to

$$\frac{1}{n}\left[\frac{k}{n} - \left(\frac{k}{n}\right)^2\right] = \frac{k}{n^2} - \frac{k^2}{n^3}.$$

Hence the area of Region A is approximately equal to

$$\text{Area } A_n = \left(\frac{1}{n^2} - \frac{1^2}{n^3}\right) + \left(\frac{2}{n^2} - \frac{2^2}{n^3}\right) + \cdots + \left(\frac{n}{n^2} - \frac{n^2}{n^3}\right)$$

$$= \left(\frac{1}{n^2} + \frac{2}{n^2} + \cdots + \frac{n}{n^2}\right) - \left(\frac{1^2}{n^3} + \frac{2^2}{n^3} + \cdots + \frac{n^2}{n^3}\right)$$

$$= \frac{1}{n^2}(1 + 2 + \cdots + n) - \frac{1}{n^3}(1^2 + 2^2 + \cdots + n^2)$$

$$= \frac{1}{n^2} \cdot \frac{n(n+1)}{2} - \frac{1}{n^3} \cdot \frac{n(n+1)(2n+1)}{6}$$

$$= \frac{1}{2} \cdot \frac{n+1}{n} - \frac{1}{6} \cdot \frac{n+1}{n} \cdot \frac{2n+1}{n}$$

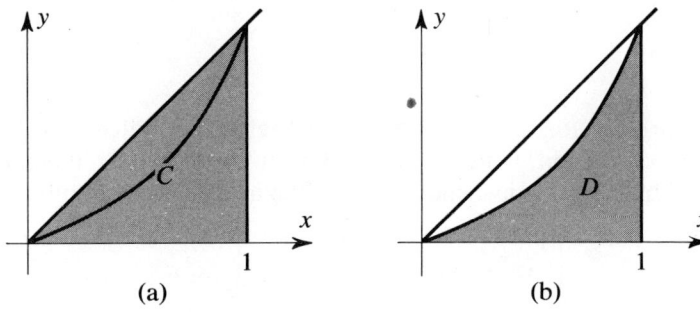

FIGURE 1.25

$$= \frac{1}{2}\left(1 + \frac{1}{n}\right) - \frac{1}{6}\left(1 + \frac{1}{n}\right)\left(2 + \frac{1}{n}\right),$$

which tends to $\frac{1}{2} - \frac{1}{6} \cdot 2 = \frac{1}{2} - \frac{1}{3} = \frac{1}{6}$ as n tends to infinity.

To compute the area of Region B, we can subtract the area of Region D from the area of Region C in Figure 1.25.

From Section 1 we know that Area $D = \frac{1}{3}$ and Area $C = \frac{1}{2}$, and hence

$$\text{Area } B = \text{Area } C - \text{Area } D$$
$$= \frac{1}{2} - \frac{1}{3}$$
$$= \frac{1}{6}.$$

We have shown that the area under the graph of $y = x - x^2$ equals the area between the graphs of $y = x$ and $y = x^2$. The reason why these regions have equal areas is simple. The area of each region is approximated by a sum of areas of rectangles. The kth rectangle in each case has area equal to

$$\frac{1}{n}\left[\frac{k}{n} - \left(\frac{k}{n}\right)^2\right],$$

as in Figure 1.26.

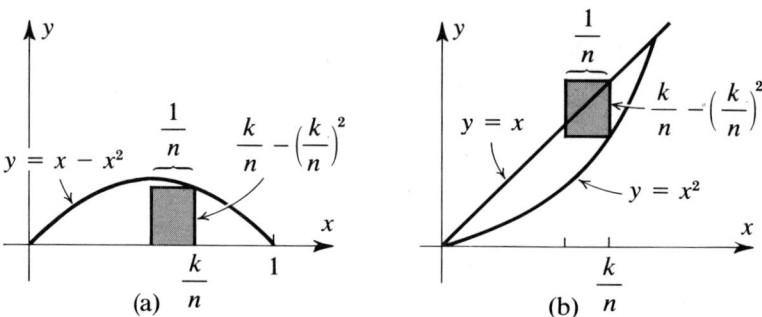

FIGURE 1.26

The area of Region A in Figure 1.23 equals the integral $\int_0^1 (x - x^2)\,dx$, and the area of Region B in Figure 1.23 is a difference of integrals $\int_0^1 x\,dx - \int_0^1 x^2\,dx$. What we have just shown is that the areas are equal and hence that

$$\int_0^1 (x - x^2)\,dx = \int_0^1 x\,dx - \int_0^1 x^2\,dx.$$

This is a special case of the following theorem.

SECTION 6: THE INTEGRAL OF A SUM   49

THEOREM 3. If f and g are continuous on $[a, b]$ and A, B are any real numbers, then

(14) $$\int_a^b [Af(x) \pm Bg(x)] \, dx = A \int_a^b f(x) \, dx \pm B \int_a^b g(x) \, dx.$$

Sketch of proof. Subdivide the interval $[a, b]$ into n subintervals, each of length $(b - a)/n$. Consider the area of the kth rectangle in each of the three figures 1.27, 1.28, and 1.29. The area of the kth rectangle in Figure 1.29 is

$$[f(x_k) + g(x_k)] \frac{b - a}{n} = f(x_k) \frac{b - a}{n} + g(x_k) \frac{b - a}{n},$$

which equals the sum of the areas of the kth rectangles in Figures 1.27 and 1.28. It follows that

$$\int_a^b (f + g) = \int_a^b f + \int_a^b g.$$

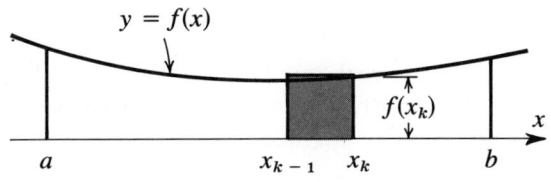

FIGURE 1.27

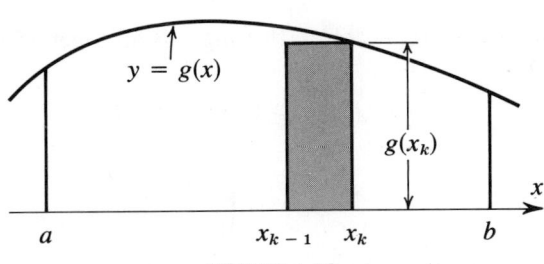

FIGURE 1.28

50 CHAPTER 1: AREA AND INTEGRATION

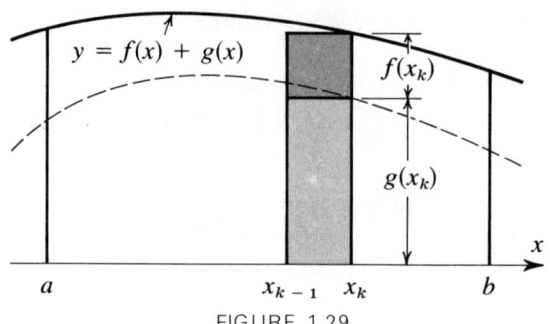

FIGURE 1.29

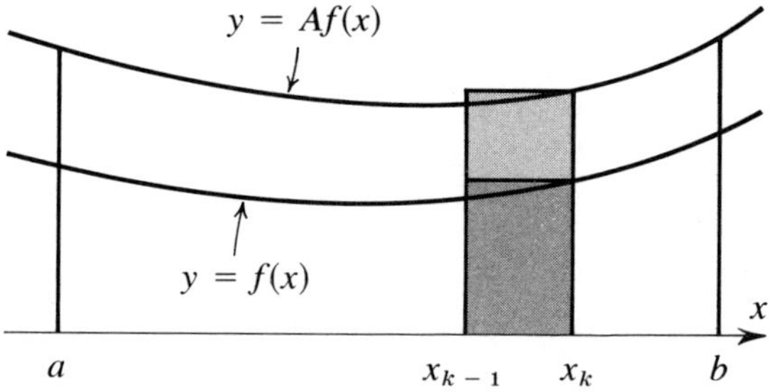

FIGURE 1.30

Now let A be some real number. In order to compute the integral $\int_a^b Af$, the interval $[a, b]$ is subdivided into n equal parts, as in Figure 1.30. The area of the kth rectangle equals

$$[Af(x_k)]\frac{b-a}{n} = A\left[f(x_k)\frac{b-a}{n}\right].$$

But this result is A times the area of the kth rectangle in Figure 1.27. Thus

$$\int_a^b Af = A\int_a^b f.$$

By the same reasoning, $\int_a^b Bg = B\int_a^b g$. Putting this all together gives

$$\int_a^b (Af + Bg) = \int_a^b Af + \int_a^b Bg$$
$$= A\int_a^b f + B\int_a^b g.$$

The fact that $+$ can be replaced by $-$ will be left to the reader.

SECTION 6: THE INTEGRAL OF A SUM

Repeated use of Theorem 3 easily shows that

$$\int_a^b (f_1 + f_2 + f_3 + \cdots + f_n) = \int_a^b f_1 + \int_a^b f_2 + \int_a^b f_3 + \cdots + \int_a^b f_n.$$

This last theorem is also useful in computing integrals for large classes of functions once and for all. In particular, the following example shows how to integrate an arbitrary polynomial function of degree 2.

EXAMPLE 27. Compute $\int_a^b (Ax^2 + Bx + C)\, dx$.

Solution. By Theorem 3,

$$\int_a^b (Ax^2 + Bx + C)\, dx = A\int_a^b x^2\, dx + B\int_a^b x\, dx + C\int_a^b 1\, dx.$$

Now

$$\int_a^b x^2\, dx = \lim_{n\to\infty} \sum_{k=1}^n \left(a + k\frac{b-a}{n}\right)^2 \frac{b-a}{n}$$

$$= \lim_{n\to\infty} \left[\frac{a^2(b-a)}{n}\sum_{k=1}^n 1 + \frac{2a(b-a)^2}{n^2}\sum_{k=1}^n k + \frac{(b-a)^3}{n^3}\sum_{k=1}^n k^2\right]$$

$$= \lim_{n\to\infty} \left[\frac{a^2(b-a)}{n}\, n + \frac{2a(b-a)^2}{n^2}\frac{n(n+1)}{2} + \frac{(b-a)^3}{n^3}\frac{n(n+1)(2n+1)}{6}\right]$$

$$= a^2(b-a) + a(b-a)^2 + \frac{1}{3}(b-a)^3$$

$$= a^2 b - a^3 + ab^2 - 2a^2 b + a^3 + \frac{1}{3}b^3 - b^2 a + ba^2 - \frac{1}{3}a^3$$

$$= \frac{1}{3}(b^3 - a^3).$$

Similarly,

$$\int_a^b x\, dx = \frac{1}{2}(b^2 - a^2)$$

and

$$\int_a^b dx = b - a.$$

Putting this all together, we have

(15) $$\int_a^b (Ax^2 + Bx + C)\, dx = \frac{A}{3}(b^3 - a^3) + \frac{B}{2}(b^2 - a^2) + C(b - a).$$

52 CHAPTER 1: AREA AND INTEGRATION

Notice that this formula does take care of *all* possible integrals of *all* possible quadratic functions and (letting $A = 0$) of all possible linear functions.

EXAMPLE 28. Compute $\int_0^2 (5x^2 + 4x + 1)\,dx$.

Solution. From Example 27,

$$\int_0^2 (5x^2 + 4x + 1)\,dx = 5\int_0^2 x^2\,dx + 4\int_0^2 x\,dx + \int_0^2 dx$$

$$= \frac{5}{3}(2^3 - 0^3) + \frac{4}{2}(2^2 - 0^2) + (2 - 0)$$

$$= \frac{70}{3}$$

$$\approx 23.33.$$

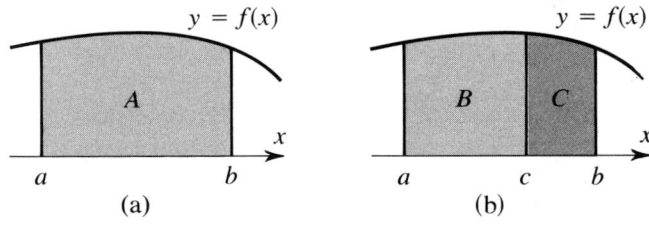

FIGURE 1.31

Consider Region A of Figure 1.31. If c is some number between a and b, then the area of Region A is equal to the area of Region B plus the area of Region C. Restated in terms of integrals, this is the statement

(16) $$\int_a^b f = \int_a^c f + \int_c^b f.$$

This formula is valid for any function f that is continuous on $[a, b]$ and not just for nonnegative valued functions. In fact, c need not lie between a and b, for if f is continuous on any interval that contains the three points a, b, and c, then

$$\int_a^b f = \int_a^c f + \int_c^b f.$$

Notice that this formula is consistent with the definitions $\int_a^a f = 0$ and $\int_a^b f = -\int_b^a f$ because $\int_a^b f = \int_a^a f + \int_a^b f$ and $\int_a^b f = \int_a^b f + \int_b^a f$.

SECTION 6: THE INTEGRAL OF A SUM

If f is a function, then we can define a new function F by

$$F(x) = \int_a^x f(t)\,dt.$$

For example, if $f(t) = 2t + 1$ and $a = 0$, then

$$F(x) = \int_0^x (2t + 1)\,dt$$

$$= \frac{2}{2}(x^2 - 0^2) + (x - 0)$$

$$= x^2 + x.$$

Also,

$$F(x + h) = \int_0^{x+h} f(t)\,dt$$

$$= \int_0^x f(t)\,dt + \int_x^{x+h} f(t)\,dt$$

$$= F(x) + \int_x^{x+h} f(t)\,dt,$$

so that

$$\int_x^{x+h} f(t)\,dt = F(x + h) - F(x).$$

PROBLEMS

Compute the following integrals. [Use Formula (15) where possible.]

1. $\int_0^2 (3x^2 + 4x + 5)\,dx$

2. $\int_{-1}^x (5t + 1)\,dt$

3. $\int_1^{3t} (x + 3x^2)\,dx$

4. $\int_0^1 (x^3 + x)\,dx$

5. $\int_1^{4t+1} (3x^2 + 2x)\,dx$

6. $\int_{4b}^{7a} (3x^2 + 2x)\,dx$

7. $\int_1^7 (3x^2 + x)\,dx$

8. $\int_{-2}^2 (x^2 + 2)\,dx$

9. Suppose that the function f is defined by $f(t) = \int_0^t (3x^2 + 5x + 1)\,dx$. Compute $f(t)$.

10. Compute $g(z)$ if $g(z) = \int_z^{z^2} (x + 5x^2)\,dx$.

11. Draw some appropriate pictures and argue that Theorem 3 is valid if $\pm$ is interpreted as $-$.

54   CHAPTER 1: AREA AND INTEGRATION

12. An embryo grows at a rate of approximately $0.005t^2 + 0.01t + 0.2$ pound per week during a certain stage of development, where t is time in weeks. If the embryo weighs 2 pounds at $t = 10$ weeks, how much does it weigh at $t = 14$ weeks? [*Hint:* During the eleventh week, the weight gain is approximately $0.005(11)^2 + 0.01(11) + 0.2$; and during the twelfth week, the gain is approximately $0.005(12)^2 + 0.01(12) + 0.2$. Altogether, the weight gain is roughly $\sum_{k=11}^{14} (0.005k^2 + 0.01k + 0.2)$, which is a Riemann sum for the integral $\int_{10}^{14} (0.005t^2 + 0.01t + 0.2)\, dt$. Now use (15) and add 2 pounds to the answer.]

13. During a 15-week term, a student works calculus problems at the rate of $t^2 - 12t + 36$ problems per week during week number t. (By looking at the graph of the function $t \to t^2 - 12t + 36$, you may suspect that the student starts out enthusiastically, gets interested in girls partway through the term, then starts worrying about the final examination.) How many problems does the student work altogether during the term?

REVIEW SECTION

EXAMPLE 29.   A section of land (one mile by one mile) is bounded on one side by a lake with a straight shoreline. Shoreline property is worth $8000 per acre, while land a mile away is only worth $1000. In fact, land x miles from the lake is worth $8000 - 7000x$ dollars per acre, where x is any number between 0 and 1. What is the total value of this section of land?

Solution.   Divide the land into a large number of thin strips, each parallel to the shoreline and of width $1/n$, as in Figure 1.32. The value of the kth strip

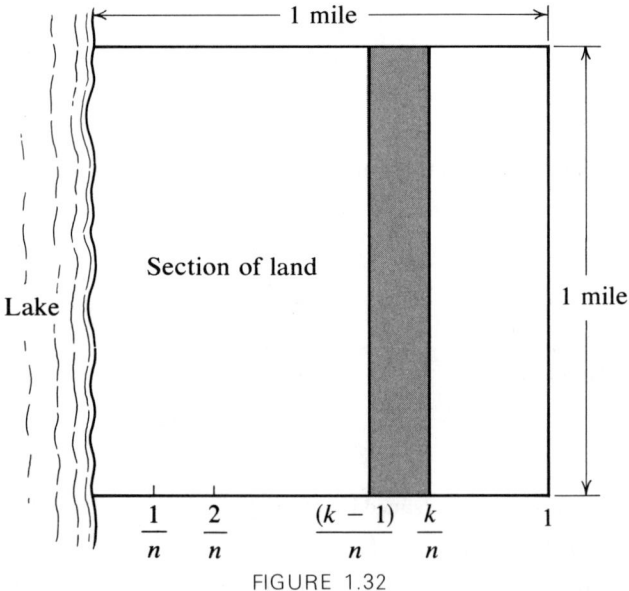

FIGURE 1.32

of land is approximately equal to $(8000 - 7000k/n)$ dollars per acre, and there are 640 acres in one square mile. It follows that the total value of the kth strip is approximately equal to $(8000 - 7000k/n)(640/n)$ dollars. Thus the section of land has value approximately equal to

$$\sum_{k=1}^{n}\left(8000 - 7000\frac{k}{n}\right)\left(\frac{640}{n}\right) = 5{,}120{,}000 \frac{1}{n}\sum_{k=1}^{n} 1 - 4{,}480{,}000 \frac{1}{n^2}\sum_{k=1}^{n} k$$

$$= 5{,}120{,}000 - 4{,}480{,}000 \frac{1}{n^2}\frac{n(n+1)}{2}$$

$$= 5{,}120{,}000 - 2{,}240{,}000\left(1 + \frac{1}{n}\right)$$

dollars. By taking the limit as n tends to infinity, we see that the land is worth

$$\$5{,}120{,}000 - \$2{,}240{,}000 = \$2{,}880{,}000.$$

We can argue intuitively that this is the right answer. If lakeshore property is worth \$8000 per acre and land a mile away is worth \$1000 per acre, then the average value of the land should be $(8000 + 1000)/2 = 4500$ dollars per acre. Multiplying, we get

$$4500 \text{ dollars/acre} \times 640 \text{ acres} = \$2{,}880{,}000.$$

EXAMPLE 30. If the cross-sectional area of a length of copper tubing is A and the *velocity* (speed in a given direction) of a fluid flowing through this tubing is v, the *flow* (volume of fluid per unit of time) is given by $F = Av$. However, the velocity close to the tube wall is retarded by the fluid "rubbing" against the wall. In fact, the velocity at a point is given by

$$v(r) = \frac{P}{4cL}(R^2 - r^2),$$

where R is the radius of the tubing, r is the distance from the point to the center of the tubing, P is the pressure difference between the ends of the tube, L is the length of the tube, and c is the viscosity of the fluid.

To find the flow, we subdivide the interval $[0, R]$ into n subintervals, each of width R/n, and label as in Figure 1.33. If n is large, then the velocity is nearly constant throughout each washer-shaped region. Hence the flow through the shaded washer-shaped region is approximately $F_k \approx v(r_k)A_k$, where A_k is the area of the shaded region and $v(r_k)$ is the velocity at a point a distance r_k from the center. Since $r_k = kR/n$, it follows that

$$v(r_k) = \frac{P}{vcL}\left[R^2 - \left(\frac{kR}{n}\right)^2\right]$$

$$= \frac{PR^2}{4cL}\left(1 - \frac{k^2}{n^2}\right),$$

56 CHAPTER 1: AREA AND INTEGRATION

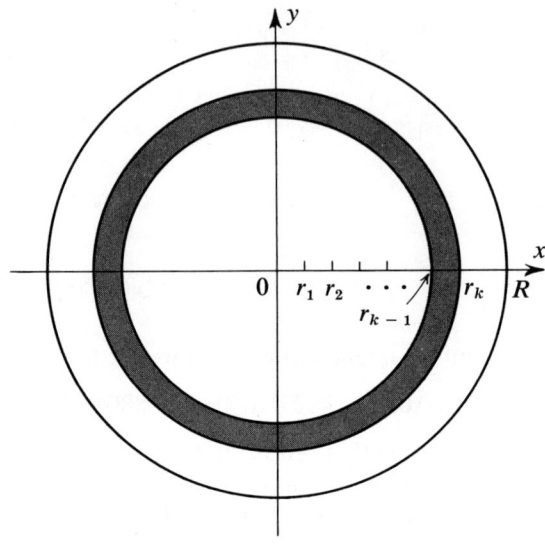

FIGURE 1.33

and A_k is the difference between a circle of radius kR/n and one of radius $(k-1)R/n$; that is,

$$A_k = \pi\left(\frac{kR}{n}\right)^2 - \pi\left[\frac{(k-1)R}{n}\right]^2$$

$$= \frac{\pi R^2}{n^2}\left[k^2 - (k-1)^2\right]$$

$$= \frac{\pi R^2}{n^2}(2k-1).$$

It follows that

$$F_k \approx \frac{PR^2}{4cL}\left(1 - \frac{k^2}{n^2}\right)\frac{\pi R^2}{n^2}(2k-1)$$

$$= \frac{P\pi R^4}{4cLn^4}(n^2 - k^2)(2k-1)$$

$$= \frac{P\pi R^4}{4cLn^4}(2kn^2 - 2k^3 - n^2 + k^2).$$

The total flow is the sum of the flows through the washer-shaped regions, so that

$$F \approx \sum_{k=1}^{n} \frac{P\pi R^4}{4cLn^4}(2kn^2 - 2k^3 - n^2 + k^2)$$

$$= \frac{P\pi R^4}{4cLn^4}\left(2n^2\sum_{k=1}^{n}k - 2\sum_{k=1}^{n}k^3 - n^2\sum_{k=1}^{n}1 + \sum_{k=1}^{n}k^2\right)$$

$$= \frac{P\pi R^4}{4cLn^4}\left[2n^2\frac{n(n+1)}{2} - 2\frac{n^2(n+1)^2}{4} - n^3 + \frac{n(n+1)(2n+1)}{6}\right]$$

$$= \frac{P\pi R^4}{4cL}\left[\frac{n+1}{n} - \frac{(n+1)^2}{2n^2} - \frac{1}{n} + \frac{(n+1)(2n+1)}{6n^3}\right].$$

Taking the limit as n tends to infinity, we see that

$$F = \lim_{n\to\infty} \frac{P\pi R^4}{4cL}\left[\frac{n+1}{n} - \frac{(n+1)^2}{2n^2} - \frac{1}{n} + \frac{(n+1)(2n+1)}{6n^3}\right]$$

$$= \frac{P\pi R^4}{4cL}\left[\lim_{n\to\infty}\frac{n+1}{n} - \lim_{n\to\infty}\frac{(n+1)^2}{2n^2} - \lim_{n\to\infty}\frac{1}{n} + \lim_{n\to\infty}\frac{(n+1)(2n+1)}{6n^3}\right]$$

$$= \frac{P\pi R^4}{4cL}\left(1 - \frac{1}{2} - 0 + 0\right)$$

$$= \frac{P\pi R^4}{8cL}$$

Since the flow depends on the fourth power of the radius, a relatively small change in the radius translates into a relatively large change in the flow (see Problem 22). This point is significant, for example, when treating a heart patient who has restrictions in some major arteries.

REVIEW PROBLEMS

Compute the areas of the regions bounded by $y = f(x)$, $x = a$, $x = b$, and $y = 0$ in Problems 1 through 4.

1. $f(x) = x^2$, $a = 0$, $b = 2$
2. $f(x) = 2x$, $a = 0$, $b = 2$
3. $f(x) = x^3$, $a = 0$, $b = 1$
4. $f(x) = x - x^2$, $a = 0$, $b = 1$

Compute the following integrals.

5. $\int_{-1}^{1} x^2\, dx$
6. $\int_{-1}^{1} x^3\, dx$
7. $\int_{0}^{1} x^4\, dx$
8. $\int_{1}^{2} (x-1)^4\, dx$
9. Evaluate $\sum_{i=100}^{200} i^2$.
10. Evaluate $\sum_{i=1}^{10} (1 + i + i^2)$.
11. Compute $\lim_{n\to\infty} \frac{n^2 - 3}{2n^2 + 1}$.
12. Compute $\lim_{n\to\infty} \left(\frac{n^2+1}{n} - \frac{n^2-1}{n}\right)$.
13. Compute $\lim_{n\to\infty} \left(\sqrt{n} - \sqrt{n^2+1-1}\right)$.

14. Show that $\sum_{i=0}^{n} r^i = \dfrac{1 - r^{n+1}}{1 - r}$. $\left[\textit{Hint:} \text{ Show that } (1 - r)\left(\sum_{i=0}^{n} r^i\right) \text{ is a collapsing sum.} \right]$

15. Compute $\lim_{n \to \infty} \left[\sum_{i=0}^{n} \left(\dfrac{1}{2}\right)^i \right]$. [*Hint:* Use Problem 14.]

16. Compute $\sum_{n=1}^{k} \dfrac{2n + 1}{n^2(n + 1)^2}$. [*Hint:* Rewrite $(2n + 1)/n^2(n + 1)^2$ in the form $A/n^2 + B/(n + 1)^2$ and get a collapsing sum.]

17. Use Problem 16 to compute $\lim_{k \to \infty} \sum_{n=1}^{k} \dfrac{2n + 1}{n^2(n + 1)^2}$.

18. Compute $\lim_{n \to \infty} \sum_{k=1}^{n} 5\left(2 + \dfrac{3k}{n}\right)^2 \dfrac{3}{n}$. [*Hint:* Rewrite the limit as an integral.]

19. Linda borrows $4000 and agrees to pay back $100 on the principal at the end of each month, at which time interest on the principal outstanding during the month is also paid at the rate of 7% per year. Find the total amount paid in discharging the debt. [*Hint:* Use Section 3.]

20. In Example 29, show that the value of the section of land is given by the integral $\int_0^1 640(8000 - 7000x)\, dx$.

21. In Example 30, show that the flow is given by the integral $\int_0^R v(r)2\pi r\, dr$ and compute the integral.

22. If it takes 2 hours to water a section of a lawn with a $\frac{1}{2}$-inch hose, how long should it take with a $\frac{5}{8}$-inch hose? (Use the results of Example 30 and assume that P, L, and c remain the same in both cases.)

CHAPTER 2

DIFFERENTIATION AND THE FUNDAMENTAL THEOREM

Everyone expects the cost of living to increase each year because of inflation. The critical question, however, is: How rapidly is the cost of living increasing? A rise of 2% per year in the cost of living is quite reasonable, but a rise of 10% per year is considered extreme. The rate at which something is increasing (or decreasing) is called a rate of change. Finding rates of change can be done by a process known as differentiation, the second basic concept in calculus (the first was integration). Although both concepts have been recognized for centuries, the fact that these concepts are closely related was first formalized by Gottfried Wilhelm Leibnitz (1646–1716) and Sir Isaac Newton (1642–1727). This relationship is the content of the Fundamental Theorem of Calculus.

As you would expect from its name, the Fundamental Theorem is of central significance in that the theorem gives a method that allows many integrals to be evaluated easily. As we have seen, the methods given in Chapter 1 for computing integrals often require a great deal of hard work, and easier methods will prove a welcome relief. There will always remain, however, many integrals that can only be approximated by using methods like those outlined in Chapter 1.

60   CHAPTER 2: DIFFERENTIATION

1. RATES OF CHANGE

Let $f(t)$ denote some quantity (population of New York City, price of a share of Bankruptcy, Inc., mass of an electron, or the temperature at the North Pole) at time t. The **change** in f between time t and time $t + h$ is the number $f(t + h) - f(t)$, where h can be either positive or negative. The **average rate of change** in f between times t and $t + h$ is the quotient

$$(1) \qquad \frac{f(t + h) - f(t)}{h}.$$

If this quotient is positive, then an interpretation might be that the population of New York City is increasing, that the price of a share of Bankruptcy, Inc. is going up, that the mass of an electron is increasing, or that it is getting hotter at the North Pole.

The average rate of change indicates trends but does not say what is happening at a particular instant. For example, the price of Bankruptcy, Inc. might be $50 in January and $60 in June, which would yield an average rate of change of $2 per month over the 5-month period. However, the price may have fluctuated wildly between January and June, perhaps $100 in February, $40 in March, $10 in April, and $70 in May. In order to become a millionaire by age 30, it is necessary to analyze the ups and downs a bit more closely.

By taking h to be close (but not equal) to zero, we can find the average rate of change over a very short period of time.

EXAMPLE 1. The volume of air inside a balloon is given by $V(t) = 1000/(t + 10)$, where t is measured in hours and $V(t)$ is given in cubic inches. How fast is the volume of air inside the balloon changing after 10 hours?

Solution. Let h be a nonzero number. The average rate of change between $t = 10$ and $t = 10 + h$ is given by

$$\frac{V(10 + h) - V(10)}{h} = \frac{1}{h}\left(\frac{1000}{10 + h + 10} - \frac{1000}{10 + 10}\right)$$

$$= \frac{1000}{h}\left(\frac{1}{20 + h} - \frac{1}{20}\right)$$

$$= \frac{1000}{h}\left[\frac{20 - (20 + h)}{20(20 + h)}\right]$$

$$= -\frac{1000}{20(20 + h)}.$$

If h is very close to zero, then the average rate of change is very close to $-1000/400 = -2.5$, where the minus sign refers to the fact that the balloon is losing air. Thus after 10 hours the balloon is shrinking at a rate of 2.5 cubic inches per hour.

A number L is called the ***instantaneous rate of change*** of the function f at t if the average rate of change $[f(t+h)-f(t)]/h$ between t and $t+h$ can be made as close to L as desired by requiring that h be sufficiently close to zero. In this case, we write

(2) $$L = \lim_{h \to 0} \frac{f(t+h) - f(t)}{h}.$$

In Example 1, the instantaneous rate of change of volume of air was -2.5 cubic inches per hour at $t = 10$.

EXAMPLE 2. The temperature at the North Pole is

$$T(t) = \frac{1}{18} t^3 - t^2,$$

where t is the number of hours after 12:00 noon. How fast is the temperature rising or falling at 6 P.M.? At 12 midnight?

Solution. We must compute the instantaneous rate of change at $t=6$ and $t=12$. The average rate of change between t and $t+h$ is

$$\frac{T(t+h) - T(t)}{h} = \frac{(1/18)(t+h)^3 - (t+h)^2 - [(1/18)t^3 - t^2]}{h}$$

$$= \frac{1}{h}\left[\frac{1}{18}(t^3 + 3t^2h + 3th^2 + h^3) - (t^2 + 2th + h^2) - \frac{1}{18}t^3 + t^2\right]$$

$$= \frac{1}{h}\left[\frac{1}{18}(3t^2h + 3th^2 + h^3) - (2th + h^2)\right]$$

$$= \frac{1}{18}(3t^2 + 3th + h^2) - (2t + h).$$

If h is close to zero, then $[T(t+h) - T(t)]/h$ is close to $\frac{1}{18}(3t^2) - 2t = \frac{1}{6}t^2 - 2t$, and hence the instantaneous rate of change is

$$\lim_{h \to 0} \frac{T(t+h) - T(t)}{h} = \frac{1}{6}t^2 - 2t.$$

At $t = 6$, the instantaneous rate of change is

$$\frac{1}{6} \cdot 6^2 - 2 \cdot 6 = -6,$$

where the minus sign indicates that the temperature is dropping. Thus at 6 P.M. the temperature is dropping at the rate of 6 degrees per hour. At midnight the instantaneous rate of change is

$$\frac{1}{6} \cdot 12^2 - 2 \cdot 12 = 0,$$

which means that the temperature is neither rising nor falling.

PROBLEMS

Find the instantaneous rate of change of each of the following functions at the indicated time.

1. $f(t) = 2t$, $t = 5$
2. $f(t) = t^2$, $t = 3$
3. $V(t) = \frac{1}{t}$, $t = 1$
4. $h(t) = t - t^2$, $t = 0$
5. $f(t) = 5$, $t = 3$
6. $S(t) = -16t^2$, $t = c$

2. SLOPE OF THE TANGENT LINE

Given a circle and a point on that circle, it is always possible to construct a line tangent to the circle at the given point. This tangent line has several interesting properties. It touches the circle in exactly one point; in fact, any line that touches the circle in exactly one point is necessarily tangent to the circle. The line tangent at a point on a circle is perpendicular to the line segment joining the given point and the center. A tangent line lies entirely on one side of the circle. The tangent line through P seems to lie in the same direction as the circle does at P; that is, if a very small section containing P is examined through a powerful microscope, the tangent line and the section of the circle will look about the same. Thus the tangent line is a "good approximation" to the circle in a small region about P. This idea is illustrated in Figure 2.1.

We wish to extend the concept of tangent line so that we can consider lines that are tangent to the graph of a function. Some of the preceding properties will certainly not hold in general, but some of them will.

Since it works for circles, one possible definition would be that a line is tangent to a graph if the line touches the graph exactly once. This would not be a good definition, however, because the line in Figure 2.2 appears to be tangent although it touches the graph three times, and the line in Figure 2.3 should not be considered tangent to the graph. (If such lines were called tangent, then there could be many tangent lines at a point, and somehow our intuitive concept tells us that tangent lines should be unique.)

SECTION 2: SLOPE OF THE TANGENT LINE   63

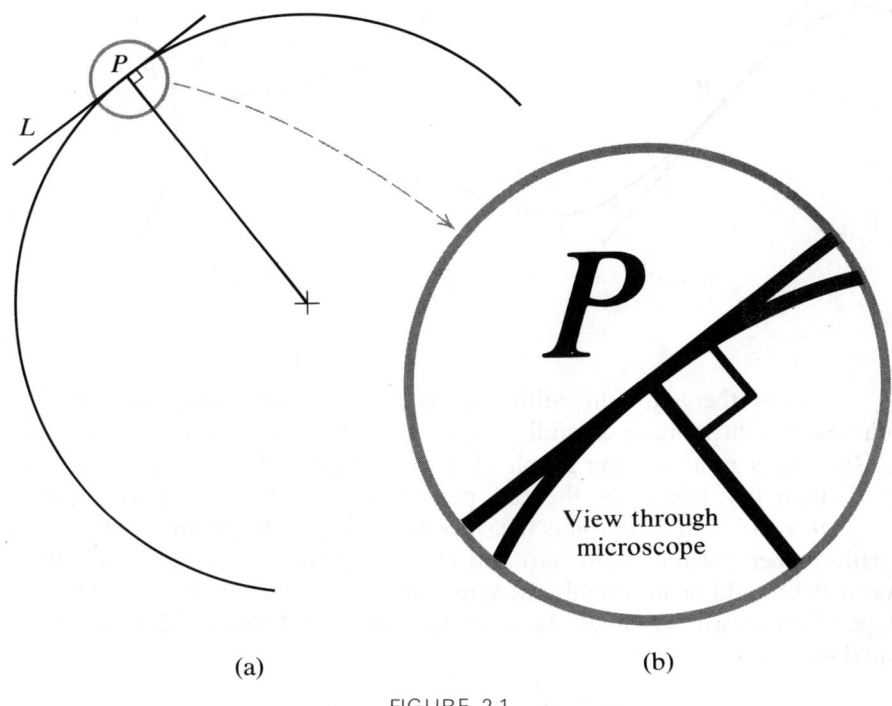

(a)   (b)

FIGURE 2.1

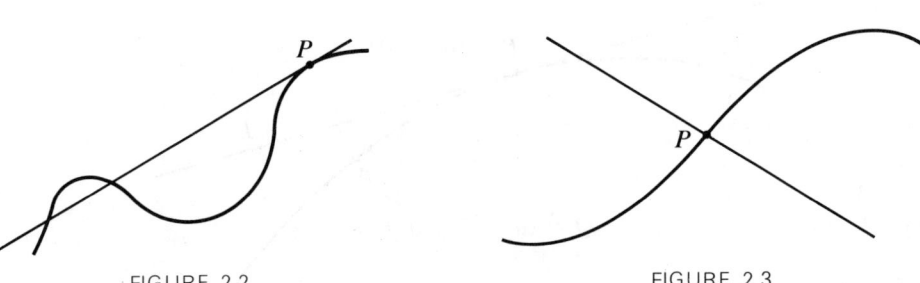

FIGURE 2.2   FIGURE 2.3

We do not wish to require a tangent line to lie entirely on one side of the graph because, intuitively, the lines in Figures 2.2 and 2.4 should be tangent, whereas the line in Figure 2.5 should not be tangent. (Again, tangent lines should be uniquely determined.)

64 CHAPTER 2: DIFFERENTIATION

FIGURE 2.4   FIGURE 2.5

Thus there are difficulties. Perhaps we should consider the idea of "direction" a little more carefully. Suppose that $y = f(x)$ is a function, and let $(a, f(a))$ be a point on the graph of f, as in Figure 2.6. If L is to be tangent at P, then (by means of the old point-slope trick) L must have equation $y - f(a) = m(x - a)$, where m is the slope of the line L. What must m be? If h is a small number different from zero and Q is the point $(a + h, f(a + h))$, then the chord PQ should lie in roughly the same direction as the tangent line; that is, the slope of the chord PQ should be approximately equal to m. The slope of PQ is equal to

$$\frac{f(a+h) - f(a)}{(a+h) - a} = \frac{f(a+h) - f(a)}{h},$$

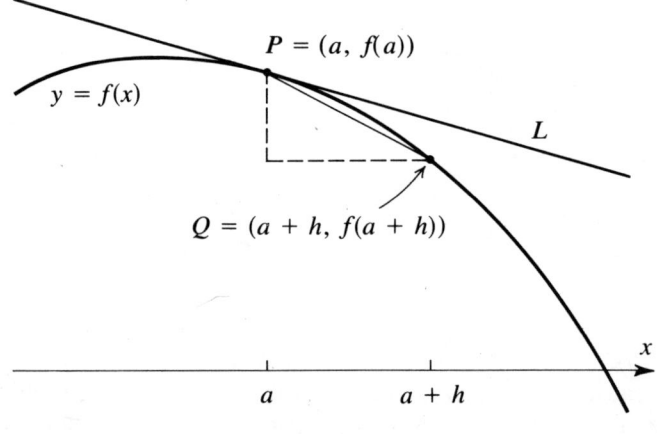

FIGURE 2.6

and hence it seems reasonable to expect that $[f(a + h) - f(a)]/h$ should be very close to m whenever h is close to zero. We write

(3) $$\lim_{h \to 0} \frac{f(a + h) - f(a)}{h} = m$$

if $[f(a + h) - f(a)]/h$ can be made as close to m as desired simply by requiring that h be sufficiently close to zero. We *define* the tangent line so that it exists if and only if the preceding limit exists.

If f is a function and if $\lim_{h \to 0} \{[f(a + h) - f(a)]/h\}$ exists, then the line through the point $(a, f(a))$ with slope $m = \lim_{h \to 0} \{[f(a + h) - f(a)]/h\}$ is called the **tangent line** to the graph of f at the point $(a, f(a))$. Notice that the point-slope form of the tangent line is given by

$$y - f(a) = m(x - a),$$

where

$$m = \lim_{h \to 0} \frac{f(a + h) - f(a)}{h}.$$

The expression $\lim_{h \to 0} \{[f(a + h) - f(a)]/h\}$ is equivalent to the expression for the instantaneous rate of change developed in Section 1. In fact, the slope of the tangent line can be thought of as the rate of change in the height of the graph.

EXAMPLE 3. Find the line tangent to $y = 3x^2 + 2$ at $(1, 5)$.

Solution. Let $f(x) = 3x^2 + 2$ and compute

$$m = \lim_{h \to 0} \frac{f(1 + h) - f(1)}{h}$$

$$= \lim_{h \to 0} \frac{3(1 + h)^2 + 2 - [3(1)^2 + 2]}{h}$$

$$= \lim_{h \to 0} \frac{3(1 + 2h + h^2) - 3}{h}$$

$$= \lim_{h \to 0} 3(2 + h)$$

$$= 6.$$

It follows that the tangent line is

$$y - 5 = 6(x - 1),$$

which simplifies to

$$y = 6x - 1.$$

The graph is given in Figure 2.7.

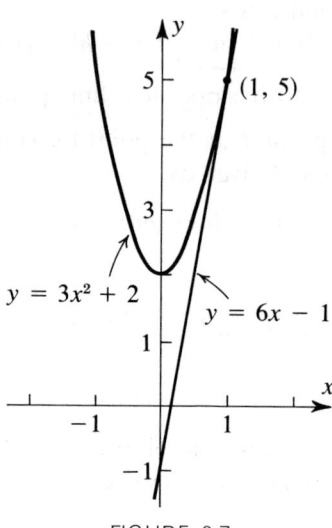

FIGURE 2.7

EXAMPLE 4. Find the line tangent to $y = f(x) = |x|$ at the point $(0, 0)$.

Solution. We get

$$m = \lim_{h \to 0} \frac{f(0 + h) - f(0)}{h}$$

$$= \lim_{h \to 0} \frac{|h| - |0|}{h}.$$

But

$$\frac{|h|}{h} = \begin{cases} 1 & \text{if } h > 0 \\ -1 & \text{if } h < 0 \end{cases},$$

so $\lim_{h \to 0} |h|/h$ does not exist because no choice of m can be close to both 1 and -1 at the same time. Thus a tangent line does not exist at $(0, 0)$. By looking at Figure 2.8, we can see that the graph does not appear to have a well-defined direction at $(0, 0)$.

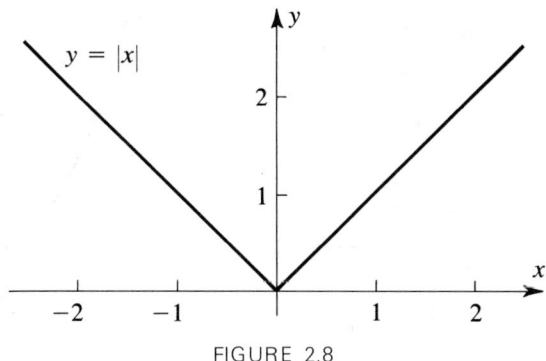

FIGURE 2.8

EXAMPLE 5. Sketch the graph of $y = x^2$.

Solution. We first note that the graph passes through the points $(-2, 4)$, $(-1, 1)$, $(0, 0)$, $(1, 1)$, and $(2, 4)$. The slope m of the tangent line at $(a, f(a))$ is given by

$$m = \lim_{h \to 0} \frac{f(a+h) - f(a)}{h}$$

$$= \lim_{h \to 0} \frac{(a+h)^2 - a^2}{h}$$

$$= \lim_{h \to 0} \frac{a^2 + 2ah + h^2 - a^2}{h}$$

$$= \lim_{h \to 0} (2a + h)$$

$$= 2a.$$

In particular, the tangent line through $(-2, 4)$ is given by

$$y - 4 = 2(-2)[x - (-2)] = -4(x + 2).$$

This is a line through $(-2, 4)$ with slope -4. Similarly, the tangent line through $(-1, 1)$ has slope -2, the tangent line at $(0, 0)$ has slope 0, the tangent line at $(1, 1)$ has slope 2, and the tangent line at $(2, 4)$ has slope 4. These tangents are sketched in Figure 2.9.

The graph in Figure 2.10 is then obtained by carefully connecting these five points by a smooth curve that seems to lie in the same direction as the tangent lines at the points $(-2, 4), (-1, 1), (0, 0), (1, 1)$, and $(2, 4)$.

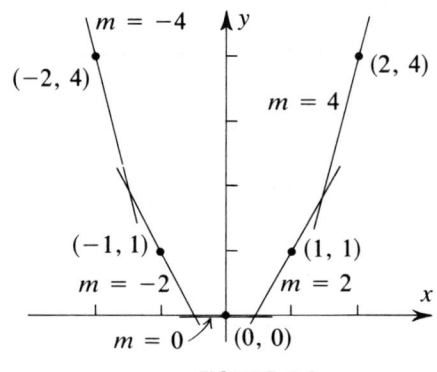

FIGURE 2.9

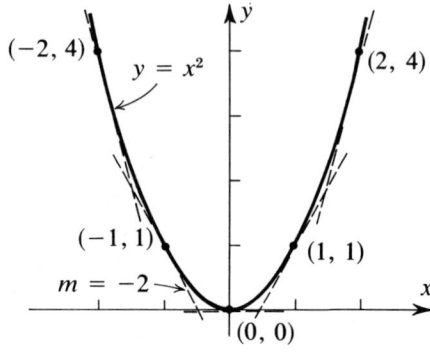

FIGURE 2.10

PROBLEMS

Find the equation of the tangent line at the point P. By first plotting a few points, attempt to sketch the graph.

1. $f(x) = 4x^2 - 2x + 1$, $P = (0, 1)$
2. $f(x) = 3x^2 + 2x - 1$, $P = (-1, 0)$
3. $f(x) = \sqrt{x}$, $P = (4, 2)$. [*Hint:* Multiply and divide by $\sqrt{4+h} + 2$.]
4. $f(x) = 1/x$, $P = (1, 1)$
5. $f(x) = 4x + 3$, $P = (-1, -1)$
6. $f(x) = 1/(x^2 + 1)$, $P = (1, 1/2)$
7. $f(x) = x^4$, $P = (-1, 1)$
8. $f(x) = x + \sqrt{x}$, $P = (4, 6)$
9. $f(x) = (x^2 + 1)^2$, $P = (1, 4)$
10. $f(x) = x^3 + 2x^2$, $P = (2, 16)$

3. LIMITS

In the previous two sections we encountered numerous limits of the form
$$\lim_{h \to 0} \frac{f(x+h) - f(x)}{h}.$$

In general,

(4) $$\lim_{t \to a} g(t) = L$$

means that $g(t)$ is as close to L as desired whenever t is sufficiently close (but not equal) to a. The symbol $\lim_{t \to a} g(t)$ is read "The limit as t tends to a of $g(t)$."

EXAMPLE 6. Compute $\lim_{t \to 2} (t - 3)$.

Solution. As t tends to 2, $t-3$ gets close to -1. Thus $\lim_{t \to 2}(t-3) = -1$. This situation is illustrated in Figure 2.11.

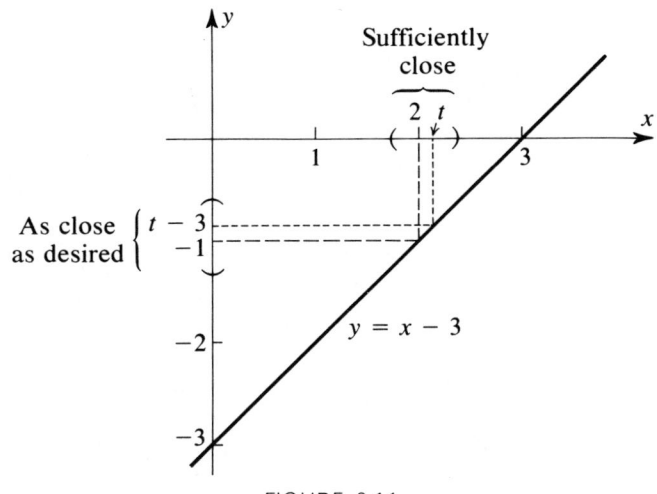

FIGURE 2.11

EXAMPLE 7. Compute $\lim_{t \to 3}(t^3 + 2t)$.

Solution. As t tends to 3, t^3 gets close to 27, $2t$ gets close to 6, and hence $t^3 + 2t$ gets close to $27 + 6 = 33$. It follows that $\lim_{t \to 3}(t^3 + 2t) = 33$ (see Figure 2.12).

EXAMPLE 8. Compute $\lim_{t \to 1} 1/(1-t)$.

Solution. For t close to 1, $1-t$ is close to zero and the quotient $1/(1-t)$ can be made arbitrarily large by picking t sufficiently close to 1. The quotient $1/(1-t)$ cannot possibly stay close to any fixed number L, and hence the limit does not exist.

EXAMPLE 9. Compute $\lim_{t \to 0} t/|t|$.

Solution. For $t > 0$,

$$\frac{t}{|t|} = \frac{t}{t} = 1$$

70   CHAPTER 2: DIFFERENTIATION

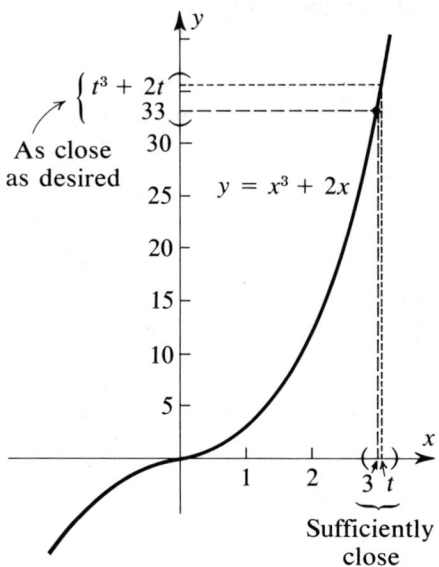

FIGURE 2.12

and for $t < 0$,

$$\frac{t}{|t|} = \frac{t}{-t} = -1.$$

No matter what L we pick as a possible limit, it is always possible to find t_1 and t_2 as close to zero as desired such that $t_1/|t_1| = 1$ and $t_2/|t_2| = -1$, so at least one of these quotients is not close to L. It follows that $\lim_{t \to 0} t/|t|$ does not exist (see Figure 2.13).

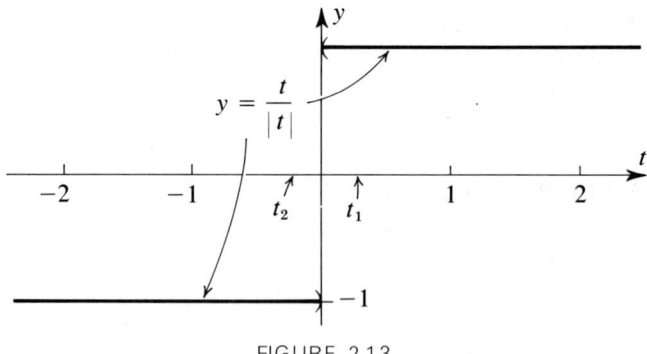

FIGURE 2.13

In the preceding example, if we compute the limit of $t/|t|$ as t assumes only positive values, then this one-sided limit exists and is equal to 1. We express this fact by writing $\lim_{t \to 0^+} t/|t| = 1$. Similarly, the symbolism $\lim_{t \to 0^-} t/|t| = -1$ means that $t/|t|$ tends to -1 as t tends to zero by assuming only negative values. In general,

$$\lim_{t \to a^+} f(t) \qquad (5)$$

is the limit of $f(t)$ as t tends to a from the right; that is, $t > a$, and

$$\lim_{t \to a^-} f(t) \qquad (6)$$

is the limit of $f(t)$ as t tends to a from the left ($t < a$).

The following test for limits is sometimes useful.

THEOREM 1. $\lim_{t \to a} f(t) = L$ if and only if $\lim_{t \to a^+} f(t) = L$ and $\lim_{t \to a^-} f(t) = L$.

EXAMPLE 10. The limit $\lim_{t \to 0} t/|t|$ does not exist because $\lim_{t \to 0^+} t/|t| = 1$ while $\lim_{t \to 0^-} t/|t| = -1$ (see Figure 2.13).

It is sometimes useful to think of limits in terms of error and the control of error. The expression $|g(t) - L|$, which is the distance between $g(t)$ and L, is called the *error*, while the distance $|t - a|$ is called the *control*. The definition of limit can be rephrased to read

$$\lim_{t \to a} g(t) = L$$

if "the error can be controlled"; that is, if the error $|g(t) - L|$ can be made as small as desired by finding a sufficiently small control $|t - a|$.

EXAMPLE 11. Show that $\lim_{t \to 3} (t^3 + 2t) = 33$ by showing that the error can be controlled.

Solution. We need a bit of algebraic trickery to write the error $|(t^3 + 2t) - 33|$ in terms of the control $|t - 3|$. This manipulation is needed to show that the error is small whenever the control is small. Now

$$|(t^3 + 2t) - 33| = |t^3 + 2t - 27 - 6|$$
$$= |t^3 - 3^3 + 2t - 2 \cdot 3|$$

$$= |(t-3)(t^2+3t+9) + 2(t-3)|$$
$$= |t-3| \cdot |t^2+3t+9+2|$$
$$= |t-3| \cdot |t^2+3t+11|.$$

If t is going to be close to 3, then we may as well assume that $2 < t < 4$, in which case $|t^2+3t+11| < 4^2+4\cdot 3+11 = 39$. Thus

$$|(t^3+2t) - 33| = |t-3| \cdot |t^2+3t+11|$$
$$< 39|t-3|,$$

which can be made as small as desired by requiring the control $|t-3|$ to be sufficiently small. For example, the error is less than 10^{-7} whenever the control is less than $10^{-7}/39$.

We tacitly assumed in Example 7 that since t^3 gets close to 27 and $2t$ gets close to 6, it follows that $t^3 + 2t$ gets close to 33. Another way to write this is

$$\lim_{t \to 3} (t^3 + 2t) = \lim_{t \to 3} t^3 + \lim_{t \to 3} 2t.$$

That this is indeed a valid statement is shown in the following theorem. The theorem will also simplify the task of computing other limits. There should be nothing surprising in this theorem; it is very similar to Theorem 1 of Chapter 1.

THEOREM 2. Suppose that $\lim_{x \to a} f(x)$ and $\lim_{x \to a} g(x)$ both exist. Then

(i) $\lim_{x \to a} [f(x) + g(x)] = \lim_{x \to a} f(x) + \lim_{x \to a} g(x)$; that is, the limit of a sum is the sum of the limits.

(ii) $\lim_{x \to a} [f(x) \cdot g(x)] = \lim_{x \to a} f(x) \cdot \lim_{x \to a} g(x)$; that is, the limit of a product is the product of the limits.

(iii) If $\lim_{x \to a} g(x) \neq 0$, then

$$\lim_{x \to a} \frac{f(x)}{g(x)} = \frac{\lim_{x \to a} f(x)}{\lim_{x \to a} g(x)};$$

that is, the limit of a quotient is the quotient of the limits.

(iv) $\lim_{x \to a} [c \cdot f(x)] = c \cdot \lim_{x \to a} f(x)$ for any number c.

(v) If f is continuous on $[a, b]$ and $a < c < b$, then $\lim_{x \to c} f(x) = f(c)$.

(vi) If $\lim_{x \to d} g(x) = c$ and f is continuous on $[a, b]$, $a < c < b$, then $\lim_{x \to d} f(g(x)) = f(c) = f(\lim_{x \to d} g(x))$.

So far we have only considered functions that are continuous on an interval. Motivated by part (v) of Theorem 2, we say that a function f is **continuous at the point** $x = c$ if $\lim_{x \to c} f(x) = f(c)$. If $\lim_{x \to c} f(x) \neq f(c)$, if $\lim_{x \to c} f(x)$ does not exist, or if $f(c)$ is undefined, then f has a **discontinuity** at $x = c$. It is a deep result in analysis that continuity on $[a, b]$ is equivalent to being continuous at each point of $[a, b]$.

Part (vi) of Theorem 2 can now be restated:

(vi') If $\lim_{x \to d} g(x) = c$ and f is continuous at c, then

$$\lim_{x \to d} f(g(x)) = f(c) = f(\lim_{x \to d} g(x)).$$

EXAMPLE 12. Compute

$$\lim_{h \to 0} \frac{\sqrt{1+h} - 1}{h}.$$

Solution. Notice that part (iii) of Theorem 2 does not apply directly. In order to evaluate this limit, we must first simplify the quotient. Multiplying both numerator and denominator by $\sqrt{1+h} + 1$, we obtain

$$\frac{\sqrt{1+h} - 1}{h} = \frac{\sqrt{1+h} - 1}{h} \cdot \frac{\sqrt{1+h} + 1}{\sqrt{1+h} + 1}$$

$$= \frac{(1+h) - 1}{h(\sqrt{1+h} + 1)}$$

$$= \frac{h}{h(\sqrt{1+h} + 1)}$$

$$= \frac{1}{\sqrt{1+h} + 1}.$$

As h tends to zero, $\sqrt{1+h}$ tends to 1, so that

$$\lim_{h \to 0} \frac{\sqrt{1+h} - 1}{h} = \lim_{h \to 0} \frac{1}{\sqrt{1+h} + 1}$$

$$= \frac{1}{1 + 1}$$

$$= 0.5.$$

In previous algebra courses you were taught to rationalize the

CHAPTER 2: DIFFERENTIATION

denominator. As this example illustrates, it is sometimes necessary to rationalize the numerator in order to compute some limits.

PROBLEMS

Compute the following limits. Use appropriate parts of Theorem 2 to justify each step.

1. $\lim\limits_{x \to 1} (2x^2 + x + 5)$

2. $\lim\limits_{x \to -1} \dfrac{x+2}{x-1}$

3. $\lim\limits_{t \to 0} \dfrac{(a+t)^3 - a^3}{t}$

4. $\lim\limits_{x \to 2} \dfrac{x^2 - 4}{x - 2}$

5. $\lim\limits_{h \to a} \dfrac{a^2 - h^2}{a - h}$

6. $\lim\limits_{h \to 0} \left(\dfrac{1}{h} - \dfrac{h^2 + 1}{h} \right)$

7. $\lim\limits_{h \to 0} \dfrac{\sqrt{(x+h)^2 + 1} - \sqrt{x^2 + 1}}{h}$

8. $\lim\limits_{t \to 5} \dfrac{2t^2 - 11t + 5}{5 - t}$

9. $\lim\limits_{x \to \sqrt{2}} \dfrac{x^4 - 4}{x^2 - 2}$

10. $\lim\limits_{h \to 0} \dfrac{1}{h} \left(\dfrac{1}{x + h} - \dfrac{1}{x} \right)$

11. $\lim\limits_{x \to 1} \dfrac{\sqrt{x} - 1}{x - 1}$

12. $\lim\limits_{x \to 1} \dfrac{x + 2}{x - 1}$

13. In Problem 1, how close must x be to 1 in order to guarantee that the error is less than 10^{-3}?

14. In Problem 2, how close must x be to -1 in order to guarantee that the error is less than 10^{-2}?

15. Suppose that E is a permissible error. In Problem 1, how small must the control be in order to guarantee that the error is less than the permissible error E?

16. Suppose that E is a permissible error. In Problem 2, how close must x be to -1 in order that

$$\left| \dfrac{x+2}{x-1} - \left(-\dfrac{1}{2} \right) \right| < E?$$

THE BINOMIAL THEOREM (OPTIONAL)

In order to differentiate x^n, we will need to expand expressions of the form $(a + b)^n$, where n is some positive integer. This step is easy for small n. For example,

$$(a + b)^1 = a + b,$$

$$(a + b)^2 = (a + b)(a + b) = a^2 + 2ab + b^2,$$

and $(a + b)^3 = (a + b)(a^2 + 2ab + b^2) = a^3 + 3a^2b + 3ab^2 + b^3$.

THE BINOMIAL THEOREM   75

Writing out a few more, it is apparent that some sort of pattern is developing:

$$(a + b)^4 = a^4 + 4a^3b + 6a^2b^2 + 4ab^3 + b^4,$$
$$(a + b)^5 = a^5 + 5a^4b + 10a^3b^2 + 10a^2b^3 + 5ab^4 + b^5,$$

and $\quad (a + b)^6 = a^6 + 6a^5b + 15a^4b^2 + 20a^3b^3 + 15a^2b^4 + 6ab^5 + b^6.$

Somehow the coefficient of $a^i b^{n-i}$ should depend on n (the row) and i (the position in the row). After a few hours of looking at the coefficients and after scribbling on both sides of several sheets of scratch paper, it is possible to notice that

$$3 = \frac{3 \cdot 2 \cdot 1}{1 \cdot (2 \cdot 1)},$$

$$6 = \frac{4 \cdot 3 \cdot 2 \cdot 1}{(2 \cdot 1) \cdot (2 \cdot 1)},$$

$$10 = \frac{5 \cdot 4 \cdot 3 \cdot 2 \cdot 1}{(3 \cdot 2 \cdot 1) \cdot (2 \cdot 1)},$$

and $\quad 15 = \dfrac{6 \cdot 5 \cdot 4 \cdot 3 \cdot 2 \cdot 1}{(4 \cdot 3 \cdot 2 \cdot 1) \cdot (2 \cdot 1)}.$

In each case, the coefficient of $a^i b^{n-i}$ equals

$$\frac{n(n-1)(n-2)\cdots 3 \cdot 2 \cdot 1}{i(i-1)(i-2)\cdots 2 \cdot 1 \cdot (n-i)(n-i-1)\cdots 2 \cdot 1}.$$

The three dots indicate terms that are not written but that continue in the same pattern. Whenever three dots are used, enough information should be given to establish an obvious pattern so that the missing terms can be found.

To simplify notation, we write $k!$, read "k-factorial," instead of $k(k-1)(k-2)(k-3)\cdots 3 \cdot 2 \cdot 1$. Notice that

$$1! = 1,$$
$$2! = 2,$$
$$3! = 3 \cdot 2$$
$$= 6,$$
$$4! = 4 \cdot 3 \cdot 2 \cdot 1$$
$$= 24,$$

and $\quad 5! = 5 \cdot 4 \cdot 3 \cdot 2 \cdot 1$
$$= 120.$$

We define $0! = 1$. This may seem strange, but it does allow many formulas to be

CHAPTER 2: DIFFERENTIATION

stated in a simple form. In particular, the coefficient of $a^i b^{n-i}$ can now be written in the form

$$\frac{n!}{i!(n-i)!},$$

and hence

(7) $$(a+b)^n = \sum_{i=0}^{n} \frac{n!}{i!(n-i)!} a^i b^{n-i}.$$

Formula (7) is commonly known as the **Binomial Theorem**.

It should be noted that the preceding remarks do *not* constitute a proof of the Binomial Theorem. However, this theorem can be proved using mathematical induction. (See Problem 13.)

PROBLEMS

Simplify each of the following expressions.

1. $6!$
2. $\dfrac{5!}{2!3!}$
3. $\dfrac{6!}{0!6!}$
4. $\dfrac{4!}{2!2!}$
5. $\dfrac{100!}{99!}$
6. $100! - 100(99!)$

Write out each of the following expressions using the Binomial Theorem.

7. $(5+2)^3$
8. $(1+1)^4$
9. $(x+h)^5$
10. $(x+h)^6$
11. $(x+h)^n$, where n is a positive integer (first three terms only).
12. $(a+3)^4$

13. Use mathematical induction to prove the Binomial Theorem. $\Big[$*Hint:* Let $P(n)$ denote

$$(a+b)^n = \sum_{i=0}^{n} \frac{n!}{i!(n-i)!} a^i b^{n-i}.\Big]$$

4. THE DERIVATIVE

Let f be any function defined on an interval $[a, b]$ and suppose that $a < x < b$. Then f is **differentiable** at x if

$$\lim_{h \to 0} \frac{f(x+h) - f(x)}{h}$$

exists, in which case we write

(8) $$f'(x) = \lim_{h \to 0} \frac{f(x+h) - f(x)}{h}.$$

The function f' (read "f prime") is called the **derivative** of f. Notice that instantaneous rates of change and slopes of tangent lines are examples of derivatives.

We will be mainly interested in functions that are differentiable at each point of an interval. Intuitively, a function is differentiable on an interval if its graph is smooth—that is, smooth enough to attach a tangent line at each point on the graph.

It seems that such functions should at least be continuous. Indeed, if

$$L = \lim_{h \to 0} \frac{f(x+h) - f(x)}{h},$$

then, multiplying both sides by $0 = \lim_{h \to 0} h$, we get

$$0 = \left[\lim_{h \to 0} h\right]\left[\lim_{h \to 0} \frac{f(x+h) - f(x)}{h}\right]$$
$$= \lim_{h \to 0} [f(x+h) - f(x)],$$

using the fact that the limit of the product is the product of the limits if both limits exist. But this result is equivalent to $\lim_{h \to 0} f(x+h) = f(x)$, which implies that f is continuous at x. [Recall that f is continuous at x if $\lim_{t \to x} f(t) = f(x)$. By replacing t by $x + h$, this equation reads $\lim_{h \to 0} f(x+h) = f(x)$.]

We have verified the following theorem.

THEOREM 3. *If f is differentiable at x, then f is continuous at x.*

The converse of Theorem 3 is false; that is, there are functions f that are continuous at some point x_0 but that are not differentiable at x_0. For example, the function $f(x) = |x|$ is continuous everywhere, but we have already seen that f is not differentiable at $x = 0$ (see Example 4).

As with the integral, the derivative has many applications and interpretations. The derivative will first be used as a tool to help evaluate integrals. In Chapter 4 the derivative serves as a powerful aid in sketching the graph of a function. In Chapter 5 the derivative is interpreted as a rate of change.

Before developing these applications and interpretations further, we need to become more proficient at computing derivatives. We first look at a few basic functions.

EXAMPLE 13. Let $F(x) = 5$ for all x. Then

$$F'(x) = \lim_{h \to 0} \frac{F(x+h) - F(x)}{h}$$

$$= \lim_{h \to 0} \frac{5 - 5}{h}$$

$$= \lim_{h \to 0} 0$$

$$= 0.$$

This result should not be surprising. If a function is constant, then the rate of change must certainly be zero.

EXAMPLE 14. Let $g(x) = 2x$. Then

$$g'(x) = \lim_{h \to 0} \frac{2(x+h) - 2x}{h}$$

$$= \lim_{h \to 0} \frac{2h}{h}$$

$$= \lim_{h \to 0} 2$$

$$= 2.$$

Again, not surprising. The fact that the derivative is 2 simply comes from the plausible statement that $2x$ changes twice as fast as x.

EXAMPLE 15. Let $f(x) = x^2$. Then

$$f'(x) = \lim_{h \to 0} \frac{f(x+h) - f(x)}{h}$$

$$= \lim_{h \to 0} \frac{(x+h)^2 - x^2}{h}$$

$$= \lim_{h \to 0} \frac{x^2 + 2xh + h^2 - x^2}{h}$$

$$= \lim_{h \to 0} (2x + h)$$

$$= 2x.$$

EXAMPLE 16. Let $h(x) = ax^2 + bx + c$. Then

$$h'(x) = \lim_{t \to 0} \frac{h(x+t) - h(x)}{t}$$

$$= \lim_{t \to 0} \frac{a(x+t)^2 + b(x+t) + c - (ax^2 + bx + c)}{t}$$

$$= \lim_{t \to 0} \frac{ax^2 + 2axt + at^2 + bx + bt + c - ax^2 - bx - c}{t}$$

$$= \lim_{t \to 0} (2ax + at + b)$$

$$= 2ax + b.$$

In particular, if $h(x) = x^2 + 2x$, then $h'(x) = 2x + 2$. In this case, notice that $h'(x) = f'(x) + g'(x)$, where f and g are given in Examples 15 and 14, respectively.

Suppose that $y = f(x)$. There are many symbols that are sometimes used instead of $f'(x)$. They include

$$\frac{dy}{dx}, \quad \frac{d}{dx}[f(x)], \quad \frac{df}{dx}, \quad Dy, \quad Df(x), \quad D_x f(x), \quad y', \quad \text{and} \quad \dot{y}.$$

The symbol $\dot{y}$ is often used if $y = f(t)$, where t denotes time. This notation was first introduced by Newton. Leibnitz invented the notation dy/dx, which is still very popular today. Although the symbol dy/dx is not a fraction, it is very suggestive of a limit of a fraction

$$\frac{f(x + \Delta x) - f(x)}{\Delta x} = \frac{\Delta y}{\Delta x},$$

the change in y divided by the change in x.

EXAMPLE 17. Let $y = \sqrt{x}$. Then

$$\frac{dy}{dx} = \lim_{s \to 0} \frac{\sqrt{x+s} - \sqrt{x}}{s}$$

$$= \lim_{s \to 0} \frac{\sqrt{x+s} - \sqrt{x}}{s} \cdot \frac{\sqrt{x+s} + \sqrt{x}}{\sqrt{x+s} + \sqrt{x}}$$

$$= \lim_{s \to 0} \frac{x+s-x}{s(\sqrt{x+s} + \sqrt{x})}$$

$$= \lim_{s \to 0} \frac{1}{\sqrt{x+s} + \sqrt{x}}$$

$$= \frac{1}{2\sqrt{x}}.$$

CHAPTER 2: DIFFERENTIATION

THEOREM 4. If $f(x) = x^n$, where n is a positive integer, then $f'(x) = nx^{n-1}$.

Proof. Recall (see the previous section) that

$$(x + h)^n = \sum_{i=0}^{n} \frac{n!}{i!(n-i)!} x^{n-i} h^i$$

by the Binomial Theorem. Thus

$$\frac{d}{dx} f(x) = \lim_{h \to 0} \frac{(x+h)^n - x^n}{h}$$

$$= \lim_{h \to 0} \frac{x^n + nx^{n-1}h + [n(n-1)/2]x^{n-2}h^2 + \cdots + h^n - x^n}{h}$$

$$= \lim_{h \to 0} \left\{ nx^{n-1} + h \left[\frac{n(n-1)}{2} x^{n-2} + \cdots + h^{n-2} \right] \right\}$$

$$= nx^{n-1}.$$

Notice that Example 15 is the special case $n = 2$.

EXAMPLE 18. Let $g(t) = t^9$. Then

$$g'(t) = 9t^8$$

by Theorem 4.

EXAMPLE 19. Let $w(x) = \dfrac{1}{x}$. Then

$$w'(x) = \lim_{t \to 0} \frac{1}{t} \left(\frac{1}{x+t} - \frac{1}{x} \right)$$

$$= \lim_{t \to 0} \frac{1}{t} \left[\frac{x - x - t}{x(x+t)} \right]$$

$$= \lim_{t \to 0} \frac{-t}{tx(x+t)}$$

$$= \lim_{t \to 0} \frac{-1}{x(x+t)}$$

$$= -\frac{1}{x^2}.$$

SECTION 4: THE DERIVATIVE

To summarize, the following is a list of differentiation formulas that should be memorized:

(9) $$\frac{d}{dx}(k) = 0 \quad (k \text{ is any constant})$$

(10) $$\frac{d}{dx}(x) = 1$$

(11) $$\frac{d}{dx}(x^n) = nx^{n-1} \quad (n \text{ is any positive integer})$$

(12) $$\frac{d}{dx}\left(\frac{1}{x}\right) = -\frac{1}{x^2}$$

(13) $$\frac{d}{dx}(\sqrt{x}) = \frac{1}{2\sqrt{x}}$$

PROBLEMS

Compute the derivative of each of the following functions, using the definition of derivative or Formulas (9) to (13) above.

1. $f(x) = x + 1$
2. $t \to 3t + 2$
3. $s(t) = t^2 + 2t + 3$
4. $f(t) = \frac{1}{t^2}$
5. $y = \frac{x}{x+1}$
6. $w = z^3 + 5$
7. $f(w) = \frac{1}{w^2 + 1}$
8. $f(s) = 16s^2 - s$
9. $x \to \frac{x^2 + 1}{x}$
10. $f(t) = \sqrt{2t}$
11. $y = 2\sqrt{x}$
12. $s(t) = \sqrt{t^2 + 1}$
13. $y = \sqrt{x + 2}$
14. $f(x) = \frac{1}{\sqrt{x}}$
15. $f(w) = \frac{1}{\sqrt{w+2}}$
16. $r(t) = (t+1)^2$
17. $f(z) = z^2 + 2z + 1$
18. $f(x) = x^{3/2}$
19. $f(x) = \frac{1}{x^2}$
20. $f(t) = \frac{1}{(t+2)^2}$
21. $f(x) = (x+2)^3$
22. $f(x) = (\sqrt{x} + 1)^2$
23. $f(x) = \sqrt[3]{x^3 + 1}$
24. $s \to s^4$

25. The temperature in Asphalt City is approximately $T(t) = 91 + 6t - t^2$ degrees Fahrenheit, where t is the number of hours since noon. How fast is the temperature rising (or falling) at 6:00 P.M.?

26. The number of miles driven by a bus is given by $s(t) = 65t$, where t is the number of hours since 9:00 A.M., $0 \le t \le 3$. At what speed is the bus traveling at 11:00 A.M.?

27. The weight (in pounds) of a t-month-old calf is $W(t) = 20t + 60\sqrt{t} + 20$. How fast (in pounds per month) is the calf gaining weight at age 4 months?

5. THE MEAN VALUE THEOREM FOR INTEGRALS

In this section we consider a theorem that may appear to be of little interest in itself but that turns out to be the key to finding an easier method for computing integrals. This theorem is called the **Mean Value Theorem for Integrals.**

We have repeatedly used a large collection of thin rectangles to approximate areas. Is it possible to get by with just one rectangle? More precisely, is there a rectangle with area equal to the area of the region in Figure 2.14 and with base $[a, b]$? In an effort to answer this question, we might try drawing some rectangles and would probably decide that the rectangle in Figure 2.15 is too large and that the rectangle in Figure 2.16 is too small. By moving the top edge of the rectangle up and down, it seems reasonable that some rectangle should have the correct area, perhaps the rectangle in Figure 2.17.

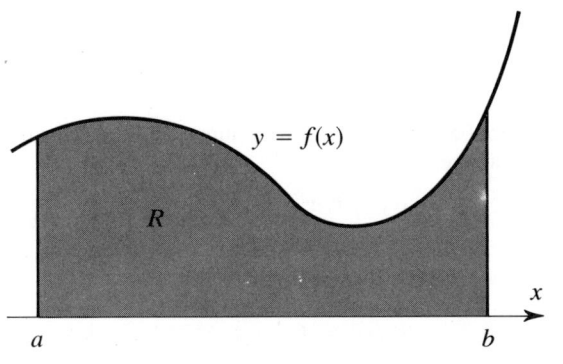

FIGURE 2.14

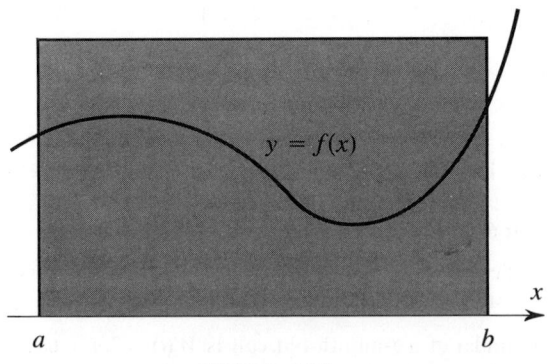

FIGURE 2.15

SECTION 5: THE MEAN VALUE THEOREM FOR INTEGRALS 83

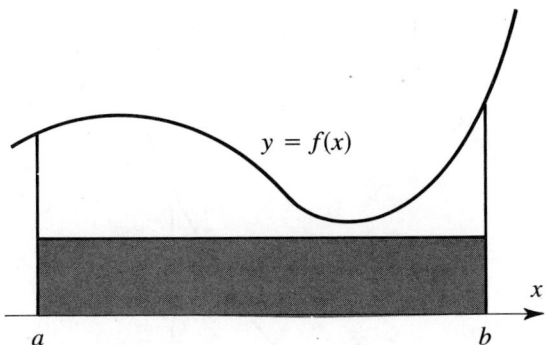

FIGURE 2.16

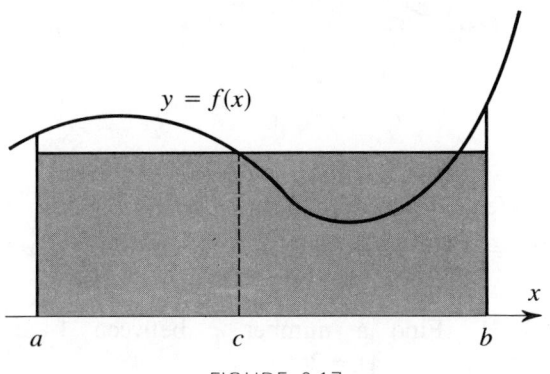

FIGURE 2.17

Notice that if the rectangle is correctly drawn, then

$$\text{Area } R = \int_a^b f(x)\,dx$$

$= $ the area of the rectangle in Figure 2.17

$= f(c)(b - a).$

That such a rectangle exists is the content of the following theorem.

THEOREM 5. (Mean Value Theorem for Integrals)   If f is continuous on $[a, b]$, then there exists a number c between a and b such that

(14) $$\int_a^b f(x)\,dx = f(c)(b - a).$$

EXAMPLE 20. Find a number c between 0 and 1 such that
$$\int_0^1 x^2 \, dx = c^2(1-0) = c^2.$$

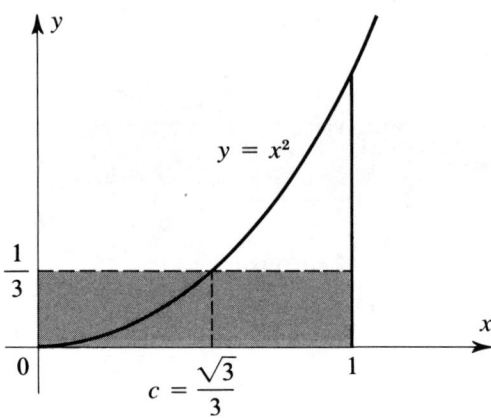

FIGURE 2.18

Solution. We found earlier that $\int_0^1 x^2 \, dx = \tfrac{1}{3}$. Thus if we let $c = \sqrt{\tfrac{1}{3}}$, then $c^2 = \tfrac{1}{3}$. Checking, we see that $(\sqrt{\tfrac{1}{3}})^2(1-0) = \tfrac{1}{3} = \int_0^1 x^2 \, dx$. (See Figure 2.18.)

EXAMPLE 21. Find a number c between 1 and 3 such that $\int_1^3 (x^2 - 3x) \, dx = (c^2 - 3c)(3-1) = 2c^2 - 6c$.

Solution. We have
$$\int_1^3 (x^2 - 3x) \, dx = \int_1^3 x^2 \, dx - 3 \int_1^3 x \, dx$$
$$= \frac{3^3 - 1^3}{3} - 3 \frac{3^2 - 1^2}{2}$$
$$= \frac{26}{3} - 12$$
$$= -\frac{10}{3}$$

by Example 27 of Chapter 1. If c satisfies $-10/3 = 2c^2 - 6c$, then $3c^2 - 9c + 5 = 0$. Using the quadratic formula, we see that $c = (9 \pm \sqrt{21})/6$. Since $(9 - \sqrt{21})/6$ is not between 1 and 3, we must choose $(9 + \sqrt{21})/6$. The location of $(9 + \sqrt{21})/6$ is shown in Figure 2.19.

SECTION 5: THE MEAN VALUE THEOREM FOR INTEGRALS

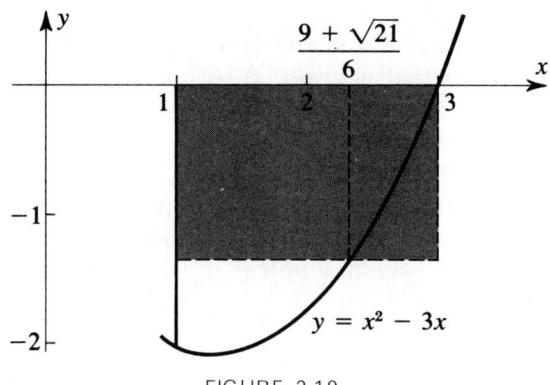

FIGURE 2.19

The number

$$f(c) = \frac{1}{b-a}\int_a^b f(x)\,dx$$

is called the *average value of f on* $[a,b]$. This is a calculus analog of the concept of the average or mean $(a_1 + a_2 + \cdots + a_n)/n$ of a finite collection of numbers $a_1, a_2, \ldots, a_n$.

PROBLEMS

In Problems 1 through 6, compute a number c between a and b that satisfies the Mean Value Theorem for Integrals.

1. $f(x) = x + 1$, $a = 1$, $b = 3$
2. $f(x) = x^2$, $a = 1$, $b = 2$
3. $f(x) = x^2 + x + 1$, $a = -1$, $b = 3$
4. $f(x) = x^2 + x$, $a = 0$, $b = 1$
5. $f(x) = x^2 + 2$, $a = -2$, $b = 2$
6. $f(x) = 3x^2 + 2x$, $a = 1$, $b = 7$
7. Justify your choice of c in Problem 1 by using some geometry you learned in high school (or earlier).
8. Find the average value of $f(x) = x^3$ on $[0, 1]$.
9. Find the average value of $f(x) = 1 + x + x^2$ on $[1, 3]$.
10. Find the average value of $f(x) = x^2 - 1$ on $[-1, 1]$.
11. A garment worker can sew buttons on shirts at the rate of $b(t) = 100 - 2t - 0.3t^2$ buttons per hour, where t is the number of hours since the worker began a shift. What is the average number of buttons per hour if the worker works an 8-hour shift? If the worker works a 12-hour shift?

86 CHAPTER 2: DIFFERENTIATION

6. FINDING THE FUNCTION IF AREA IS KNOWN

Consider the case of the absent-minded student. (Professors are not the only people who can be absent-minded.) After finding that

$$\int_1^x f(t)\,dt = x^3 - 1$$

for each x between 1 and 2, he could not remember what function he had started with. In an attempt to reconstruct f, he drew a picture as in Figure 2.20. Since he did not know what the graph of f looks like, he drew the graph of some arbitrary function.

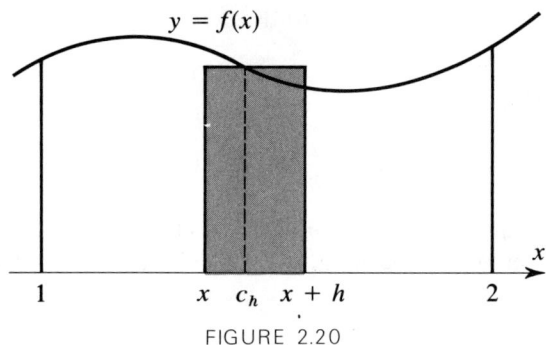

FIGURE 2.20

By the Mean Value Theorem for Integrals,

$$\int_x^{x+h} f(t)\,dt = f(c_h)(x+h-x)$$
$$= f(c_h) \cdot h$$

for some number c_h between x and $x + h$. Thus

$$f(c_h) = \frac{\int_x^{x+h} f(t)\,dt}{h}$$

$$= \frac{\int_1^{x+h} f(t)\,dt - \int_1^x f(t)\,dt}{h}$$

$$= \frac{[(x+h)^3 - 1] - (x^3 - 1)}{h}$$

$$= \frac{1}{h}(x^3 + 3x^2h + 3xh^2 + h^3 - x^3)$$

$$= 3x^2 + 3xh + h^2$$

for every $h \neq 0$. The number c_h must be close to x for h close to zero. The student remembered that the function f was continuous, which meant that since c_h was close to x, then $f(c_h)$ must be close to $f(x)$. But $f(c_h) = 3x^2 + 3xh + h^2$ tends to $3x^2$ as h tends to zero, so the student concluded that $f(x) = 3x^2$. "Oh yes, I remember now," sighed the student, "I started with $f(x) = 3x^2$."

Let us analyze the preceding solution more carefully. By knowing enough about the area under a graph, the student was able to find the equation of the graph.

Suppose that f is continuous on $[a, b]$ and assume that $\int_a^x f(t)\,dt$ is known for each x between a and b, say $\int_a^x f(t)\,dt = F(x)$. Then for each number $h \neq 0$, it follows that $\int_x^{x+h} f(t)\,dt = f(c_h)(x + h - x) = f(c_h) \cdot h$ for some c_h between x and $x + h$ by the Mean Value Theorem for Integrals. Thus dividing both sides by h,

$$f(c_h) = \frac{\int_x^{x+h} f(t)\,dt}{h}$$

$$= \frac{\int_a^{x+h} f(t)\,dt - \int_a^x f(t)\,dt}{h}$$

$$= \frac{F(x+h) - F(x)}{h}.$$

Now $\lim_{h \to 0} f(c_h) = f(x)$ because f is continuous at x, and

$$\lim_{h \to 0} \frac{F(x+h) - F(x)}{h} = F'(x)$$

by the definition of derivative. Taking the limit on both sides of the equation

$$f(c_h) = \frac{F(x+h) - F(x)}{h}$$

as h tends to zero, we get

$$f(x) = \lim_{h \to 0} f(c_h)$$

$$= \lim_{h \to 0} \frac{F(x+h) - F(x)}{h}$$

$$= F'(x).$$

We can now state the following theorem.

THEOREM 6. If f is continuous on $[a, b]$, then

(15)
$$\frac{d}{dx} \int_a^x f(t)\, dt = f(x)$$

for each x between a and b.

Since every differentiable function is continuous, it follows that the function $F(x) = \int_a^x f(t)\, dt$ is continuous.

EXAMPLE 22. Water is flowing out of a tank. After t minutes, $60t - t^2$ gallons have poured out of the tank, $0 \le t \le 30$. It is known from physics that $60t - t^2 = \int_0^t f(x)\, dx$, where $f(t)$ is the rate of flow in gallons per minute. What is the rate of flow?

Solution. By Theorem 6,

$$f(t) = \frac{d}{dt} \int_0^t f(x)\, dx$$

$$= \frac{d}{dt}(60t - t^2)$$

$$= 60 - 2t.$$

Notice that at $t = 0$, the water starts flowing out of the tank at a rate of 60 gallons per minute; and at $t = 30$, the rate of flow is zero. A physical interpretation might be that the water pressure at the outlet depends on the depth of water in the tank, and after 30 minutes the tank is empty.

PROBLEMS

Find the function f in Problems 1 through 12.

1. $\int_0^x f(t)\, dt = x$

2. $\int_0^a f(x)\, dx = a^2$

3. $\int_0^x f(t)\, dt = x + x^2$

4. $\int_a^z f(x)\, dx = z^3 + 3z - a^3 - 3a$

5. $\int_0^z f(y)\, dy = z^3 + z^2$

6. $\int_b^t f(z)\, dz = \frac{1}{t} - \frac{1}{b}$

7. $\int_0^x f(t)\, dt = x^4$

8. $\int_1^x f(t)\, dt = \sqrt{x - 1}$

9. $\displaystyle\int_a^x f(t)\,dt = \dfrac{1}{x+1} - \dfrac{1}{a+1}$

10. $\displaystyle\int_0^t f(x)\,dx = \sqrt{t+1} - 1$

11. $\displaystyle\int_1^a f(x)\,dx = \dfrac{a^2+1}{a} - 2$

12. $\displaystyle\int_0^x f(t)\,dt = x^5$

13. Check your answer to Problem 3 above by computing the integral.

14. Check your answer to Problem 5 above by computing the integral.

15. Compute $\int_0^x (5 + t + t^2)\,dt$; then use the method developed in this section to recover the function $f(t) = 5 + t + t^2$.

16. Compute $\int_0^x t^3\,dt$; then recover the function $f(t) = t^3$.

7. THE FUNDAMENTAL THEOREM OF CALCULUS

Suppose that f and g are functions that differ by a constant; that is, $f(x) = g(x) + C$ for all x. If f is differentiable,

$$f'(x) = \lim_{h \to 0} \frac{f(x+h) - f(x)}{h}$$

$$= \lim_{h \to 0} \frac{g(x+h) + C - [g(x) + C]}{h}$$

$$= \lim_{h \to 0} \frac{g(x+h) - g(x)}{h}$$

$$= g'(x),$$

and hence g is differentiable and $g' = f'$. We have proved the following lemma.

LEMMA 1. If f is differentiable on $[a, b]$ and $f(x) = g(x) + C$ for all x in $[a, b]$, then g is differentiable on $[a, b]$ and $g' = f'$.

The following converse is also true, but we shall not prove it here. (See Problem 14, Section 7 of Chapter 3.)

LEMMA 2. If f and g are differentiable on $[a, b]$ with $f' = g'$, then there is a real number C such that $f(x) = g(x) + C$ for all x between a and b.

Let $F_a(x) = \int_a^x f(t)\,dt$. (If $f(t) \geq 0$ on $[a, b]$, then $F_a(x)$ is simply the area under the graph of f between a and x, as in Figure 2.21.) We have seen that

90 CHAPTER 2: DIFFERENTIATION

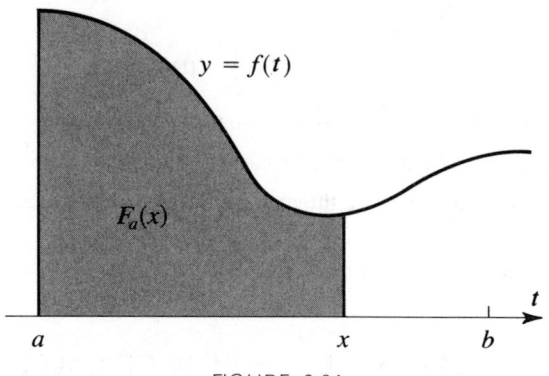

FIGURE 2.21

$F'_a(x) = f(x)$. If F is any other function such that $F' = f$, then by Lemma 2, $F(x) = F_a(x) + C$ for some constant C, and hence

$$\int_a^b f(x)\,dx = \int_a^b f(x)\,dx - \int_a^a f(x)\,dx$$
$$= F_a(b) - F_a(a)$$
$$= F_a(b) + C - [F_a(a) + C]$$
$$= F(b) - F(a).$$

We have just proved the following theorem.

THEOREM 7. (Fundamental Theorem of Calculus) Suppose that f is continuous on $[a, b]$ and let F be any function such that $F' = f$. Then

(16) $$\int_a^b f(t)\,dt = F(b) - F(a).$$

We will use the standard notation

(17) $$F(x)\big|_a^b = F(b) - F(a),$$

which reads "$F(x)$ evaluated from a to b equals $F(b)$ minus $F(a)$." For example,

$$(x^2 + \sqrt{x})\big|_1^4 = (4^2 + \sqrt{4}) - (1^2 + \sqrt{1})$$
$$= 16.$$

The following examples illustrate the significance of the Fundamental Theorem. Derivatives are relatively easy to compute, and the Fundamental Theorem allows us to compute integrals in case we can recognize functions as being derivatives of known functions.

SECTION 7: THE FUNDAMENTAL THEOREM OF CALCULUS

EXAMPLE 23. Compute $\int_2^4 x^2 \, dx$.

Solution. $\dfrac{d}{dx}(x^3) = 3x^2$, and hence

$$\int_2^4 x^2 \, dx = \frac{1}{3} \int_2^4 3x^2 \, dx$$

$$= \frac{1}{3} x^3 \bigg|_2^4$$

$$= \frac{56}{3}.$$

EXAMPLE 24. Compute $\int_a^b (Ax^2 + Bx + C) \, dx$.

Solution. By Theorem 3 of Chapter 1,

$$\int_a^b (Ax^2 + Bx + C) \, dx = A \int_a^b x^2 \, dx + B \int_a^b x \, dx + C \int_a^b dx.$$

Now by Theorem 4 of Section 4, we know that

$$\frac{d}{dx}(x^3) = 3x^2, \quad \frac{d}{dx}(x^2) = 2x, \quad \text{and} \quad \frac{d}{dx}(x) = 1,$$

so by the Fundamental Theorem of Calculus,

$$\int_a^b (Ax^2 + Bx + C) \, dx = \frac{A}{3} \int_a^b 3x^2 \, dx + \frac{B}{2} \int_a^b 2x \, dx + C \int_a^b dx$$

$$= \frac{A}{3} x^3 \bigg|_a^b + \frac{B}{2} x^2 \bigg|_a^b + Cx \bigg|_a^b$$

$$= \frac{A}{3}(b^3 - a^3) + \frac{B}{2}(b^2 - a^2) + C(b - a),$$

which agrees with Example 27 of Chapter 1.

EXAMPLE 25. Compute $\int_a^b x^n \, dx$, where n is a positive integer.

Solution. From Theorem 4 of Section 4,

$$\frac{d}{dx}(x^{n+1}) = (n+1)x^n.$$

Hence
$$\int_a^b x^n\,dx = \frac{1}{n+1}\int_a^b (n+1)x^n\,dx$$
$$= \frac{1}{n+1} x^{n+1}\Big|_a^b$$
$$= \frac{1}{n+1}(b^{n+1} - a^{n+1}).$$

EXAMPLE 26. Find the area of the region bounded by $x = 2$, $x = 5$, the x-axis, and the graph of $y = 1/\sqrt{x}$.

Solution. The area is equal to the integral $\int_2^5 dx/\sqrt{x}$. Since $d/dx(\sqrt{x}) = 1/(2\sqrt{x})$ by Example 17, we get

$$\int_2^5 \frac{dx}{\sqrt{x}} = 2\int_2^5 \frac{dx}{2\sqrt{x}}$$
$$= 2\sqrt{x}\Big|_2^5$$
$$= 2(\sqrt{5} - \sqrt{2}).$$
$$\approx 1.6437.$$

EXAMPLE 27. Compute $\int_1^2 \left(-\frac{1}{x^2}\right) dx$.

Solution. We know that
$$\frac{d}{dx}\left(\frac{1}{x}\right) = -\frac{1}{x^2}$$

by Example 19, and hence

$$\int_1^2 \left(-\frac{1}{x^2}\right) dx = \frac{1}{x}\Big|_1^2$$
$$= \frac{1}{2} - 1$$
$$= -\frac{1}{2}.$$

SECTION 7: THE FUNDAMENTAL THEOREM OF CALCULUS

EXAMPLE 28. Evaluate $\int_{-2}^{0} (2x + 2) \, dx$.

Solution. Since $D(x^2 + 2x) = 2x + 2$, it follows that

$$\int_{-2}^{0} (2x + 2) \, dx = (x^2 + 2x) \Big|_{-2}^{0}$$

$$= 0 - [(-2)^2 + 2(-2)]$$

$$= 0.$$

PROBLEMS

Use the Fundamental Theorem of Calculus to evaluate the following integrals.

1. $\int_{0}^{2} (4x^3 + 3x^2 + 2x) \, dx$

2. $\int_{-1}^{2} \frac{1}{x^2} \, dx$

3. $\int_{0}^{1} x^{10} \, dx$

4. $\int_{1}^{4} \frac{1}{\sqrt{x}} \, dx$

5. $\int_{5}^{4} (x^2 - 10x) \, dx$

6. $\int_{a}^{b} x^3 \, dx$

7. $\int_{0}^{x} t^2 \, dt$

8. $\int_{-1}^{1} (x + 4x^3) \, dx$

9. $\int_{1}^{3} \frac{1}{x^2} \, dx$

10. $\int_{1}^{2} \frac{x^3 + 1}{x^2} \, dx$

11. $\int_{0}^{2} 5 \, dx$

12. $\int_{x}^{x^2} (t^3 + 2t) \, dt$

13. $\int_{-1}^{-2} (3x^2 + 2x) \, dx$

14. $\int_{1}^{\sqrt{2}} 2x \, dx$

15. $\int_{5}^{8} \frac{2x^3 + 3}{x^2} \, dx$

16. $\int_{a}^{b} (b + a) \, dx$ (a and b constant)

17. $\int_{1}^{-1} \left(x + \frac{1}{\sqrt{x}} \right) dx$

18. $\int_{0}^{1} (4t^3 + 5t) \, dt$

19. $\int_{x}^{x^2+1} \frac{dt}{\sqrt{t}}$

20. $\int_{0}^{10} 10x^9 \, dx$

21. Charley Brown moves to a new town and starts making friends at the rate of $10/\sqrt{t+1}$ friends per day, where t is the number of days since he moved into town. How many friends has he made at the end of 5 weeks? At the end of 360 days?

$\left[\text{Hint: Argue that the number of friends in 5 weeks is} \right.$

$$\int_{0}^{35} \frac{10}{\sqrt{t+1}} \, dt$$

and show that

$$\frac{d}{dt}(20\sqrt{t+1}) = \frac{10}{\sqrt{t+1}}.$$

22. After sprouting, a certain plant grows at a rate of approximately $0.00003t^2 + 0.005t + 0.2$ inch per day for 100 days, where t is the number of days since sprouting. How tall should the plant be after 100 days? [*Answer:* 45 inches.]

8. THE ANTIDERIVATIVE

We have seen that integration and differentiation are very closely related. In fact, Theorem 6 states that

(18) $$\frac{d}{dx}\int_a^x f(t)\,dt = f(x),$$

or in words, "The derivative of the integral of a function is that function back again." (This is actually one form of the Fundamental Theorem of Calculus.)

If f and F are functions such that $F' = f$, then F is called an **antiderivative** of f. Thus $x \to x^3 + 5$ and $x \to x^3 - 3$ are both antiderivatives of $x \to 3x^2$.

If F and G are both antiderivatives of f, then by Lemma 2 of Section 7, $G(x) = F(x) + C$ for some constant C. It therefore follows that to get an arbitrary antiderivative of f, it is sufficient to find one antiderivative $F(x)$ and add a constant C. The **indefinite integral** of f is given by $\int f(x)\,dx = F(x) + C$, where F is an antiderivative of f and C is an unspecified constant. In particular, $\int 3x^2\,dx = x^3 + C$, where C remains unspecified.

Notice that the definite (or Riemann) integral $\int_a^b f(x)\,dx$ is a number, whereas the indefinite integral $\int f(x)\,dx$ is a collection of functions. In particular, $\int_0^1 x\,dx$ is the number $\frac{1}{2}$, and $\int x\,dx$ is the collection of functions $x \to \frac{1}{2}x^2 + C$, one for each choice of C.

EXAMPLE 29. Compute $\int x^3\,dx$.

Solution. Since

$$\frac{d}{dx}\left(\frac{1}{4}x^4\right) = x^3,$$

we get

$$\int x^3\,dx = \frac{1}{4}x^4 + C.$$

Many additional integrals can now be computed by simply *guessing*. In order to evaluate an integral $\int_a^b f(x)\,dx$, a thoughtful guess is made of a possible antiderivative $F(x)$. If $F'(x) = f(x)$, the guess is correct and the Fundamental Theorem can be used to evaluate the integral. Otherwise a second guess is made. This process may need to be repeated several times, and it *may* not work at all!

EXAMPLE 30. Compute $\int_1^2 \dfrac{1}{x^3}\,dx$.

Solution. Let $f(x) = x^{-3}$. As a first guess, take $F(x) = x^{-2}$. [Since $D(1/x) = -1/x^2$, possibly $D(1/x^2) = \pm 1/x^3$.] Computing, we have

$$D(F(x)) = \lim_{h \to 0} \frac{(x+h)^{-2} - x^{-2}}{h}$$

$$= \lim_{h \to 0} \frac{x^2 - x^2 - 2xh - h^2}{x^2(x+h)^2 h}$$

$$= \lim_{h \to 0} \frac{-2xh - h^2}{x^2(x+h)^2 h}$$

$$= \lim_{h \to 0} \frac{-2x - h}{x^2(x+h)^2}$$

$$= \frac{-2x}{x^2 \cdot x^2}$$

$$= \frac{-2}{x^3}$$

$$\neq \frac{1}{x^3}.$$

Thus the guess $F(x) = 1/x^2$ is off by a factor of -2. As a second guess, take $F(x) = -1/(2x^2)$. Computations similar to those above show that indeed

$$\frac{d}{dx}\left(\frac{-1}{2x^2}\right) = \frac{1}{x^3}.$$

Therefore

$$\int_1^2 \frac{1}{x^3}\,dx = \left.\frac{-1}{2x^2}\right|_1^2$$

$$= -\frac{1}{8} + \frac{1}{2}$$

$$= \frac{3}{8}.$$

CHAPTER 2: DIFFERENTIATION

PROBLEMS

Compute the following indefinite integrals.

1. $\int \dfrac{1}{x^2}\,dx$
2. $\int (7x^6 + 5x^4 + 3x^2)\,dx$
3. $\int \dfrac{3}{\sqrt{x}}\,dx$
4. $\int x^{99}\,dx$
5. $\int \dfrac{1}{t^3}\,dt$
6. $\int 5\,dx$
7. $\int t^2\,dt$
8. $\int \left(x^3 + \dfrac{1}{\sqrt{x}}\right)dx$

Find antiderivatives of the following functions.

9. $f(x) = \dfrac{3x^3 + 4}{x^2}$
10. $s(t) = \dfrac{1}{\sqrt{2t}}$
11. $y = \dfrac{1}{x} - \dfrac{x+1}{x}$
12. $f(x) = \dfrac{1}{(x+1)^3}$
13. $v(t) = 64 - 32t$
14. $g(x) = (x+1)^2$

REVIEW SECTION

For convenience, we repeat a few basic ideas and results that were developed in this chapter.

The **derivative** of a function f is given by

(a) $$f'(x) = \lim_{h \to 0} \dfrac{f(x+h) - f(x)}{h}$$

for all x where this limit exists. If $f'(x)$ exists, then f is said to be **differentiable** at x.

The **tangent line** to the graph of f at the point $(a, f(a))$ is given by

(b) $$y - f(a) = f'(a)(x - a)$$

if $f'(a)$ exists.

If f is differentiable at x, then f is continuous at x.
If f is continuous on $[a, b]$, then

(c) $$\dfrac{d}{dx}\int_a^x f(t)\,dt = f(x)$$

for all x between a and b.

If f is continuous on $[a, b]$ and F is any antiderivative of f, then

(d) $$\int_a^b f(x)\,dx = F(b) - F(a).$$

The following differentiation and integration formulas should be memorized:

$$\frac{d}{dx}(k) = 0 \quad \text{(where } k \text{ is any constant)}$$

$$\frac{d}{dx}(x) = 1 \qquad \int dx = x + C$$

$$\frac{d}{dx}(x^n) = nx^{n-1} \qquad \int x^n \, dx = \frac{1}{n+1} x^{n+1} + C$$

$$\frac{d}{dx}(\sqrt{x}) = \frac{1}{2\sqrt{x}} \qquad \int \frac{dx}{2\sqrt{x}} = \sqrt{x} + C$$

$$\frac{d}{dx}\left(\frac{1}{x}\right) = -\frac{1}{x^2} \qquad \int \frac{dx}{x^2} = -\frac{1}{x} + C$$

REVIEW PROBLEMS

Use the definition of derivative to compute each of the following derivatives.

1. $\dfrac{d}{dx}\sqrt{x+1}$

2. $D_x\left(\dfrac{1}{\sqrt{x+2}}\right)$

3. $D\left(\dfrac{1}{x^2+1}\right)$

4. $D_x(x^{1/3})$
   [Hint: Show that $(a-b)(a^2 + ab + b^2) = a^3 - b^3$ and solve for $b - a$.]

5. $\dfrac{d}{dx}\left(\dfrac{1}{x^3}\right)$

6. $\dfrac{d}{dx}(x^{2/3})$

Compute the following limits.

7. $\lim\limits_{x \to 1} \dfrac{x^2 - 1}{x - 1}$

8. $\lim\limits_{x \to a} \dfrac{x^2 - a^2}{x - a}$

9. $\lim\limits_{x \to 2} \dfrac{x^3 + 2x - 1}{x^2 + 5x - 13}$

10. $\lim\limits_{x \to a} \left(\dfrac{1}{\sqrt{x}} - \dfrac{1}{\sqrt{a}}\right)$

11. $\lim\limits_{x \to a} \dfrac{1}{x - a}\left(\dfrac{1}{\sqrt{x}} - \dfrac{1}{\sqrt{a}}\right)$

12. $\lim\limits_{t \to b} \dfrac{\sqrt{t} - \sqrt{b}}{t - b}$

Compute derivatives of the following functions.

13. $f(x) = \dfrac{1}{x^2 + x}$

14. $x \to \sqrt{2x + 3}$

15. $f(x) = (x^2 + 2)^2$

16. $g(t) = (t + 1)^3$

17. $g(z) = \dfrac{1}{z^2} - \dfrac{1}{z}$

18. $x \to \dfrac{1}{\sqrt{2x}}$

19. Let $C(x)$ be the total cost of producing a quantity x of a certain product. Assume that x can be any nonnegative real number (this is a common assumption if, for example, the product is clean water, wheat flour, or steel). The **marginal cost** is defined to be the derivative $C'(x)$. What is the marginal cost of producing 400 tons of cement if the total cost of producing x tons of cement is $C(x) = 5x + 15\sqrt{x}$ dollars?

20. A large manufacturing company is considering the installation of a data processing system to handle its bookkeeping. One of the criteria for the selection of such a system is the average daily cost of operation. Two systems are considered with daily costs

$$f(x) = \frac{x^2}{1000} - x + 300 \text{ for system } A$$

and

$$g(x) = \frac{2x^2}{1000} - 1.6x + 360 \text{ for system } B,$$

where x is the number of items handled daily by such a system.

a. Which system should be used if the average number x of items is approximately 400 and does not vary greatly from day to day?

b. Which should be chosen if the daily requirements varied widely from 100 to 700 items daily (each requirement equally likely)? [*Hint:* Compute the average values of the functions f and g on $[100, 700]$.]

CHAPTER 3

ELEMENTARY TECHNIQUES OF DIFFERENTIATION AND INTEGRATION

By applying the definition of derivative, we have been able to differentiate such functions as x^2 and x^3. Suppose that we let $f(x) = x^2 + x^3$. Is there some easy way to differentiate f without going all the way back to the definition of derivative? There is indeed, as we shall soon discover.

Addition is just one of the standard methods of obtaining new functions from old functions. Other methods discussed in Chapter 0 include products of functions, quotients of functions, and even functions of functions. In each case, we will find a rule that will allow us to differentiate the new, more complicated functions in terms of the derivatives of the old functions. The Fundamental Theorem of Calculus can then be used to derive some corresponding rules for evaluating integrals.

1. THE DERIVATIVE OF A SUM

Suppose that $f = u + v$, where u and v are both differentiable. Then

$$\frac{f(x+h) - f(x)}{h} = \frac{u(x+h) + v(x+h) - u(x) - v(x)}{h}$$

$$= \frac{u(x+h) - u(x)}{h} + \frac{v(x+h) - v(x)}{h}$$

and hence

$$f'(x) = \lim_{h \to 0} \frac{f(x+h) - f(x)}{h}$$

$$= \lim_{h \to 0} \left[\frac{u(x+h) - u(x)}{h} + \frac{v(x+h) - v(x)}{h} \right]$$

$$= \lim_{h \to 0} \frac{u(x+h) - u(x)}{h} + \lim_{h \to 0} \frac{v(x+h) - v(x)}{h}$$

$$= u'(x) + v'(x).$$

This result gives us the following rule:

(1) $$(u + v)' = u' + v',$$

or in words, "The derivative of a sum equals the sum of the derivatives." [More precisely, if any two of the functions u, v, and $u + v$ are differentiable at x, then so is the third, in which case $(u + v)'(x) = u'(x) + v'(x)$.]

EXAMPLE 1. $D(x^2 + x^3) = Dx^2 + Dx^3 = 2x + 3x^2.$

Now let f be a differentiable function and let c be some real number. Let cf be the function defined by $(cf)(x) = c \cdot f(x)$. Then

$$D[cf(x)] = \lim_{h \to 0} \frac{(cf)(x+h) - (cf)(x)}{h}$$

$$= \lim_{h \to 0} \frac{c \cdot f(x+h) - c \cdot f(x)}{h}$$

$$= c \cdot \lim_{h \to 0} \frac{f(x+h) - f(x)}{h}$$

$$= c \cdot f'(x).$$

This equation yields a second rule:

(2) $$(cf)' = cf'.$$

Restated in words, this rule becomes "The derivative of a constant times a function equals the constant times the derivative of the function."

If we let $c = -1$ in Rule (2), we get

$$(-f)' = [(-1) \cdot f]' = (-1)f' = -f'.$$

Thus
$$(f - g)' = [f + (-g)]' = f' + (-g)' = f' - g'.$$
Combining this with (1) and (2), it follows that

(3) $$(af \pm bg)' = af' \pm bg',$$

where a and b are any real numbers and f and g are any differentiable functions. The symbol $\pm$ in Rule (3) means $(af + bg)' = af' + bg'$ and $(af - bg)' = af' - bg'$.

EXAMPLE 2. Differentiate $f(x) = 3x + 5x^3$.

Solution.
$$\begin{aligned} f'(x) &= D(3x + 5x^3) \\ &= 3Dx + 5Dx^3 \\ &= 3 \cdot 1 + 5 \cdot 3x^2 \\ &= 3 + 15x^2. \end{aligned}$$

EXAMPLE 3. Differentiate the function
$$t \to \frac{3t^2 - 1}{t}.$$

Solution. Rewriting gives
$$\frac{3t^2 - 1}{t} = 3t - \frac{1}{t}.$$
Hence recalling that $D(1/t) = -1/(t^2)$, we have
$$\begin{aligned} D\left(\frac{3t^2 - 1}{t}\right) &= D\left(3t - \frac{1}{t}\right) \\ &= 3D(t) - D\left(\frac{1}{t}\right) \\ &= 3 + \frac{1}{t^2} \\ &= \frac{3t^2 + 1}{t^2}. \end{aligned}$$

EXAMPLE 4. Differentiate $H(x) = x + \sqrt{x} + \frac{1}{x}$.

Solution. Applying Rule (1) two times, we get

$$H'(x) = D(x + \sqrt{x}) + D\left(\frac{1}{x}\right)$$

$$= Dx + D\sqrt{x} + D\left(\frac{1}{x}\right)$$

$$= 1 + \frac{1}{2\sqrt{x}} + \left(-\frac{1}{x^2}\right)$$

$$= 1 + \frac{1}{2\sqrt{x}} - \frac{1}{x^2}.$$

PROBLEMS

Differentiate the following functions.

1. $f(x) = 3x^2 + 2x + 5$
2. $g(t) = t^3 + \dfrac{1}{t}$
3. $h(z) = 3\sqrt{z} + 29z^2$
4. $f(t) = at^2 + bt + c$
5. $f(x) = \dfrac{x + 1}{x}$
6. $s(t) = \frac{1}{4}t^4 + \frac{1}{3}t^3 + \frac{1}{2}t^2 + t$
7. $y = ax + b$
8. $y = x^2 + 2$
9. $f(x) = 3x^4 + 4x^3$
10. $g(t) = \sqrt{t} + t^3$
11. $f(x) = (x^2 + 1)^3$
12. $y = (x^3 + 2)^2$
13. $h(s) = \dfrac{s^2 + 1}{s}$
14. $t \to (t^2 + t + 1)^2$
15. $g(t) = (t + 1)(t + 2)$
16. $f(x) = (x^2 + 1)(x - 1)$
17. $G(t) = \dfrac{\sqrt{t^3 + 1}}{t}$
18. $L(x) = (x + 2)^4$
    [*Hint:* Use the Binomial Theorem.]
19. $w \to \dfrac{1}{w} - w^5$
20. $f(x) = 2x + 4 - \sqrt{x}$
21. $g(t) = (2t + 3t^2)^2$
22. $h(z) = \dfrac{1}{z} - (z + 1)^2$
23. $K(s) = s^4 - (s + 1)^4$
24. $W(t) = 37t^3 + \sqrt{t}$

2. THE INTEGRAL OF A SUM

Rule (3) can be used to help evaluate integrals, as the following example illustrates.

SECTION 2: THE INTEGRAL OF A SUM

EXAMPLE 5. Compute $\int_1^3 (1 + 2x)\,dx$.

Solution. Since $D(x) = 1$ and $D(x^2) = 2x$, it follows from Rule (3) and the Fundamental Theorem of Calculus that

$$\int_1^3 (1 + 2x)\,dx = (x + x^2)\Big|_1^3$$
$$= (3 + 3^2) - (1 + 1^2)$$
$$= 10.$$

Suppose that $F' = f$ and $G' = g$. Then $(AF \pm BG)' = AF' \pm BG' = Af \pm Bg$ for any real numbers A and B, and hence $AF \pm BG$ is an antiderivative of $Af \pm Bg$. Thus

$$\int_a^b [Af(x) \pm Bg(x)]\,dx = [AF(x) \pm BG(x)]\Big|_a^b$$
$$= [AF(b) \pm BG(b)] - [AF(a) \pm BG(a)]$$
$$= A[F(b) - F(a)] \pm B[G(b) - G(a)]$$
$$= A\int_a^b f(x)\,dx \pm B\int_a^b g(x)\,dx.$$

This result gives the following rule for integrals:

(4) $$\int_a^b (Af \pm Bg) = A\int_a^b f \pm B\int_a^b g$$

whenever all these integrals exist. More precisely, if any two of the integrals exist, then so does the third, in which case (4) holds.

Recall that this rule was obtained earlier by entirely different methods (see Theorem 3 of Chapter 1).

EXAMPLE 6. Evaluate $\int_1^2 \left(5x + \dfrac{3}{2\sqrt{x}}\right) dx$.

Solution. An antiderivative of x is $x^2/2$ and an antiderivative of $1/(2\sqrt{x})$ is $\sqrt{x}$. Hence

$$\int_1^2 \left(5x + \frac{3}{2\sqrt{x}}\right) dx = 5\int_1^2 x\,dx + 3\int_1^2 \frac{1}{2\sqrt{x}}\,dx$$
$$= 5 \cdot \frac{1}{2}x^2\Big|_1^2 + 3\sqrt{x}\Big|_1^2$$

$$= \frac{5}{2}(4-1) + 3(\sqrt{2}-1)$$

$$= \frac{9}{2} + 3\sqrt{2}.$$

EXAMPLE 7. Evaluate $\int_{-1}^{1} \left(\frac{1}{x} + 3x^2\right) dx.$

Solution. The function $f(x) = 1/x$ is not continuous on the interval $[-1, 1]$ because it is not defined at zero. The Fundamental Theorem does not apply, and we are unable to evaluate the integral without additional techniques. Such integrals will be investigated in more detail in Chapter 8, where methods will be given that could show that this particular integral does not exist.

PROBLEMS

Compute the following integrals.

1. $\int_{1}^{2} \left(\frac{1}{x^2} + 3x^2\right) dx$

2. $\int_{1}^{3} \frac{x^2 + 1}{x^2} dx$

3. $\int_{-1}^{1} (4x^3 + 2x) dx$

4. $\int \left(\frac{1}{2\sqrt{z}} + z\right) dz$

5. $\int_{0}^{1} (ax^2 + bx + c) dx$

6. $\int_{4}^{1} (1 + t + t^2) dt$

7. $\int \left(\frac{1}{\sqrt{x}} + \frac{1}{x^2}\right) dx$

8. $\int_{0}^{1} (10x^9 + 9x^8) dx$

9. $\int_{1}^{4} \left(\frac{1}{\sqrt{x}} - 4x\right) dx$

10. $\int_{0}^{1} (x^2 + 1)^2 6x \, dx$

[*Hint:* Expand $(x^2 + 1)^2 6x$.]

11. $\int_{1}^{2} \frac{x^3 + 1}{x^2} dx$

12. $\int_{-1}^{1} (ax^3 + bx) dx$

13. $\int_{-1}^{1} (ax^4 + bx^2 + c) dx$

14. $\int \frac{5x^4 + 4x^2 + 1}{x^2} dx$

15. $\int_{a}^{b} (x^3 + ax) dx$

16. $\int_{0}^{t} (5x^4 + 3x^2) dx$

3. THE DERIVATIVE OF A PRODUCT

Suppose that we wish to differentiate $f(x) = x^{3/2} = x\sqrt{x}$. We know how to differentiate x and $\sqrt{x}$. Can we use this information to help differentiate

SECTION 3: THE DERIVATIVE OF A PRODUCT

the product $x\sqrt{x}$? Since it works for sums, we might be tempted to assume that the derivative of the product equals the product of the derivatives. However, this assumption would lead to nothing but trouble. For example,

$$\frac{d}{dx}(x^2) \cdot \frac{d}{dx}(x) = 2x \cdot 1 = 2x,$$

whereas

$$\frac{d}{dx}(x^3) = 3x^2.$$

The derivative of a product will require a bit more effort. Still, it would be useful if we could find a formula for $f'(x)$ in terms of the derivatives $u'(x)$ and $v'(x)$, where $f(x) = u(x)v(x)$. We start with the difference quotient

$$\frac{f(x+h) - f(x)}{h} = \frac{u(x+h)v(x+h) - u(x)v(x)}{h}.$$

If we are to write $f'(x)$ in terms of $u'(x)$ and $v'(x)$, it will be necessary to look for the difference quotients

$$\frac{u(x+h) - u(x)}{h} \quad \text{and} \quad \frac{v(x+h) - v(x)}{h}.$$

To do so, we add and subtract the term $u(x+h)v(x)$. This step gives

$$\frac{u(x+h)v(x+h) - u(x+h)v(x) + u(x+h)v(x) - u(x)v(x)}{h}$$

$$= \frac{u(x+h)v(x+h) - u(x+h)v(x)}{h} + \frac{u(x+h)v(x) - u(x)v(x)}{h}$$

$$= u(x+h)\frac{v(x+h) - v(x)}{h} + v(x)\frac{u(x+h) - u(x)}{h}.$$

Taking limits on both sides as h approaches zero, we get

$$f'(x) = \lim_{h \to 0} \frac{f(x+h) - f(x)}{h}$$

$$= \lim_{h \to 0} u(x+h) \lim_{h \to 0} \frac{v(x+h) - v(x)}{h}$$

$$+ \lim_{h \to 0} v(x) \lim_{h \to 0} \frac{u(x+h) - u(x)}{h}$$

$$= u(x)v'(x) + v(x)u'(x).$$

[Recall that every differentiable function is continuous. In particular, $\lim_{h \to 0} u(x+h) = u(x)$ because u is differentiable at x.]

CHAPTER 3: DIFFERENTIATION AND INTEGRATION

We have derived the following rule:

(5) $$(uv)' = uv' + vu'.$$

In words, "The derivative of the product equals the first times the derivative of the second plus the second times the derivative of the first."

EXAMPLE 8. Differentiate $f(x) = x^{3/2} = x\sqrt{x}$.

Solution.
$$f'(x) = x\frac{d}{dx}(\sqrt{x}) + \sqrt{x}\frac{d}{dx}(x)$$
$$= x\frac{1}{2\sqrt{x}} + \sqrt{x} \cdot 1$$
$$= \frac{1}{2} \cdot \sqrt{x} + \sqrt{x}$$
$$= \frac{3}{2} \cdot \sqrt{x}$$
$$= \frac{3}{2} x^{1/2}.$$

EXAMPLE 9. Differentiate $w(z) = (z^2 + 2)(3z^3 + z)$.

Solution. We will use two methods. First, using the rule for differentiating a product, we obtain

$$w'(z) = (z^2 + 2)\frac{d}{dz}(3z^3 + z) + (3z^3 + z)\frac{d}{dz}(z^2 + 2)$$
$$= (z^2 + 2)(9z^2 + 1) + (3z^3 + z)(2z)$$
$$= 9z^4 + 19z^2 + 2 + 6z^4 + 2z^2$$
$$= 15z^4 + 21z^2 + 2.$$

We may also multiply out the two terms and then differentiate term by term:

$$w'(z) = \frac{d}{dz}(3z^5 + 7z^3 + 2z)$$
$$= 15z^4 + 21z^2 + 2.$$

PROBLEMS

Differentiate the following functions.
1. $f(x) = (x^3 + 2)^2$
    $= (x^3 + 2)(x^3 + 2)$
2. $g(x) = (x^2 + 1)(x - 1)$
3. $f(x) = x^{5/2}$
4. $h(t) = (t + 1)(t + 2)$
5. $k(x) = \sqrt{x}(x + 2)$
6. $f(x) = x^2(x + 1)(x - 2)$
7. $g(x) = x^{n+0.5}$
   (n is a positive integer.)
8. $h(x) = (x + 2)^4$
9. $k(x) = \dfrac{1}{x} \cdot \sqrt{x} + x^3$
10. $f(x) = \dfrac{1}{x^2} = \dfrac{1}{x} \cdot \dfrac{1}{x}$
11. $g(x) = (ax + b)^2$
12. $f(x) = (x^2 + 3x + 4)(x^3 + 1)$
13. $f(x) = 5x^2(x^3 + 1)$
14. $h(z) = (\sqrt{z} + 1)(z^2 + 1)$
15. $f(z) = (z^2 + 1)^3$
16. $g(t) = (t^2 - 1)(t^2 + 1)$

4. THE DERIVATIVE OF A QUOTIENT

We can use the product rule for derivatives to obtain a rule for differentiating quotients. Let $f(x) = u(x)/v(x)$ and assume that $f'(x)$, $u'(x)$, and $v'(x)$ all exist and $v'(x) \neq 0$. Then $u(x) = f(x)v(x)$, and hence

$$u'(x) = \frac{d}{dx}[f(x)v(x)]$$
$$= f(x)v'(x) + v(x)f'(x).$$

Solving for $f'(x)$, we get

$$f'(x) = \frac{u'(x) - f(x)v'(x)}{v(x)}$$
$$= \frac{u'(x) - [u(x)/v(x)]v'(x)}{v(x)}$$
$$= \frac{v(x)u'(x) - u(x)v'(x)}{[v(x)]^2}.$$

We have derived the *quotient rule* for derivatives:

(6) $$\left(\frac{u}{v}\right)' = \frac{vu' - uv'}{v^2}.$$

In words, "The derivative of the quotient equals the denominator times the derivative of the numerator, minus the numerator times the derivative of the denominator, all divided by the square of the denominator."

108   CHAPTER 3: DIFFERENTIATION AND INTEGRATION

As a special case, let $u(x) = 1$. Then

$$\frac{d}{dx}\left[\frac{1}{v(x)}\right] = \frac{v(x)u'(x) - u(x)v'(x)}{[v(x)]^2}$$

$$= \frac{v(x) \cdot 0 - 1 \cdot v'(x)}{[v(x)]^2}$$

$$= -\frac{v'(x)}{[v(x)]^2}.$$

Hence we have the *reciprocal rule*

(7)
$$\left(\frac{1}{v}\right)' = -\frac{v'}{v^2}.$$

EXAMPLE 10.   Differentiate $f(x) = \dfrac{x^2 + 1}{x - 1}$.

Solution. $f'(x) = \dfrac{(x - 1)D(x^2 + 1) - (x^2 + 1)D(x - 1)}{(x - 1)^2}$

$$= \frac{2x(x - 1) - (x^2 + 1) \cdot 1}{(x - 1)^2}$$

$$= \frac{x^2 - 2x - 1}{(x - 1)^2}.$$

EXAMPLE 11.   Differentiate $f(x) = x^{-3/2} = \dfrac{1}{x^{3/2}}$.

Solution.   We have already seen that

$$\frac{d}{dx}(x^{3/2}) = \frac{3}{2}x^{1/2},$$

so

$$f'(x) = -\frac{(3/2)x^{1/2}}{(x^{3/2})^2}$$

$$= -\frac{3}{2}x^{-5/2}.$$

We have computed $D(x^a)$ for several choices of the exponent a, including $a = 1, 2, 3, 1/2, 3/2$, and $-3/2$. In each case, we observed that $D(x^a) = ax^{a-1}$.

SECTION 4: THE DERIVATIVE OF A QUOTIENT

This statement is true for any numerical exponent a. We will assume the following rule without proof.

For any real number a,

(8) $$\frac{d}{dx} x^a = a x^{a-1}.$$

EXAMPLE 12. The derivative of $f(x) = x^{-\pi}$ is given by $f'(x) = -\pi x^{-\pi - 1}$, where $\pi = 3.14159\ldots$ is the ratio of the circumference of a circle to its diameter.

EXAMPLE 13. $\dfrac{d}{dx}(x^{\sqrt{2}}) = \sqrt{2} x^{\sqrt{2} - 1}.$

PROBLEMS

Differentiate the following functions.

1. $f(x) = \dfrac{1}{x^3}$

2. $g(t) = \dfrac{\sqrt{t}}{t^2 + 1}$

3. $f(t) = \dfrac{t^2 + 1}{t^2 - 1}$

4. $h(x) = \dfrac{x^2 + 3x + 2}{\sqrt{x}}$

5. $f(t) = (t^3 + 1)^{-2} = \dfrac{1}{(t^3 + 1)} \cdot \dfrac{1}{(t^3 + 1)}$

6. $k(t) = (3t^{-5} + 1)(t - 1)$

7. $f(x) = x^{\sqrt{2} - 1}$

8. $s(t) = \dfrac{3t^2 + 2t + 1}{t^3 + 3t + 5}$

9. $h(x) = x^{-5/2}$

10. $f(x) = \dfrac{1}{x^2 + 3x + 2}$

11. $g(z) = (z^2 + 2) \cdot (z^{-2} + 1)$

12. $h(x) = \left(\dfrac{1}{x} + 1\right)\sqrt{x}$

13. $f(x) = \dfrac{3}{x^5 + 2x}$

14. $g(x) = \dfrac{x^2}{x^2 - 1}$

15. $h(z) = \dfrac{1}{\sqrt{z}}$

16. $f(x) = \dfrac{x^3 + 2x - 1}{x^3 - 2x + 1}$

17. $f(x) = \dfrac{1/x}{(1/x) - 1}$

18. $k(x) = \dfrac{ax + b}{cx + d}$

19. $f(x) = \dfrac{3x^2 + 2}{x^3 + 2x}$

20. $f(x) = \dfrac{x^{1/3} - 1}{x^{2/3}}$

21. $f(x) = \dfrac{x^3 + 2x}{3x^2 + 2}$

22. $g(x) = (x^2 + 1)^{-2}$

23. $h(x) = \dfrac{\sqrt{x} + x}{\sqrt{x} - x}$
24. $t \to (t^2 + 2t + 1)^{-2}$

5. THE DERIVATIVE OF A COMPOSITION

Consider the function $y = \sqrt{x^3 + 2}$. We can write it as a composition of simpler functions by letting $z = x^3 + 2$ so that $y = \sqrt{z}$. (See Chapter 0 for a discussion of the composition of functions.) We know that $dy/dz = 1/(2\sqrt{z})$ and $dz/dx = 3x^2$. Is there some way to use this information directly to compute dy/dx? There is indeed, as the following computations show. From the definition of derivative,

$$\begin{aligned}
\frac{dy}{dx} &= \lim_{h \to 0} \frac{\sqrt{(x+h)^3 + 2} - \sqrt{x^3 + 2}}{h} \\
&= \lim_{h \to 0} \frac{\sqrt{(x+h)^3 + 2} - \sqrt{x^3 + 2}}{h} \cdot \frac{\sqrt{(x+h)^3 + 2} + \sqrt{x^3 + 2}}{\sqrt{(x+h)^3 + 2} + \sqrt{x^3 + 2}} \\
&= \lim_{h \to 0} \frac{x^3 + 3x^2 h + 3xh^2 + h^3 + 2 - x^3 - 2}{h[\sqrt{(x+h)^3 + 2} + \sqrt{x^3 + 2}]} \\
&= \lim_{h \to 0} \frac{3x^2 + 3xh + h^2}{\sqrt{(x+h)^3 + 2} + \sqrt{x^3 + 2}} \\
&= \frac{3x^2}{2\sqrt{x^3 + 2}}.
\end{aligned}$$

Since $z = x^3 + 2$, this equation can be rewritten

$$\begin{aligned}
\frac{dy}{dx} &= \frac{3x^2}{2\sqrt{z}} \\
&= \frac{1}{2\sqrt{z}} \cdot 3x^2 \\
&= \frac{dy}{dz} \frac{dz}{dx}.
\end{aligned}$$

We see in this example, at least, that the derivative dy/dx can indeed be written in terms of dy/dz and dz/dx; namely, dy/dx is the *product* of dy/dz with dz/dx.

The fact that

$$\frac{dy}{dx} = \frac{dy}{dz} \frac{dz}{dx}$$

in this example is no accident. Recall that a derivative can be interpreted as a rate of change of a function. If

$$\frac{dy}{dz} = 4 \quad \text{and} \quad \frac{dz}{dx} = 3,$$

then, roughly speaking, y is changing four times as fast as z, z is changing three times as fast as x, so y should be changing 12 times as fast as x; that is,

$$\frac{dy}{dx} = 12 = 4 \cdot 3 = \frac{dy}{dz}\frac{dz}{dx}.$$

The following *composition rule* or *chain rule* is presented without proof. If f, g, and h are functions such that $f(x) = h(g(x))$, then

(9) $$f'(x) = h'(g(x))g'(x)$$

for all x such that $h'(g(x))$ and $g'(x)$ both exist. If we let $y = h(z)$ and $z = g(x)$, then this rule reads

$$\frac{dy}{dx} = \frac{dy}{dz}\frac{dz}{dx}.$$

Rule (9) is easy to remember in the latter form because it looks like an elementary cancellation rule for fractions, although these expressions are *not* to be confused with ordinary fractions.

EXAMPLE 14. Differentiate $h(x) = \sqrt{x^2 + 1}$.

Solution. Let $y = \sqrt{x^2 + 1}$ and $z = x^2 + 1$. Then $y = \sqrt{z}$ so that

$$\frac{dy}{dz} = \frac{1}{2\sqrt{z}}, \quad \frac{dz}{dx} = 2x,$$

and hence

$$\frac{dy}{dx} = \frac{dy}{dz}\frac{dz}{dx}$$

$$= \frac{1}{2\sqrt{z}} 2x$$

$$= \frac{x}{\sqrt{x^2 + 1}}$$

since $z = x^2 + 1$. [When differentiating a function $y = f(x)$, the derivative $dy/dx = f'(x)$ should be expressed entirely in terms of x.]

112   CHAPTER 3: DIFFERENTIATION AND INTEGRATION

EXAMPLE 15. Differentiate $y = (2x^3 + 3)^9$.

Solution. It would be possible to expand the polynomial $(2x^3 + 3)^9$ and then differentiate. However, it is much simpler to use the composition rule. Let $u = 2x^3 + 3$. Then $y = u^9$, and hence

$$\frac{dy}{dx} = \frac{dy}{du}\frac{du}{dx}$$

$$= 9u^8 6x^2$$

$$= 9(2x^3 + 3)^8 6x^2$$

$$= 54x^2(2x^3 + 3)^8.$$

EXAMPLE 16. Differentiate $f(x) = \dfrac{1}{x^3 + 2x + 1}$.

Solution. Let $g(x) = x^3 + 2x + 1$ and $h(z) = 1/z$. Then

$$f(x) = \frac{1}{x^3 + 2x + 1} = \frac{1}{g(x)} = h(g(x)).$$

Now $h'(z) = -1/z^2$ and

$$f'(x) = -\frac{1}{(g(x))^2}(3x^2 + 2)$$

$$= -\frac{3x^2 + 2}{(x^3 + 2x + 1)^2}.$$

PROBLEMS

Differentiate the following functions.

1. $f(x) = (3x^2 + 4)^3$
2. $f(x) = \sqrt{3x + 2}$
3. $g(z) = (\sqrt{z^2 + 1} + z)^{99}$
4. $s(t) = \sqrt{(5t + 2)^3 + 4}$
5. $f(t) = (t^3 - 1)^{1/3}$
6. $H(x) = \dfrac{1}{x^3} + x^3$
7. $F(w) = (w^9 + 2w + 1)^{15}$
8. $u(x) = \sqrt{x^3 + x + 1} + (x^3 + x + 1)^5$
9. $f(x) = \sqrt{x^2 + \sqrt{x^2 + 1}}$
10. $g(x) = (x^3 + 4x + 1)^{-3}$
11. $h(t) = \sqrt{t + \sqrt{t + \sqrt{t}}}$
12. $f(x) = [x^2 + (x^2 + 1)^2]^2$
13. $g(z) = \dfrac{1}{z + 1/(z + 1)}$
14. $x \to (x^8 + 1)^9$

15. $L(s) = (s^3 + 4)^5(s^2 + 5)^7$

16. $f(x) = \dfrac{x}{1 + x/(1 + x)}$

17. $T(x) = (x + 2)^9(x - 1)^8(x - 3)^7$

18. $r(s) = (s^2 + 1)^4(s^2 - 1)^5(s^3 + 2)^6$

19. $f(t) = [(t^2 + 1)^3 + (t^2 + 2)^4]^7$

20. $h(x) = \sqrt{\sqrt{x^3 - 1} + \sqrt{x^3 + 1}}$

6. INTEGRATION BY SUBSTITUTION

In this section we will see how the rule for differentiating a composition of functions yields a method for computing integrals. We first need some definitions. Let $y = f(x)$. We have remarked that the symbol dy/dx is not an ordinary fraction but is merely another expression for the derivative $f'(x)$, bequeathed to us by history. We will now use what may appear to be a "cheap trick" to enable us to treat dy/dx as if it were an ordinary fraction. Let us assume that dx is any real number different from zero. Then we define dy by the equation

(10) $$dy = f'(x)\, dx.$$

Dividing both sides by the nonzero real number dx, we get $dy/dx = f'(x)$, but now dy/dx is actually the quotient of dy by dx. The symbols dy and dx are called **differentials**.

EXAMPLE 17. Let $y = f(x) = 3x^2 + 2x$. Then $f'(x) = 6x + 2$, and hence $dy = f'(x)\, dx = (6x + 2)\, dx$. Notice that dy depends on x as well as dx. In particular, if $x = 5$ and $dx = 0.4$, then $dy = (6 \cdot 5 + 2)0.4 = 12.8$.

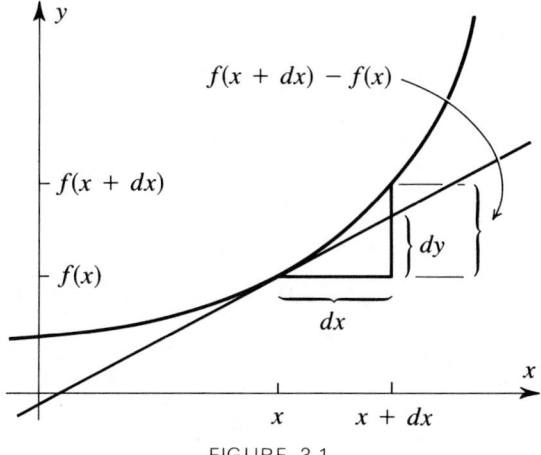

FIGURE 3.1

114   CHAPTER 3: DIFFERENTIATION AND INTEGRATION

The geometric meaning of dx and dy is shown in Figure 3.1. If dx is close to zero, then dy is a good approximation to $f(x+dx)-f(x)$. Notice that the slope of the tangent line is the quotient dy/dx.

Since $dy = f'(x)\,dx$, the approximation $dy \approx f(x+dx) - f(x)$ can be rewritten in the form

$$f(x+dx) \approx f(x) + f'(x)\,dx,$$

which has numerical applications. Suppose that we wish to estimate $\sqrt{10}$. Since we know that $\sqrt{9} = 3$, we let $x = 9$ and $dx = 1$, where $f(x) = \sqrt{x}$. Then $f'(x) = 1/(2\sqrt{x})$, so that

$$\begin{aligned}\sqrt{10} &= f(x+dx) \\ &\approx f(x) + f'(x)\,dx \\ &= 3 + \frac{1}{2\sqrt{9}} \\ &= 3 + \frac{1}{6} \\ &\approx 3.167.\end{aligned}$$

As a check, we note that $(3.167)^2 = 10.03$. A geometric interpretation is given in Figure 3.2.

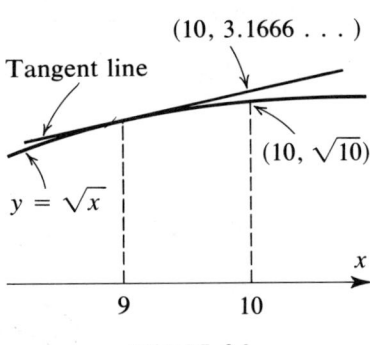

FIGURE 3.2

EXAMPLE 18.   Estimate $(1.01)^{10}$ by using differentials.

Solution.   Let $f(x) = x^{10}$. Then

$$\begin{aligned}f(x+dx) &\approx f(x) + f'(x)\,dx \\ &= x^{10} + 10x^9\,dx.\end{aligned}$$

At $x = 1$ and $dx = 0.01$,

$$(1.01)^{10} \approx 1^{10} + 10(1^9)(0.01)$$
$$= 1 + 0.1$$
$$= 1.1.$$

We have seen above how differentials can be used to make approximations. More basically, differentials are very handy objects to have around when computing integrals. In fact, the equation $dy = f'(x)\, dx$ is the basis for a technique known as **substitution**, one of the most useful techniques available for computing integrals. The following examples illustrate this technique.

EXAMPLE 19. Find $\int (x^2 + 1)^3 x\, dx$.

Solution. Let $y = x^2 + 1$. Then $dy = (dy/dx)\, dx = 2x\, dx$. We need to multiply and divide by 2 to get the factor $2x\, dx$:

$$\int (x^2 + 1)^3 x\, dx = \frac{1}{2} \int (x^2 + 1)^3 2x\, dx.$$

By replacing $x^2 + 1$ by y and $2x\, dx$ by dy in the preceding integral, we get

$$\int (x^2 + 1)^3 x\, dx = \frac{1}{2} \int y^3\, dy.$$

We know that $D_y(y^4) = 4y^3$, and hence

$$\frac{1}{2} \int y^3\, dy = \frac{1}{2} \cdot \frac{1}{4} y^4 + C$$
$$= \frac{1}{8} y^4 + C.$$

Replacing y again by $x^2 + 1$, we see that

$$\int (x^2 + 1)^3 x\, dx = \frac{1}{8}(x^2 + 1)^4 + C.$$

Checking, we obtain

$$\frac{d}{dx}\left[\frac{1}{8}(x^2 + 1)^4 + C\right] = \frac{1}{8} \cdot 4(x^2 + 1)^3 2x$$
$$= (x^2 + 1)^3 x,$$

which means that $\frac{1}{8}(x^2 + 1)^4 + C$ is indeed an antiderivative of $(x^2 + 1)^3 x$.

EXAMPLE 20. Find $\displaystyle\int \frac{(3x^2 + 4)\,dx}{(x^3 + 4x + 1)^3}$.

Solution. Let $u = x^3 + 4x + 1$. Then $du = (du/dx)\,dx = (3x^2 + 4)\,dx$, and hence

$$\int \frac{(3x^2 + 4)\,dx}{(x^3 + 4x + 1)^3} = \int \frac{du}{u^3}$$

$$= \int u^{-3}\,du$$

$$= -\frac{1}{2}u^{-2} + C$$

$$= \frac{-1}{2(x^3 + 4x + 1)^2} + C.$$

EXAMPLE 21. Compute $\displaystyle\int x\sqrt{x+1}\,dx$.

Solution. When square roots are involved, it is often useful to let the part under the radical be a perfect square. (Such substitutions are called **rationalizing** substitutions.) In this example we let $u = \sqrt{x+1}$ so that $u^2 = x + 1$. Differentiation on both sides gives $2u(du/dx) = 1$, or $2u\,du = dx$. (Remember that by using differentials du and dx, we can treat du/dx like any other fraction.) Also, $u^2 = x + 1$ implies $x = u^2 - 1$. Substituting for x in terms of u in the integral, we get

$$\int x\sqrt{x+1}\,dx = \int (u^2 - 1)u\,2u\,du$$

$$= \int (2u^4 - 2u^2)\,du$$

$$= \frac{2}{5}u^5 - \frac{2}{3}u^3 + C$$

$$= \frac{2}{5}(x+1)^{5/2} - \frac{2}{3}(x+1)^{3/2} + C.$$

EXAMPLE 22. Find $\displaystyle\int (\sqrt{x} + x)\frac{1 + 2\sqrt{x}}{2\sqrt{x}}\,dx$.

SECTION 6: INTEGRATION BY SUBSTITUTION

Solution. This operation may require a few false starts. Eventually we see that by letting $u = \sqrt{x} + x$, we get

$$du = \left(\frac{1}{2\sqrt{x}} + 1\right) dx = \frac{1 + 2\sqrt{x}}{2\sqrt{x}} dx,$$

so that

$$\int (\sqrt{x} + x) \frac{1 + 2\sqrt{x}}{2\sqrt{x}} dx = \int u \, du$$

$$= \frac{1}{2} u^2 + C$$

$$= \frac{1}{2} (\sqrt{x} + x)^2 + C.$$

EXAMPLE 23. Compute $\int_1^3 x\sqrt{x^2 - 1} \, dx$.

Solution. We use the rationalizing substitution $u = \sqrt{x^2 - 1}$ so that $u^2 = x^2 - 1$. Then $2u \, du = 2x \, dx$, or $u \, du = x \, dx$. Thus

$$\int_1^3 x\sqrt{x^2 - 1} \, dx = \int_1^3 \sqrt{x^2 - 1} \, x \, dx$$

$$= \int_{x=1}^{x=3} uu \, du$$

$$= \int_{x=1}^{x=3} u^2 \, du$$

$$= \frac{1}{3} u^3 \Big|_{x=1}^{x=3}$$

$$= \frac{1}{3} (x^2 - 1)^{3/2} \Big|_1^3$$

$$= \frac{1}{3} (8^{3/2} - 0)$$

$$= \frac{16\sqrt{2}}{3}$$

The notation $\int_{x=1}^{x=3} u^2 \, du$ denotes the fact that the limits of integration are from $x = 1$ to $x = 3$, *not* $u = 1$ to $u = 3$.

EXAMPLE 24. Find $\int_{-1}^{7} 2x \sqrt[3]{x+1}\, dx$.

Solution. If $u = \sqrt[3]{x+1}$, then $u^3 = x+1$, so that $3u^2\, du = dx$ and $x = u^3 - 1$. Thus

$$\int_{-1}^{7} 2x \sqrt[3]{x+1}\, dx = \int_{x=-1}^{x=7} 2(u^3 - 1)u 3u^2\, du$$

$$= 6 \int_{x=-1}^{x=7} (u^6 - u^3)\, du$$

$$= 6\left(\frac{1}{7}u^7 - \frac{1}{4}u^4\right)\Big|_{x=-1}^{x=7}$$

$$= \frac{6}{7}(x+1)^{7/3} - \frac{6}{4}(x+1)^{4/3}\Big|_{-1}^{7}$$

$$= \frac{600}{7}$$

$$\approx 85.714.$$

PROBLEMS

Use differentials to estimate the following numbers.

1. $\sqrt{26}$
2. $(1.02)^{50}$
3. $\dfrac{1}{99}$
4. $\sqrt[3]{26}$

5. $\sqrt{99}$
6. $\left(\dfrac{1}{99}\right)^2$
7. $(2.03)^{10}$
8. $\sqrt{15} - \sqrt[4]{15}$

9. $50^{3/2}$
10. $120^{3/2}$
11. $\sqrt{50}$
12. $\dfrac{1}{24}$

13. $63^{2/3}$

Compute the following integrals.

14. $\displaystyle\int_0^1 (x^3 + 3x)^4 (x^2 + 1)\, dx$
15. $\displaystyle\int x\sqrt{x^2 + 1}\, dx$

16. $\displaystyle\int_0^1 \frac{t^2}{\sqrt{t^3 + 1}}\, dt$
17. $\displaystyle\int_0^a x^3 \sqrt{x^2 + a^2}\, dx$

18. $\displaystyle\int \frac{1}{\sqrt{x}}(\sqrt{x} + 1)^9\, dx$
19. $\displaystyle\int x\sqrt[4]{x + 1}\, dx$

20. $\displaystyle\int_1^2 \frac{3x^4 + 5x^2 + 4}{x^2}\, dx$
21. $\displaystyle\int_0^1 \frac{x}{\sqrt{x^2 + 1}}\, dx$

22. $\displaystyle\int (x^2 + x)\sqrt{x+1}\, dx$

23. $\displaystyle\int x^5 \sqrt{x^3 + 1}\, dx$

24. $\displaystyle\int_1^2 \frac{x^2 + 1}{x^2}\, dx$

25. $\displaystyle\int \frac{dx}{(\sqrt{x} + 1)^2 \sqrt{x}}$

26. $\displaystyle\int \frac{3x^2}{\sqrt{x^3 + 1}}\, dx$

27. $\displaystyle\int_0^1 x(x+1)(x+2)\, dx$

7. THE MEAN VALUE THEOREM FOR DERIVATIVES

In the previous section we used the approximation
$$f(x + dx) \approx f(x) + f'(x)\, dx$$
for small dx. The following theorem provides a way to measure the accuracy of this approximation.

THEOREM 1. (Mean Value Theorem for Derivatives) If f is differentiable on an interval containing a and b, then

(11) $$f(b) = f(a) + f'(c)(b - a)$$

for some c between a and b.

This theorem has a geometric interpretation. The equation $f(b) = f(a) + f'(c)(b - a)$ can be rewritten as

(12) $$f'(c) = \frac{f(b) - f(a)}{b - a}.$$

The left side of this equation is the slope of the tangent line at the point $(c, f(c))$. The right-hand side is the slope of the line through the points $(a, f(a))$ and $(b, f(b))$. Two lines are parallel if and only if they have equal slopes. Hence the theorem states that there exists a point c between a and b such that the tangent line at $(c, f(c))$ is parallel to the line through the points $(a, f(a))$ and $(b, f(b))$, as in Figure 3.3.

Notice that the theorem says that there exists some number c, but it gives no hint as to how to find this number. Such a theorem is called an *existence theorem*. The theorem should be interpreted to mean that there exists *at least one* c with $f(b) = f(a) + f'(c)(b - a)$. Indeed, if the graph of f is a straight line, then *every* c between a and b satisfies $f(b) = f(a) + f'(c)(b - a)$.

By taking $x = a$ and $x + dx = b$, the theorem says that $f(x + dx) = f(x) + f'(\bar{x})\, dx$ for some $\bar{x}$ between x and $x + dx$. If f' does not change very much between x and $x + dx$, then $f'(\bar{x})$ is close to $f'(x)$, and hence $f(x + dx) = f(x) + f'(\bar{x})\, dx$ is close to $f(x) + f'(x)\, dx$.

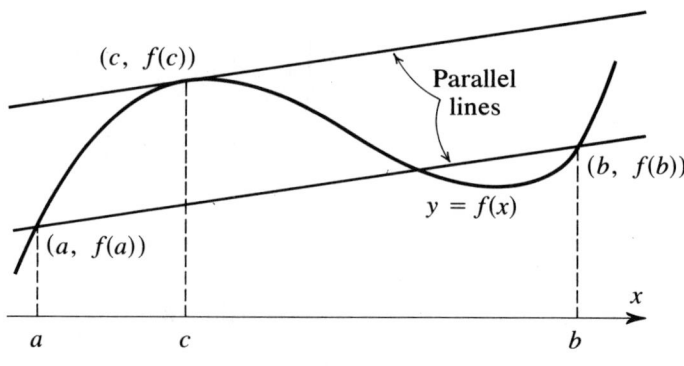

FIGURE 3.3

In particular, if $f(x) = \sqrt{x}$, then $f'(x) = 1/(2\sqrt{x})$ lies between $1/(2\sqrt{9}) = 1/6$ and $1/(2\sqrt{10})$ if x is between 9 and 10. Thus the error in the approximation $\sqrt{10} \approx 3\frac{1}{6}$ is less than

$$\left| \frac{1}{6} - \frac{1}{2\sqrt{10}} \right| \approx 0.0085528.$$

If f' is differentiable on an interval containing a and b, we can get a useful error estimate. The derivative of the derivative $(d/dx)[f'(x)]$ is called the **second derivative** of f at x and is denoted by various notations:

$$\frac{d^2}{dx^2} f(x), \quad f''(x), \quad f^{(2)}(x), \quad D_x^2 f(x), \quad \frac{d^2 y}{dx^2}, \quad \text{and} \quad y''$$

where $y = f(x)$. An **error bound** for the approximation $f(b) \approx f(a) + f'(a)(b - a)$ is given by

(13) $$|E| \leq M(b - a)^2,$$

where M is a number such that $|f''(x)| \leq M$ for all x between a and b.

In the differential notation $f(x + dx) \approx f(x) + f'(x)\,dx$, the expression for the error bound is written

(14) $$|E| \leq M\,dx^2,$$

where $|f''(t)| \leq M$ for all t between x and $x + dx$.

EXAMPLE 25. Use differentials to approximate $\sqrt[3]{26}$ and find an error bound.

Solution. Let $f(x) = \sqrt[3]{x}$. Then $f'(x) = \frac{1}{3}x^{-2/3}$ and $f''(x) = -\frac{2}{9}x^{-5/3}$. With $x = 27$ and $dx = -1$, we get

$$\sqrt[3]{26} \approx \sqrt[3]{27} + \frac{1}{3}(27)^{-2/3}(-1)$$

$$= 3 - \frac{1}{3} \cdot \frac{1}{9}$$

$$= \frac{80}{27}$$

$$\approx 2.96296.$$

For x between 26 and 27,

$$|f''(x)| = \left|\frac{2}{9x^{5/3}}\right|$$

$$= \frac{2}{9 \cdot x \cdot x^{2/3}}$$

$$\leq \frac{2}{9 \cdot 26 \cdot 4}$$

$$\approx 0.0021,$$

so we let $M = 0.0022$. An error bound is then given by

$$M\,dx^2 = M = 0.0022.$$

It follows that $\sqrt[3]{26}$ is some number between $2.96296 - 0.0022 = 2.96076$ and $2.96296 + 0.0022 = 2.96516$.

PROBLEMS

In Problems 1 through 4, find a number c between a and b such that $f(b) = f(a) + f'(c)(b - a)$.

1. $f(x) = x^3 + 4x + 1$, $a = 0$, $b = 2$
2. $f(x) = 3x^2 + 2x - 1$, $a = -1$, $b = 2$
3. $f(x) = 3x + 4$, $a = 1$, $b = 5$
4. $f(x) = \frac{1}{x} - x$, $a = 1$, $b = 3$

Use the approximation $f(b) \approx f(a) + f'(a)(b - a)$ to estimate each of the following numbers; then use Formula (13) to get an error bound.

5. $\dfrac{1}{99}$
6. $\left(\dfrac{1}{99}\right)^2$
7. $(2.03)^{10}$
8. $\sqrt{15} - \sqrt[4]{15}$
9. $50^{3/2}$
10. $120^{3/2}$
11. $\sqrt{50}$
12. $\dfrac{1}{24}$

122 CHAPTER 3: DIFFERENTIATION AND INTEGRATION

13. Use the Mean Value Theorem for Derivatives to show that if $f'(x) = 0$ for all x, then $f(x) = C$ for some C.

14. Show that if $f'(x) = g'(x)$ for every x, then $f(x) = g(x) + C$ for some constant C.

15. Derive the error bound (13). [*Hint*: Use the Mean Value Theorem for Derivatives with f replaced by f'.]

8. IMPLICIT DIFFERENTIATION

Let (a, b) and (c, d) be points in the plane. By constructing a right triangle with legs parallel to the coordinate axes, as in Figure 3.4, we see that

$$D^2 = (a - c)^2 + (b - d)^2$$

by the Pythagorean Theorem, which states that the square of the hypotenuse equals the sum of the squares of the two legs (see Appendix).

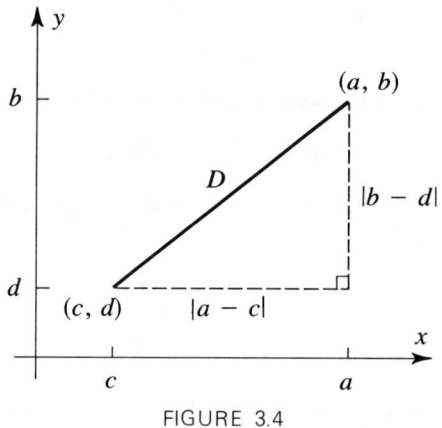

FIGURE 3.4

In particular, the **equation of a circle** with center at (h, k) and radius r is given by

(15) $$r^2 = (x - h)^2 + (y - k)^2,$$

since this circle is the locus of all points (x, y) that are a distance r from the point (h, k). Taking the positive square root of both sides, we get the **distance formula**

(16) $$r = \sqrt{(x - h)^2 + (y - k)^2}.$$

Given the equation of a circle and a point on that circle, we need a method for finding the equation of the tangent line. According to the definition of a tangent line, we must start with a function f that has a derivative. A circle is not the graph of a function, since a vertical line can intersect a circle in more than one point. We know from high school geometry, however, that circles do have tangent lines.

Suppose that we want to find a line tangent to the circle $(x - 1)^2 + (y - 2)^2 = 5^2$ at the point $(-2, 6)$, as in Figure 3.5. It is possible to solve for y in terms of x. Subtracting $(x - 1)^2$ from both sides, we get

$$(y - 2)^2 = 25 - (x - 1)^2,$$

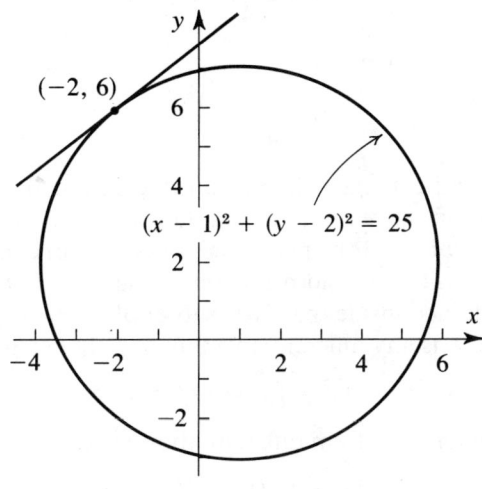

FIGURE 3.5

so that

$$y - 2 = \pm\sqrt{25 - (x - 1)^2},$$

and hence

$$y = 2 \pm \sqrt{25 - (x - 1)^2}.$$

By evaluating at $x = -2$, we see that the plus sign is needed, since the point $(-2, 6)$ is in the second quadrant. The equation

$$y = 2 + \sqrt{25 - (x - 1)^2}$$

(whose graph is the top half of the circle) defines y as a function of x; that is, each choice of x determines one and only one choice for y. Differentiating, we get

$$\frac{dy}{dx} = \frac{-2(x - 1)}{2\sqrt{25 - (x - 1)^2}}$$

$$= \frac{1-x}{\sqrt{25-(x-1)^2}},$$

so the slope of the tangent line is

$$\frac{1-(-2)}{\sqrt{25-(-2-1)^2}} = \frac{3}{\sqrt{25-9}}$$

$$= \frac{3}{4}.$$

Hence the equation of the tangent line is

$$y - 6 = \frac{3}{4}(x+2),$$

which simplifies to

$$y = \frac{3}{4}x + \frac{15}{2}.$$

In order to find a tangent line to a circle, a function was found whose graph was a subset of the circle; then a tangent line for the function was computed. Given *any* subset of the plane and a point (a, b) contained in the given subset, it is *always* possible to find a function f such that $f(a) = b$ and whose graph is entirely contained inside the given subset of the plane.

Suppose that f is any differentiable function that satisfies

$$(x-1)^2 + [f(x) - 2]^2 = 25$$

and $f(-2) = 6$. Then term-by-term differentiation yields

$$2(x-1) + 2[f(x) - 2]f'(x) = 0.$$

Solving for $f'(x)$, we get

$$f'(x) = \frac{1-x}{f(x) - 2},$$

which can be evaluated at $x = -2$ to obtain

$$f'(-2) = \frac{1-(-2)}{6-2}$$

$$= \frac{3}{4}.$$

It is comforting to end up with the same tangent line

$$y = \frac{3}{4}x + \frac{15}{2},$$

and there is good reason for it.

What was done above is known as **implicit differentiation.** We assume the existence of a function $y = f(x)$ that satisfies the given equation, but it is not necessary to know exactly what f equals. Notice that if $f(x)$ is a function, then so is $[f(x)]^2$. By the composition rule,

$$\frac{d}{dx}[f(x)^2] = 2f(x)f'(x).$$

In the preceding example it was not too difficult to solve for y in terms of x and then proceed to compute dy/dx, but even with this simple example it was easier to use implicit differentiation. The following is an example in which it would be nearly impossible to solve for y in terms of x, but yet a tangent line is easily found by using implicit differentiation.

EXAMPLE 26. Find a line tangent to the graph of $x^4 + xy^3 + y^4 = 3$, at the point $(1, 1)$.

Solution. Differentiating both sides implicitly gives

$$4x^3 + y^3 + 3xy^2 \frac{dy}{dx} + 4y^3 \frac{dy}{dx} = 0.$$

Solving for dy/dx, we get

$$\frac{dy}{dx} = -\frac{4x^3 + y^3}{3xy^2 + 4y^3}.$$

Evaluation at $x = 1, y = 1$ leads to

$$\frac{dy}{dx} = -\frac{4 + 1}{3 + 4}$$

$$= -\frac{5}{7}.$$

Hence the equation of the tangent line at $(1, 1)$ is

$$y - 1 = -\frac{5}{7}(x - 1),$$

which simplifies to

$$y = -\frac{5}{7}x + \frac{12}{7}.$$

EXAMPLE 27. Find the equation of the line tangent to the graph of $xy^2 + 4xy + 7x = 4$ at the point $(1, -1)$.

Solution. Implicit differentiation yields
$$y^2 + 2xyy' + 4y + 4xy' + 7 = 0,$$
so that
$$y' = -\frac{y^2 + 4y + 7}{2xy + 4x}.$$
Thus the slope of the tangent line at $(1, -1)$ is
$$y' = -\frac{1 - 4 + 7}{-2 + 4}$$
$$= -2,$$
so the equation of the tangent line is
$$y + 1 = -2(x - 1).$$
or
$$y = -2x + 1.$$

PROBLEMS

If $y = f(x)$, find $f'(x)$ in terms of x and y without first solving for y in terms of x in Problems 1 through 12.

1. $3x^2y + 4xy^3 = 1$
2. $y + 2xy - 1 + y^2 = 0$
3. $xy = 1$
4. $x^2 + y^2 = 4$
5. $x^2 + xy + y^2 = 1$
6. $2x + 3y + x^2 - y^2 = 4$
7. $1 + x + x^2y = 0$
8. $x^2 - 9y^2 + 2x - 8 = 0$
9. $xy = \sqrt{x^2 + y^2}$
10. $\sqrt{x} + \sqrt{y} + \sqrt{xy} = 3$
11. $\sqrt{x} + \sqrt{y} = 1$
12. $(y - 1)^2 = 4(x + 3)$
13. Find the tangent line to the graph of $4x^2 + y^2 - 8x + 4y = -3$ at the point $(2, -1)$.
14. Find the tangent line to the graph of $y^2 = 8x$ at the point $(2, 4)$.
15. Find the tangent line to the graph of $x^2 - y^2 = 1$ at the point $(\sqrt{5}, 2)$.
16. Find the tangent line to the graph of $x^2 - y^2 = 1$ at the point (a, b) on the graph.
17. Sketch the graph of $4x^2 + y^2 - 8x + 4y + 3 = 0$ by plotting points and sketching tangent lines at $(2, -1)$, $(0, -1)$, $(0, -3)$, $(2, -3)$, $(1, \sqrt{5} - 2)$, and $(1, -\sqrt{5} - 2)$.
18. What is the distance between the points $(1, -2)$ and $(-3, 2)$?
19. If Doc Holiday is 2 miles north and 1 mile west of the OK Corral, and Wyatt Erp is 3 miles south and 2 miles west of the OK Corral, how far apart are they?
20. What is the equation of the circle with center at $(1, -1)$ and with radius 3?
21. Is the point $(2, 3)$ inside or outside the circle with equation $(x + 1)^2 + (y + 3)^2 = 7^2$?

REVIEW SECTION

In this chapter the following rules were developed (a, b, c, and d are constants, whereas f and g are functions).

(a) $(f + g)' = f' + g'$

(b) $(cf)' = cf'$

(c) $(cf \pm dg)' = cf' \pm dg'$

(d) $\int_a^b (cf \pm dg) = c \int_a^b f \pm d \int_a^b g$

(e) $(fg)' = fg' + gf'$

(f) $\left(\dfrac{f}{g}\right)' = \dfrac{gf' - fg'}{g^2}$

(g) $\left(\dfrac{1}{g}\right)' = -\dfrac{g'}{g^2}$

(h) $\dfrac{d}{dx} x^a = a x^{a-1}$

(i) $\dfrac{dy}{dx} = \dfrac{dy}{dz} \dfrac{dz}{dx}$

The foregoing rules will be needed many times during the remainder of the book and should be committed to memory.

REVIEW PROBLEMS

Differentiate each of the following functions.

1. $f(x) = 3x^2 + 4$

2. $g(z) = z^4 + 3z$

3. $y = \dfrac{1}{x} - x$

4. $t \to (t + 2)^2$

5. $h(s) = \sqrt{s(s + 3)}$

6. $F(a) = (a^3 + 5)(a^7 + 5a^3 + 4)$

7. $x \to \dfrac{1}{x^2 + x + 1}$

8. $f(x) = \dfrac{x}{x^2 + x + 1}$

9. $g(x) = \dfrac{1}{\sqrt{x^2 + x + 1}}$

10. $z \to \dfrac{z}{z - 1}$

11. $C(A) = \dfrac{1}{A + 1/(A + 1)}$

12. $t \to \sqrt{t^2 + \sqrt{t^2 + 1}}$

13. $P(t) = \dfrac{t + 1}{t - 1}$

14. $f(x) = x\sqrt{1 + 2x}$

15. $R(t) = 1 + \dfrac{1}{t} + \dfrac{1}{t^2}$

16. $S(r) = (\sqrt{r} + 1)^{10}$

17. $a(z) = [1 + (z + 2)^7]^8$

18. $B(y) = y[y^2 + (y - 3)^5]^4$

19. $C(k) = k^2 + \dfrac{k - 3}{k^2 + 1}$

20. $x \to x\sqrt{(x + 1)^2}$

Compute the following integrals.

21. $\displaystyle\int_0^1 (3x + 2)\, dx$

22. $\displaystyle\int_1^2 \left(\dfrac{1}{x^2} - 3\right) dx$

23. $\displaystyle\int_0^4 \dfrac{2x\, dx}{\sqrt{x^2 + 9}}$

24. $\displaystyle\int_2^4 \dfrac{z^2 - 2z}{z^2 - 2z + 1}\, dz$

$\left[\text{Hint}: \dfrac{z^2 - 2z}{z^2 - 2z + 1} = 1 - \dfrac{1}{(z - 1)^2}.\right]$

25. $\displaystyle\int_0^1 2y(y^2 + 5)^4\, dy$

26. $\displaystyle\int_0^2 \dfrac{2x + 1}{\sqrt{x^2 + x + 1}}\, dx$

27. $\displaystyle\int_0^4 \sqrt{s(s + 3)}\, ds$

28. $\displaystyle\int_0^1 (t + 1)^5\, dt$

29. $\displaystyle\int_1^4 \dfrac{(\sqrt{r} + 1)^{10}}{\sqrt{r}}\, dr$

30. $\displaystyle\int_0^1 (x^2 + 2x + 4)(x^3 + 1)\, dx$

31. During a certain 100-day period, one type of tree will grow at a rate of approximately $0.05 + 0.0001t^{3/2} + 0.00003t^2$ inch per day, where t is the number of days since the beginning of the growing period. How many inches will the tree grow during the 100-day period?

32. An automobile travels at a rate of $100 - 10{,}000/(t + 10)^2$ feet per second, where t is the number of seconds since the car has started to move. How far does the automobile go during the first 10 seconds?

33. During the month of September 1948, potential voters registered at a rate of $500 + 10t + 3t^2$ people per day in a small Western state, where t is the number of business days since September 1. How many such people registered during the first 20 business days in September 1948?

34. During a recent year, a ski shop in Colorado made a profit at a rate of approximately $-4(t - 6)^4 + 288(t - 6)^2 - 500$ dollars per month, where t is the number of months since the beginning of the year. (Notice that during part of the summer, the shop made a negative profit; that is, expenses exceeded gross receipts and the shop took a loss.) How much profit did the ski shop make during the year?

35. The total cost of producing x liters of a product is $C(x) = 15 + 30x^{2/3}$ dollars. If marginal cost is the derivative of cost, find the marginal cost at 125-liters output. At what output is the marginal cost equal to $2?

36. Find the marginal revenue from sales of x liters of the product in Problem 35 if the revenue from selling x liters is $R(x) = 500x + 200\sqrt{x}$. (Marginal revenue is the derivative of revenue.)

CHAPTER 4
SKETCHING THE GRAPH OF A FUNCTION

One of the basic applications of differentiation is to curve sketching. Indeed we have already seen how tangent lines can be sketched at various points on the graph so that we can get some idea about which "direction" the curve lies at these points. We are now ready to develop more sophisticated techniques, based on the use of the derivative, which will yield much better sketches.

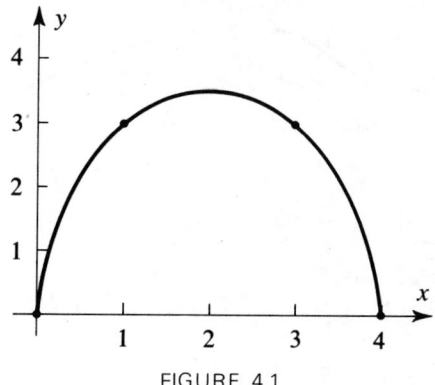

FIGURE 4.1

130   CHAPTER 4: SKETCHING THE GRAPH OF A FUNCTION

There are practical reasons for our desire to use some sophistication. For example, suppose that we wish to graph the function $f(x) = 16x - 20x^2 + 8x^3 - x^4$. The graph passes through the points $(0, 0)$, $(1, 3)$, $(3, 3)$, and $(4, 0)$, so a crude method of graphing (plot several points and connect them with a smooth curve) would yield something like Figure 4.1. However, the graph of f actually looks more like Figure 4.2. Sophistication is needed to be certain that there are no additional unexpected dips and humps.

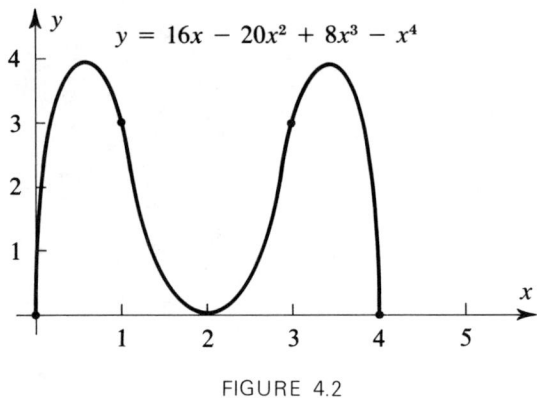

FIGURE 4.2

1. INCREASING AND DECREASING FUNCTIONS

We begin our study of graphing by analyzing the relationship between the graph of a function and its tangent lines.

A function f defined on $[a, b]$ is said to be **increasing** on $[a, b]$ if

(1) $\qquad f(x_1) < f(x_2) \quad \text{whenever} \quad a \le x_1 < x_2 \le b,$

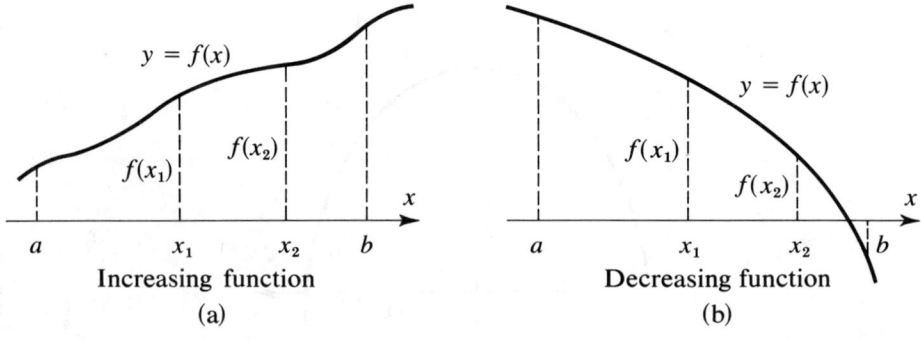

FIGURE 4.3

and *decreasing* on $[a, b]$ if

(2) $\qquad f(x_1) > f(x_2)\quad$ whenever $\quad a \le x_1 < x_2 \le b$.

Roughly speaking, an increasing function goes uphill from left to right, and a decreasing function goes downhill from left to right, as in Figure 4.3.

EXAMPLE 1. Assume that $m > 0$. Then the function $x \to mx + b$ is increasing everywhere, for if $x_1 < x_2$, then $mx_1 < mx_2$, and hence $mx_1 + b < mx_2 + b$. Similarly, if $m < 0$, then $x \to mx + b$ is decreasing. Thus the line $y = mx + b$ is increasing if and only if its slope is positive (see Figure 4.4).

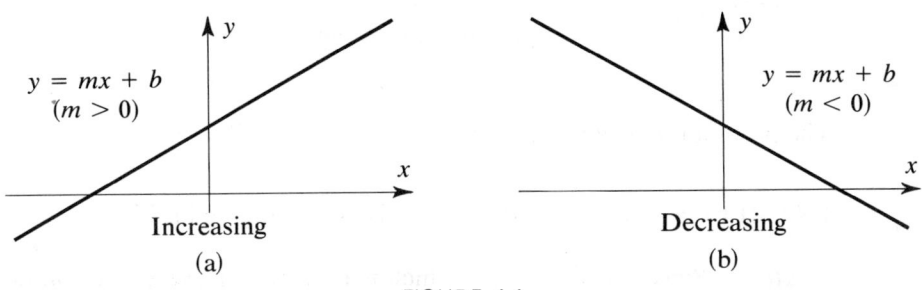

FIGURE 4.4

EXAMPLE 2. The function $f(x) = 2x^2$ is increasing on $[0, 4]$, for suppose that $0 \le x_1 < x_2 \le 4$. Then $x_1^2 < x_2^2$, and hence $2x_1^2 < 2x_2^2$; that is, $f(x_1) < f(x_2)$.

If f happens to be differentiable, then in order to test whether or not f is increasing, it is sufficient to look at f'. In fact, we have the following theorem.

THEOREM 1. Suppose that f is differentiable. If $f'(x) > 0$ for all x between a and b, then f is increasing on $[a, b]$. If $f'(x) < 0$ for all x between a and b, then f is decreasing on $[a, b]$.

Since $f'(c)$ is the slope of the tangent line at $x = c$, it follows from Theorem 1 that f is increasing if all the tangent lines are increasing between a and b (see Figure 4.5).

Proof of Theorem 1. Assume that $f'(x) > 0$ for all x between a and b and let $a \le x_1 < x_2 \le b$. By the Mean Value Theorem for Derivatives (see Chapter 3), $f(x_2) - f(x_1) = f'(c)(x_2 - x_1)$ for some c between x_1 and x_2. But $f'(c) > 0$ and $x_2 - x_1 > 0$, and hence $f(x_2) - f(x_1) > 0$; that is, $f(x_1) < f(x_2)$.

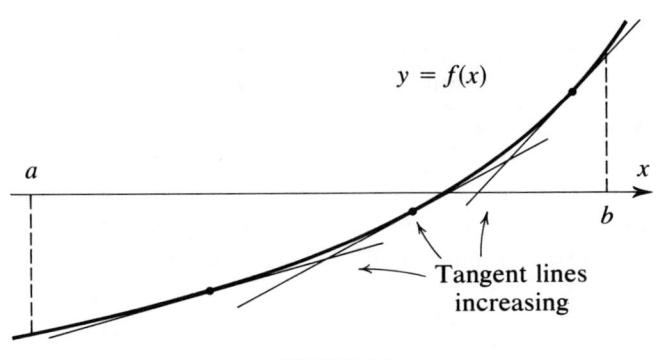

FIGURE 4.5

The proof for $f' < 0$ is similar.

EXAMPLE 3. Show that $f(x) = 1/x$ is decreasing on $[1, 2]$.

Solution. Since $f'(x) = -1/x^2$, which is negative for every x, it follows from Theorem 1 that f is decreasing.

Notice that Theorem 1 cannot be used unless f is differentiable on $[a, b]$. It is possible for a function f to be increasing on $[a, b]$ even though f is not differentiable. Indeed Figure 4.6 illustrates an example of an increasing function that is not differentiable, and Figure 4.7 is an example of a graph of an increasing function that is not even continuous. Thus no conclusion can be reached from Theorem 1 in case f is not differentiable on $[a, b]$.

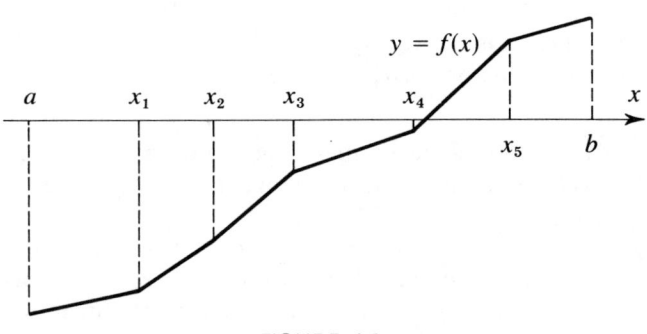

FIGURE 4.6

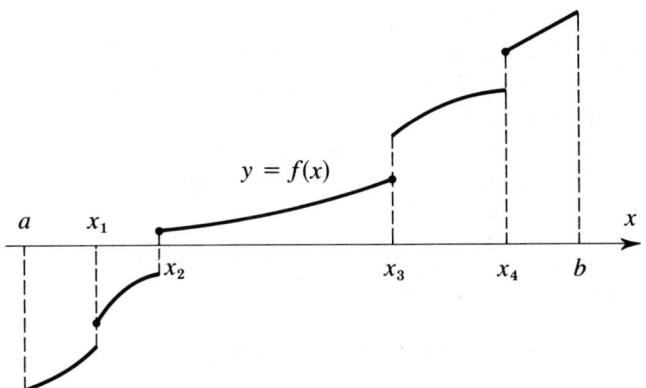

FIGURE 4.7

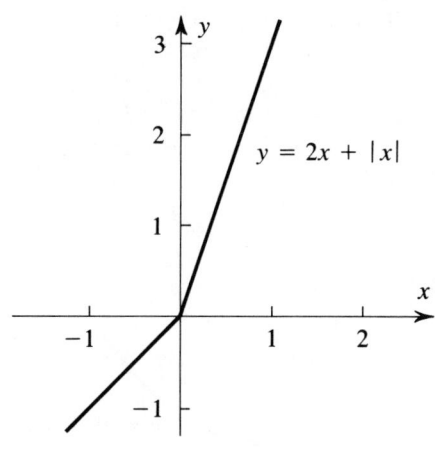

FIGURE 4.8

EXAMPLE 4. The function $f(x) = 2x + |x|$ is increasing on $[-1, 1]$, although f is not differentiable at $x = 0$ (see Figure 4.8).

EXAMPLE 5. Show that $f(x) = \sqrt{x}$ is increasing on $[0, a]$ for every positive number a.

Solution. Note that $f'(x) = 1/(2\sqrt{x}) > 0$ for $x > 0$, and hence f is increasing on every interval $[b, a]$ with $0 < b < a$. But also, $f(0) = 0 < \sqrt{a}$ if $a > 0$, and hence f is increasing on $[0, a]$.

It also makes sense to talk about a function that is increasing at a point: a function f is said to be **increasing at** $x = c$ if for all h sufficiently close to zero, $f(c) < f(c + h)$ for $h > 0$ and $f(c + h) < f(c)$ for $h < 0$. Also, f is said to be **decreasing at** $x = c$ if for all h sufficiently close to zero, $f(c) > f(c + h)$ for $h > 0$ and $f(c + h) > f(c)$ for $h < 0$.

If f is differentiable at c, then the following theorem affords an easy way to show that f is increasing or that f is decreasing at $x = c$.

THEOREM 2. *If $f'(c) > 0$, then f is increasing at $x = c$; and if $f'(c) < 0$, then f is decreasing at $x = c$.*

Proof. If $f'(c) > 0$, then the difference quotient $[f(c + h) - f(c)]/h$ is positive for h sufficiently close to zero. If $h > 0$, then necessarily $f(c + h) - f(c) > 0$, and hence $f(c) < f(c + h)$. If $h < 0$, then also $f(c + h) - f(c) < 0$, and hence $f(c + h) < f(c)$.

EXAMPLE 6. Show that $f(x) = x^3 + 1$ is increasing on $[-1, 1]$.

Solution. The derivative $f'(x) = 3x^2$ is positive except at $x = 0$. However, f is also increasing at zero, for if $a < 0 < b$, then $a^3 < 0 < b^3$, and hence $a^3 + 1 < 1 < b^3 + 1$; that is, $f(a) < f(0) < f(b)$. Hence f is increasing on all of $[-1, 1]$ (see Figure 4.9).

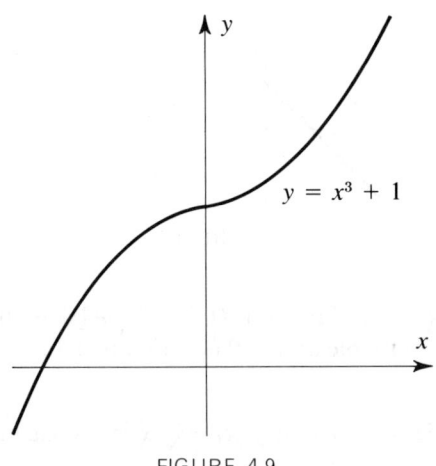

FIGURE 4.9

SECTION 1: INCREASING AND DECREASING FUNCTIONS

EXAMPLE 7. Find where the function $f(x) = x^3 + x^2$ is increasing.

Solution. Differentiating, we get $f'(x) = 3x^2 + 2x = x(3x + 2)$. Thus $f'(x) > 0$ if $x(3x + 2) > 0$, which can happen if both $x > 0$ and $3x + 2 > 0$, or if both $x < 0$ and $3x + 2 < 0$. Solving the first system $x > 0$ and $3x + 2 > 0$ of inequalities, we see that $x > 0$ and $x > -\frac{2}{3}$, so that $x > 0$. Hence f is increasing at x if $x > 0$. Solving the remaining system $x < 0$ and $3x + 2 < 0$ of inequalities, we see that $x < 0$ and $x < -\frac{2}{3}$, and hence $x < -\frac{2}{3}$. Thus f is increasing at x if either $x > 0$ or else $x < -\frac{2}{3}$, as in Figure 4.10.

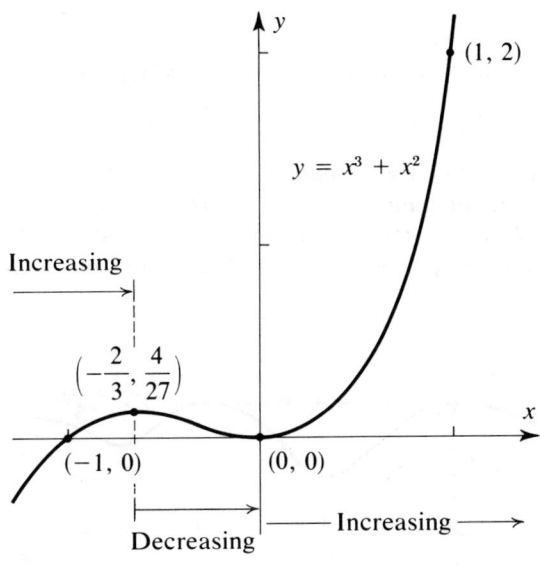

FIGURE 4.10

PROBLEMS

In each of the following problems, find where the function f is increasing and where it is decreasing.

1. $f(x) = x^2$
2. $f(x) = 3x + 4$
3. $f(x) = \dfrac{1}{x}$
4. $f(x) = \sqrt{x^2}$
5. $f(x) = x^3 - 3x$
6. $f(x) = \dfrac{1}{x^2 + 1}$
7. $f(x) = x^3$
8. $f(x) = \sqrt{x^2 + 1}$
9. $f(x) = \dfrac{x + 1}{x - 1}$
10. $f(x) = x^4 - 4x$

11. $f(x) = \dfrac{x^2 + 1}{x + 1}$

12. $f(x) = 5$

13. $f(x) = \sqrt{x + 1} - \sqrt{x}$

14. $f(x) = \dfrac{1}{x} - \dfrac{1}{x + 1}$

15. $f(x) = (x^2 + 1)^{10}$

16. $f(x) = x^3 - x^2 + x - 1$

17. Suppose that the cost of making x leather boots is $C(x)$ dollars, where $C(x) = 500 + 20x - 0.09x^2 + 0.0002x^3$. Find out where the marginal cost $C'(x)$ is (a) increasing, and (b) decreasing.

2. THE FIRST DERIVATIVE TEST FOR EXTREME VALUES

A function f has a **maximum** at c if $f(x) \leq f(c)$ for every x where f is defined, and f has a **minimum** at c if $f(x) \geq f(c)$ for every x where f is defined. Notice in Figure 4.11 that f has a maximum at c_1 and a maximum at c_2, but it has a minimum only at c_4. The function in Figure 4.12 has a minimum at c and a maximum at a.

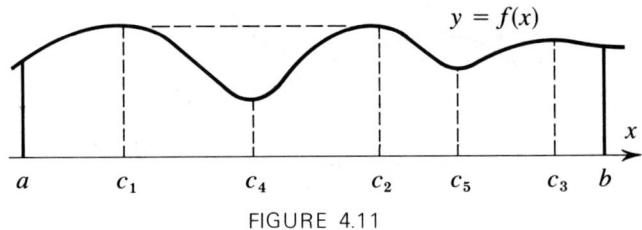

FIGURE 4.11

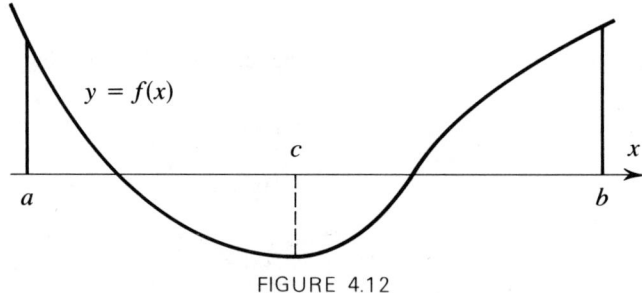

FIGURE 4.12

A function f has a **relative maximum** at c if there exists an interval $[c - h, c + h]$ such that $f(x) \leq f(c)$ for each x in the interval $[c - h, c + h]$ where

f is defined. Notice that in Figure 4.11 f has relative maxima (plural of maximum) at c_1, c_2, and c_3. In Figure 4.12 f has relative maxima at a and b.

Similarly, f has a **relative minimum** at c if there exists an interval $[c - h, c + h]$ such that $f(x) \geq f(c)$ for each x in $[c - h, c + h]$ where f is defined. In Figure 4.11 f has relative minima at a, c_4, c_5, and b, and the function in Figure 4.12 has a relative minimum at c.

We need a method for obtaining maxima, minima, relative maxima, and relative minima. The derivative gives us such a method. Suppose that f is defined on $[a, b]$ and has an **extreme value** (that is, a relative maximum or a relative minimum) at $x = c$, where $a < c < b$. If $f'(c) > 0$, then f would be increasing at $x = c$; and if $f'(c) < 0$, then f would be decreasing at c. In either case, f could not possibly have an extreme value at c. The only way out is to conclude that if f has an extreme value at c, then either $f'(c) = 0$ or else $f'(c)$ does not exist. It is also possible for f to have an extreme value at an endpoint a or b. (Such extreme values are called endpoint extreme values.) We have discovered the following useful theorem (see Figure 4.13).

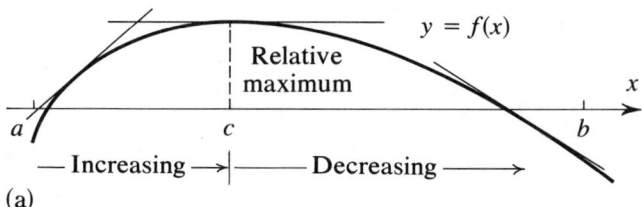

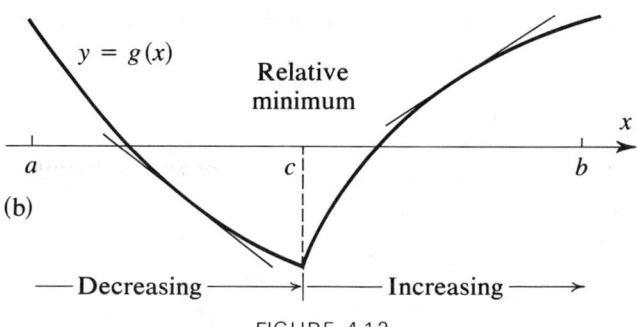

FIGURE 4.13

THEOREM 3. If f is defined on $[a, b]$ and has an extreme value at $x = c$, then $f'(c) = 0$, $f'(c)$ does not exist, $c = a$, or $c = b$.

If f is defined on $[a, b]$ and $a \leq c \leq b$, then c is called a **critical point** of f if $f'(c) = 0$, $f'(c)$ does not exist, $c = a$, or $c = b$. Thus if a function f defined on $[a, b]$ has an extreme value at $x = c$, then c is a critical point of f. In searching for extreme values, it is sufficient to check all the critical points. There may be critical points, however, that do not lead to extreme values. For example, if $f(x) = x^3 + 1$, then $f'(x) = 3x^2$, so f has a critical point at $x = 0$. However, f is increasing at zero (see Figure 4.9), and hence $f(0)$ is not an extreme value.

EXAMPLE 8. Let $f(x) = x^3 - 3x$. Then
$$f'(x) = 3x^2 - 3 = 3(x + 1)(x - 1),$$
which equals zero only when $x = 1$ or $x = -1$. Thus the only possible extreme values occur at $x = 1$ and $x = -1$. We shall see later that there is a relative maximum at $x = -1$ and a relative minimum at $x = 1$. Since f is defined for all real numbers, there are no possible endpoint extreme values.

EXAMPLE 9. Find the largest value of $\sqrt{s + 1} - \sqrt{s}$.

Solution. If $f(s) = \sqrt{s + 1} - \sqrt{s}$, then
$$f'(s) = \frac{1}{2\sqrt{s + 1}} - \frac{1}{2\sqrt{s}}$$
$$= \frac{\sqrt{s} - \sqrt{s + 1}}{2\sqrt{s}\sqrt{s + 1}}$$
$$< 0$$
because $\sqrt{s} < \sqrt{s + 1}$; thus the function is always decreasing. The function f is only defined for $s \geq 0$; hence the maximum must occur at $s = 0$, where $f(0) = \sqrt{1} - \sqrt{0} = 1$.

EXAMPLE 10. Test the function $F(x) = x^2 - 1/x^2$ for extreme values.

Solution.
$$F'(x) = 2x + \frac{2}{x^3} = \frac{2(x^4 + 1)}{x^3},$$
so the derivative is never zero and is undefined only at $x = 0$. But the original function is also undefined at $x = 0$. There are no possible endpoint extreme values, and therefore there are no extreme values of any type.

EXAMPLE 11. Helen and Sam each work on an assembly line for a company that makes electronic calculators. Helen can make $30h - \frac{1}{2}h^2$ solder

SECTION 2: THE FIRST DERIVATIVE TEST FOR EXTREME VALUES

joints and Sam can make $20h - \frac{1}{10}h^2$ solder joints, where h is the number of working hours in a day. How many hours should Helen work and how many hours should Sam work if together they work 16 hours? (Output is to be maximized.)

Solution. If Helen works h hours, then Sam works $16 - h$ hours. The total number of solder joints will be

$$S(h) = 30h - 0.5h^2 + 20(16 - h) - 0.1(16 - h)^2.$$

Differentiating, we have

$$S'(h) = 30 - h - 20 + 0.2(16 - h)$$
$$= 13.2 - 1.2h,$$

which equals zero when $h = 11$. At $h = 11$, the total output of the two workers is

$$S(11) = 367 \text{ joints.}$$

At the endpoints, $S(0) = 294.4$ and $S(16) = 352$. Thus maximum output occurs when Helen works 11 hours and Sam works 5 hours.

The preceding examples are typical of what are known as **max-min problems.** The following **First Derivative Test for Extreme Values** will be useful in solving such problems.

THEOREM 4. Suppose that f is continuous at $x = c$. If there exists $h > 0$ such that $f'(x) \geq 0$ for all x between $c - h$ and c, and $f'(x) \leq 0$ for all x between c and $c + h$, then f has a relative maximum at $x = c$. If there exists $h > 0$ such that $f'(x) \leq 0$ for all x between $c - h$ and c, and $f'(x) \geq 0$ for all x between c and $c + h$, then f has a relative minimum at $x = c$.

EXAMPLE 12. Let $f(x) = x^3 - 3x^2 + 2$. Then

$$f'(x) = 3x^2 - 6x = 3x(x - 2).$$

For $x < 0$ or $x > 2$, $f'(x) > 0$; and for $0 < x < 2$, we get $f'(x) < 0$. By the First Derivative Test, there is a relative maximum at $x = 0$. Similarly, there is a relative minimum at $x = 2$, as in Figure 4.14.

EXAMPLE 13. Test $f(x) = |x|$ for extreme values.

Solution. Since

$$f'(x) = \begin{cases} 1 & \text{if } x > 0 \\ -1 & \text{if } x < 0, \end{cases}$$

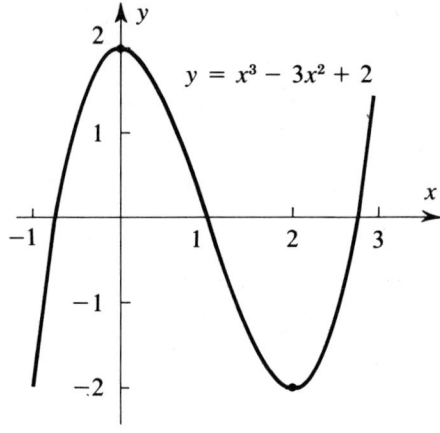

FIGURE 4.14

it follows that $f'(x) < 0$ for $x < 0$ and $f'(x) > 0$ for $x > 0$. By the First Derivative Test, f has a minimum at $x = 0$. Notice that in this example $f'(0)$ does not exist.

EXAMPLE 14. Test $f(x) = x^4 - 4x^2$ for extreme values.

Solution. Computing, we have

$$f'(x) = 4x^3 - 8x$$
$$= 4x(x^2 - 2)$$
$$= 4x(x - \sqrt{2})(x + \sqrt{2}),$$

which equals zero when $x = 0$, $x = \sqrt{2}$, or $x = -\sqrt{2}$. Notice that $f'(x) < 0$ for $x < -\sqrt{2}$, $f'(x) > 0$ for $-\sqrt{2} < x < 0$, $f'(x) < 0$ for $0 < x < \sqrt{2}$, and $f'(x) > 0$ for $x > \sqrt{2}$. From the First Derivative Test, it follows that f has a relative minimum at $x = -\sqrt{2}$, f has a relative maximum at $x = 0$, and f has a relative minimum at $x = \sqrt{2}$.

PROBLEMS

Test the following for extreme values, using the First Derivative Test.

1. $f(x) = x + \dfrac{1}{x}$

2. $x \to x^3 + 2x^2 + 1$

3. $y = \dfrac{1}{x^2 + 1}$

4. $h(w) = w^3, \ -1 \le w \le 3$

5. $s(t) = \dfrac{t-1}{t^2 + 3t}$

6. $w(z) = \sqrt{z^2 + 1}$

7. $g(u) = \sqrt{u}$

8. $H(r) = r^2 + 3r + 1, \ -3 \le r \le 3$

9. $x \to \sqrt{x} - \dfrac{1}{\sqrt{x}}$

10. $S(t) = 64t - 32t^2$

11. $f(t) = \dfrac{t^2 - 1}{t + 2}$

12. $H(x) = \dfrac{x}{x^2 + 1}, \ 0 \le x \le 10$

13. $V(t) = \dfrac{t}{t^2 - 1}$

14. $x \to \sqrt{x^2 + 1} - x$

3. THE SECOND DERIVATIVE TEST

If $f(x) = 3x^2 + 5x$, then $f'(x) = 6x + 5$ is also a function, so it makes sense to ask if this new function has a derivative. Indeed

$$\frac{d}{dx}(f'(x)) = 6.$$

Recall that the "derivative of the derivative"

$$\frac{d}{dx}(f'(x))$$

is called the **second derivative** of f and is denoted $f''(x)$.

EXAMPLE 15. Compute the second derivative of $f(x) = \sqrt{x^2 + 1}$.

Solution. The first derivative is given by

$$f'(x) = \frac{2x}{2\sqrt{x^2 + 1}}$$

$$= \frac{x}{\sqrt{x^2 + 1}}.$$

The derivative of this new function is

$$f''(x) = \frac{\sqrt{x^2 + 1} - x(x/\sqrt{x^2 + 1})}{(\sqrt{x^2 + 1})^2}$$

142 CHAPTER 4: SKETCHING THE GRAPH OF A FUNCTION

$$= \frac{x^2 + 1 - x^2}{\sqrt{x^2 + 1}(x^2 + 1)}$$

$$= \frac{1}{(x^2 + 1)^{3/2}}.$$

We now look at how the second derivative is related to some additional, more subtle aspects of sketching a graph.

The graph of a function f is called **concave upward** on $[a, b]$ if for each c between a and b, the graph of f lies above the tangent line $y = f(c) + f'(c)(x - c)$, and it is called **concave downward** on $[a, b]$ if the graph of f lies below each of its tangent lines (see Figure 4.15).

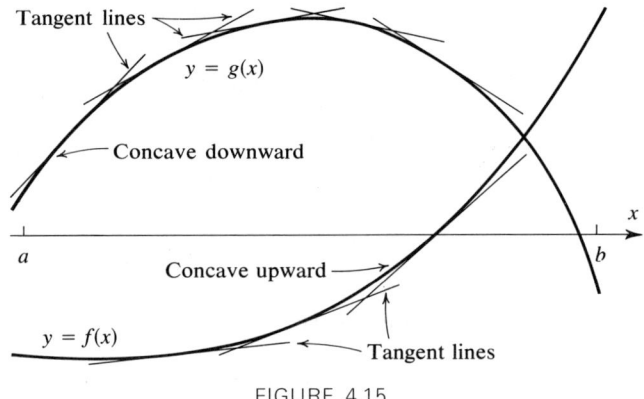

FIGURE 4.15

From a closer examination of Figure 4.15, we see that if f is concave upward on $[a, b]$ and $a \le x_1 < x_2 \le b$, then the slope of the tangent line at $x = x_1$ is less than the slope of the tangent line at $x = x_2$; that is, $f'(x_1) < f'(x_2)$. But this is exactly the statement that f' is increasing on $[a, b]$. By Theorem 1 (with f replaced by f'), in order to test whether f' is increasing, we can check to see if $f''(x) > 0$ for x between a and b. This situation is summarized in the following theorem.

THEOREM 5. Let f be continuous on $[a, b]$. If $f''(x) > 0$ for all x between a and b, then f is concave upward on $[a, b]$. If $f''(x) < 0$ for all x between a and b, then f is concave downward on $[a, b]$.

The second derivative can be used to help check for extreme values. For suppose that $f'(c) = 0$ and $f''(c) < 0$. Then the graph lies below the horizontal tangent line at $x = c$, and hence f must have a relative maximum at $x = c$, as in Figure 4.16.

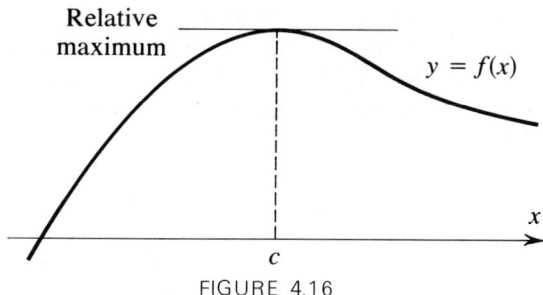

FIGURE 4.16

Similarly, if $f'(c) = 0$ and $f''(c) > 0$, then f has a relative minimum at $x = c$. We can now state the **Second Derivative Test for Extreme Values.**

THEOREM 6. If $f'(c) = 0$ and $f''(c) < 0$, then f has a relative maximum at $x = c$. If $f'(c) = 0$ and $f''(c) > 0$, then f has a relative minimum at $x = c$.

EXAMPLE 16. Test $f(x) = 2x^3 + 3x^2 - 12x$ for extreme values.

Solution. Since $f'(x) = 6x^2 + 6x - 12 = 6(x + 2)(x - 1)$, the only possible extreme values occur when $x = -2$ or when $x = 1$. But $f''(x) = 12x + 6$, so $f''(-2) < 0$ and $f''(1) > 0$. By the Second Derivative Test, there is a relative maximum at $(-2, 20)$ and a relative minimum at $(1, -7)$. The graph of f is given in Figure 4.17.

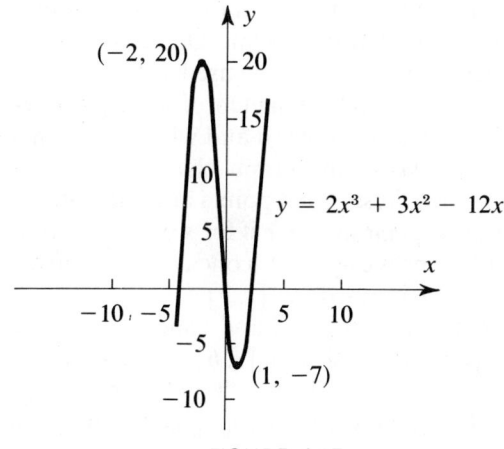

FIGURE 4.17

EXAMPLE 17. Test $f(x) = x^3$ for extreme values.

Solution. Since $f'(x) = 3x^2$, the only possible extreme value occurs when $x = 0$. But $f''(x) = 6x$, so $f''(0) = 0$ and the Second Derivative Test does not apply. However, if $x < 0$, then $f(x) = x^3 < 0$; and if $x > 0$, then $f(x) = x^3 > 0$. Consequently, f is increasing at $(0, 0)$, and so there are no extreme values, as can be seen in Figure 4.18.

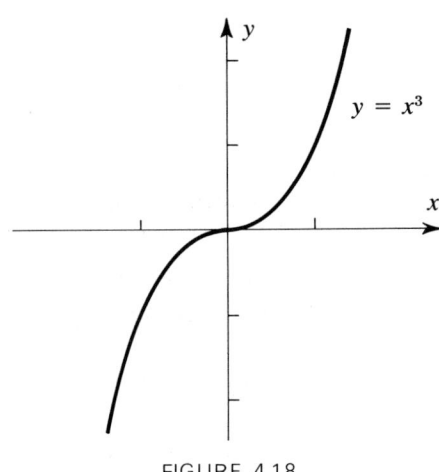

FIGURE 4.18

EXAMPLE 18. Test $f(x) = x^4$ for extreme values.

Solution. The derivative $f'(x) = 4x^3$ is equal to zero only when $x = 0$. But $f''(x) = 12x^2$, and hence $f''(0) = 0$. Here again the Second Derivative Test does not apply. When we use the First Derivative Test, $f'(x) < 0$ for $x < 0$ and $f'(x) > 0$ for $x > 0$, so there is a minimum at $(0, 0)$, as in Figure 4.19.

These examples show that the Second Derivative Test sometimes fails, in which case the First Derivative Test may be used.

In order to make an accurate sketch of a graph, it is helpful to know exactly where the graph is concave upward and where it is concave downward. In particular, we must know where the graph changes from concave upward to concave downward, and vice versa. Such points are called inflection points. More precisely, f has an **inflection point** at $x = c$ if for some a and b with $a < c < b$, f is continuous on $[a, b]$, f is concave upward (concave downward) on $[a, c]$, and f is concave downward (concave upward) on $[c, b]$.

If f is concave upward on $[a, c]$ and concave downward on $[c, b]$, then f' is increasing on $[a, c]$ and decreasing on $[c, b]$. If f' is continuous on $[a, b]$, then f' must have a maximum at $x = c$. Thus $f''(c) = 0$ or else $f''(c)$ does not exist. Similarly, if f is concave downward on $[a, c]$ and concave upward on $[c, b]$,

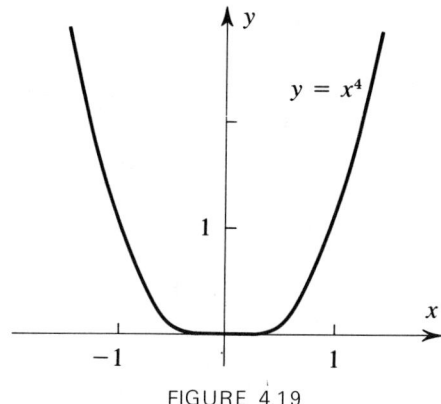

FIGURE 4.19

then $f''(c) = 0$ or else $f''(c)$ does not exist. Thus in order to check for inflection points, it suffices to find all numbers c such that $f''(c) = 0$ or $f''(c)$ does not exist.

EXAMPLE 19. Find the inflection points of $f(x) = x^4 - 6x^2 + 1$.

Solution. We need to find the extreme values of $f'(x) = 4x^3 - 12x$. But $f''(x) = 12x^2 - 12 = 12(x + 1)(x - 1)$, so the only possible extreme values of f' occur at $x = -1$ and $x = 1$. Since $f''(x)$ changes sign at both -1 and 1, there are indeed inflection points at $(-1, -4)$ and $(1, -4)$. Notice that in Figure 4.20 the graph changes from concave upward to concave downward at $(-1, -4)$ and back to concave upward at $(1, -4)$.

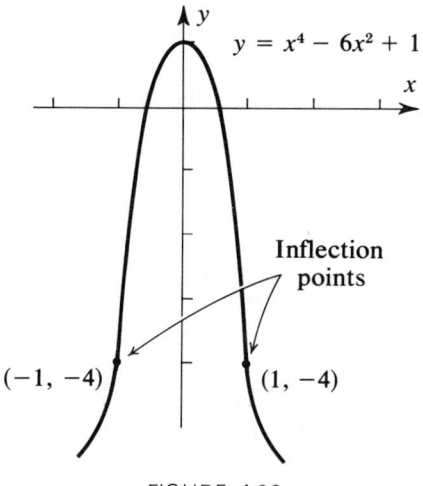

FIGURE 4.20

PROBLEMS

Find the extreme values, regions where the graph is concave upward, and all inflection points.

1. $f(x) = \dfrac{1}{x}$
2. $f(x) = 2x^3 - 3x^2 + 1$
3. $f(x) = x + \sqrt{x}$
4. $y = 3x + 4$
5. $y = \dfrac{1}{x} - \dfrac{1}{x^2}$
6. $y = \sqrt{x^2 + 1}$
7. $f(x) = 3x^4 + 4x^3$
8. $f(x) = \dfrac{x}{x+1}$
9. $f(x) = \dfrac{1}{\sqrt{x}}$
10. $f(x) = x^{3/2} + x^{1/2}$
11. $f(x) = x^4 - 2x^3 - 36x^2$
12. $y = 5x^2 + 3x - 10$
13. $y = \dfrac{x^2}{x^2 + 2}$
14. $f(x) = (x + 1)^9$
15. $f(x) = (x^2 - 1)^2$
16. $y = (x^2 + 1)^2$

17. Compute the "third derivative" $\dfrac{d}{dx} f''(x) = f'''(x)$ if $f(x) = 1/x$.

4. GETTING THE PICTURE

If $f(x) = x^4 - 6x^2 + 1$, then, in a sense, we know all about the function f. Given any real number x, we can compute the value of f at x. However, since there are infinitely many real numbers, there are always questions about f that cannot be answered by computing $f(x)$ at any finite number of points. Where is the function increasing? Where is the function concave upward? What are the inflection points? Where are the extreme values? By using techniques of calculus, we can now answer these questions easily and then use the information to sketch a graph of the function.

EXAMPLE 20. Sketch the graph of $f(x) = x^4 - 6x^2 + 1$.

Solution. We first note that $f'(x) = 4x^3 - 12x = 4x(x^2 - 3)$ and $f''(x) = 12x^2 - 12 = 12(x^2 - 1)$. We then complete Table 4.1, which at least includes all x such that $f'(x) = 0$ or $f''(x) = 0$.

Table 4.1

x	$f(x)$	$f'(x)$	$f''(x)$	Remarks
0	1	0	-12	Relative max by Second Derivative Test
$\sqrt{3}$	-8	0	24	Relative min by Second Derivative Test
$-\sqrt{3}$	-8	0	24	Relative min by Second Derivative Test
1	-4	-8	0	Inflection point
-1	-4	8	0	Inflection point
2	-7	16	36	Increasing, concave upward
-2	-7	-16	36	Decreasing, concave upward
3	28	72	96	Increasing, concave upward
-3	28	-72	96	Decreasing, concave upward

Since the entries in the $f(x)$ column range from -8 to 28 and the entries in the x column only range from -3 to 3, it is appropriate to use different scales on the x- and y-axes, as in Figure 4.21.

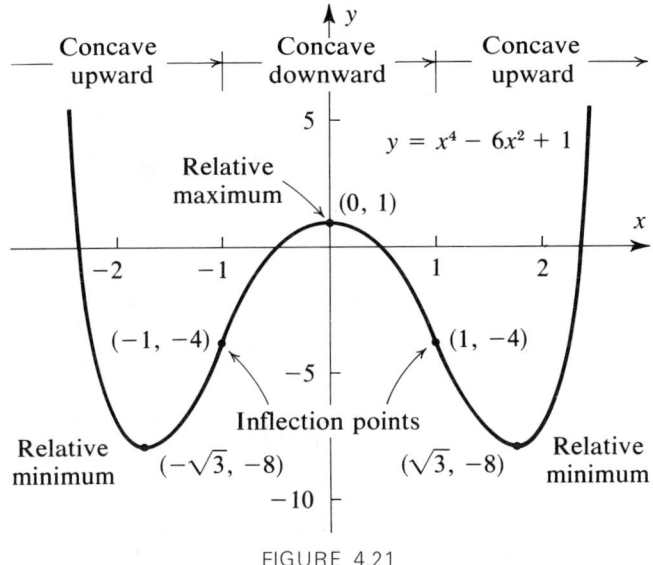

FIGURE 4.21

When we use equal scales, the graph looks like Figure 4.22.

Notice that in Figure 4.21 the graph of f for x negative looks like the mirror image of the graph of f for x positive. The reason is that $f(-x) = f(x)$ for each real number x. Such a function is called an **even function**, and its graph is called **symmetric with respect to the y-axis**.

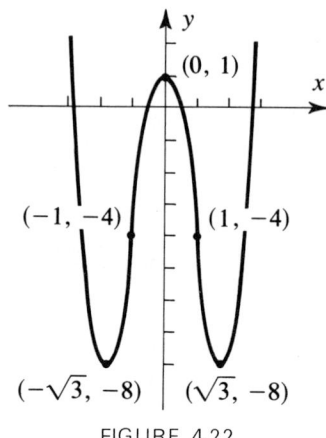

FIGURE 4.22

EXAMPLE 21. Sketch the graph of $f(x) = x^3 - 3x$.

Solution. Note that $f'(x) = 3x^2 - 3$ and $f''(x) = 6x$. We compute Table 4.2 by evaluating f, f', and f'' at several points, including all x where $f'(x) = 0$ or where $f''(x) = 0$. The graph is then sketched as in Figure 4.23.

Table 4.2

x	$f(x)$	$f'(x)$	$f''(x)$	Remarks
0	0	-3	0	Inflection point
1	-2	0	6	Relative min by Second Derivative Test
-1	2	0	-6	Relative max
2	2	9	12	Increasing, concave upward
-2	-2	9	-12	Increasing, concave downward

Notice that in this example $f(-x) = (-x)^3 - 3(-x) = -(x^3 - 3x) = -f(x)$. Functions with the property that $f(-x) = -f(x)$ are called **odd** functions. This does not mean "weird" or "strange." The reason for the names "even" and "odd" is simply that a polynomial function of the form $f(x) = x^n$ is even or odd, depending on whether n is even or odd.

The graph in Figure 4.23 has the following interesting property. If L is any line through the origin, the distance from A to 0 equals the distance from 0 to B. We say that the graph is **symmetric with respect to the origin.** In fact, the graph of any odd function is symmetric with respect to the origin. In order to sketch the graph of an odd function, it suffices to sketch the graph for x positive and then use the fact that the graph is symmetric with respect to the origin to

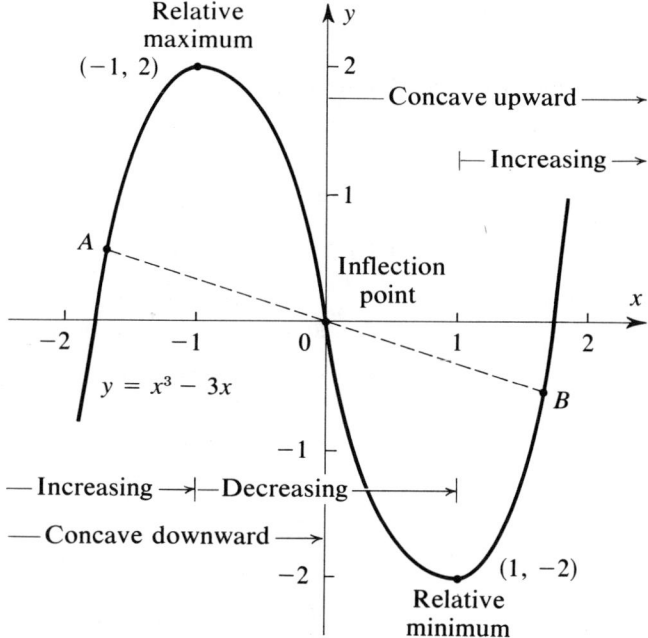

FIGURE 4.23

sketch the remainder of the graph. Similarly, in order to sketch the graph of an even function, it is sufficient to sketch the graph for x positive and then reflect it about the y-axis. Of course, most functions are neither odd nor even. The sum of an odd function with an even function is neither odd nor even. However, we do have Table 4.3, which allows us to construct new odd or even functions from old ones.

Table 4.3

Operation	Example
Odd plus odd = odd	$x + x^3$
Even plus even = even	$2 + x^2$
Odd times even = odd	$x(x^2 + x^4) = x^3 + x^5$
Odd times odd = even	$x(x + x^3) = x^2 + x^4$
Even times even = even	$x^2(3 + x^4) = 3x^2 + x^6$
Odd divided by even = odd	$\dfrac{x}{x^2 + x^4}$

EXAMPLE 22. Sketch the graph of $f(x) = x + 1/x$.

Solution. We notice that f is an odd function. We have that

$$f'(x) = 1 - \frac{1}{x^2} = \frac{x^2 - 1}{x^2} \quad \text{and} \quad f''(x) = \frac{2}{x^3}.$$

As before, we complete Table 4.4 for nonnegative x and then sketch the graph, as in Figure 4.24.

Table 4.4

x	$f(x)$	$f'(x)$	$f''(x)$	Remarks
0	Undefined	—	—	$f(x)$ gets large as $x \to 0$
1	2	0	2	Relative min by Second Derivative Test
$\frac{1}{2}$	$\frac{5}{2}$	-3	16	Decreasing, concave upward
2	$\frac{5}{2}$	$\frac{3}{4}$	$\frac{1}{4}$	Increasing, concave upward

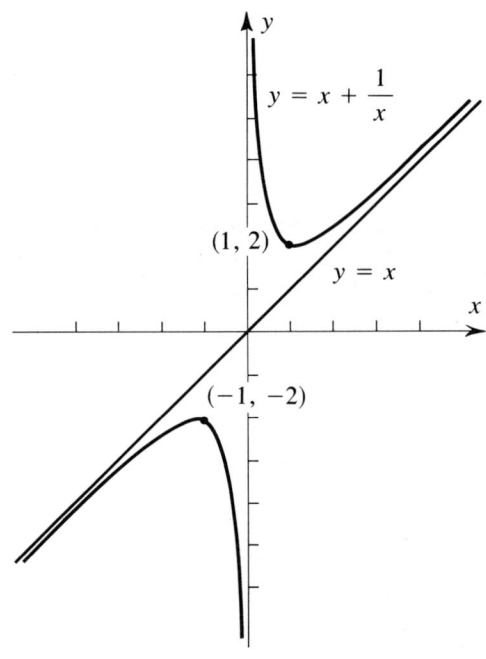

FIGURE 4.24

As x gets large, $f(x)$ also gets large. Is there anything more we can say? Suppose that we superimpose the graph of the line $y = x$ on the sketch. Since $1/x$ approaches zero as x gets large, the graph of f is very close to the graph of

$y = x$ for x large. We say that the line $y = x$ is an **asymptote** or is **asymptotic** to the graph of f. Also, the line $x = 0$ is an asymptote, usually called a **vertical asymptote**.

EXAMPLE 23. Sketch the graph of $f(x) = \dfrac{x^2 + 3}{2x^2 - 1}$.

Solution. Since

$$f(-x) = \frac{(-x)^2 + 3}{2(-x)^2 - 1} = \frac{x^2 + 3}{2x^2 - 1} = f(x),$$

it follows that f is an even function. Also,

$$f'(x) = \frac{2x(2x^2 - 1) - (x^2 + 3)4x}{(2x^2 - 1)^2}$$

$$= \frac{-14x}{(2x^2 - 1)^2},$$

and

$$f''(x) = \frac{-14(2x^2 - 1)^2 + 14x(2x^2 - 1)8x}{(2x^2 - 1)^4}$$

$$= \frac{-28x^2 + 112x^2 + 14}{(2x^2 - 1)^3}$$

$$= \frac{84x^2 + 14}{(2x^2 - 1)^3}.$$

There are vertical asymptotes at $x = \pm 1/\sqrt{2}$. Note that $f'(x) = 0$ only if $x = 0$, and by the Second Derivative Test, f has a relative maximum at $x = 0$. Furthermore, $f''(x) > 0$ if $x < -1/\sqrt{2}$, $f''(x) < 0$ if $-1/\sqrt{2} < x < 1/\sqrt{2}$, and $f''(x) > 0$ if $x > 1/\sqrt{2}$. There are no inflection points, f is concave upward when $x < -1/\sqrt{2}$ or $x > 1/\sqrt{2}$, and f is concave downward when $-1/\sqrt{2} < x < 1/\sqrt{2}$. What happens to the graph when $x \to \infty$? Dividing numerator and denominator by x^2, we get

$$f(x) = \frac{x^2 + 3}{2x^2 - 1}$$

$$= \frac{1 + 3/x^2}{2 - 1/x^2},$$

which tends to $\frac{1}{2}$ as $x \to \infty$. We complete Table 4.5 and sketch the graph as in Figure 4.25.

CHAPTER 4: SKETCHING THE GRAPH OF A FUNCTION

Table 4.5

x	$f(x)$	$f'(x)$	$f''(x)$	Remarks
0	-3	0	<0	Relative max by Second Derivative Test
$\dfrac{1}{\sqrt{2}}$	–	–	–	Vertical asymptote
1	4	-14	>0	Decreasing, concave upward
0.5	-6.5	<0	<0	Decreasing, concave downward
2	1	<0	>0	Decreasing, concave upward

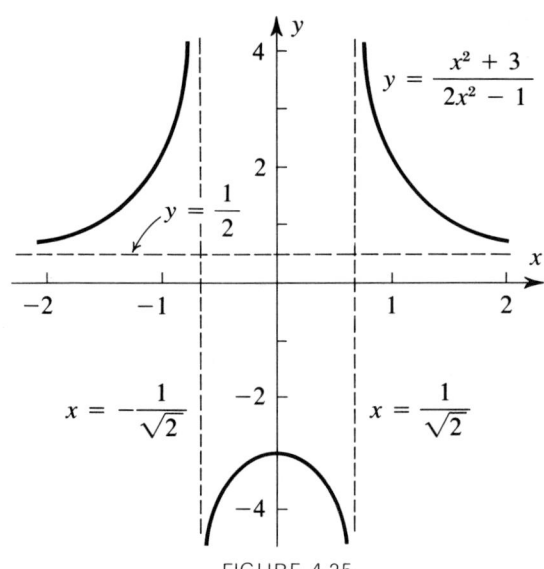

FIGURE 4.25

Notice that the line $y = \tfrac{1}{2}$ is an asymptote. In general, an asymptote of the form $y = k$ is called a **horizontal asymptote.** In checking for such asymptotes, it is sufficient to compute $\lim_{x \to +\infty} f(x)$ and $\lim_{x \to -\infty} f(x)$. In the preceding example, $\lim_{x \to +\infty} f(x) = \lim_{x \to -\infty} f(x) = \tfrac{1}{2}$.

PROBLEMS

Sketch the graph of each of the following functions. Label all extreme values, inflection points, asymptotes, regions where the function is increasing, and all regions where the function is concave upward. If the function is odd or even, state which.

1. $f(x) = x^3 + 3x^2 - 9x - 1$
2. $f(x) = \tfrac{1}{5}x^5 + \tfrac{1}{4}x^4 - \tfrac{2}{3}x^3 - 1$

3. $f(x) = x^4 - 8x^2 + 2$

4. $f(x) = \dfrac{2x^2 + 1}{x^2 - 1}$

5. $f(x) = \dfrac{1}{x}$

6. $f(x) = \sqrt{x^2}$

7. $f(x) = \dfrac{1}{x^2 + 1}$

8. $f(x) = \dfrac{x}{x^2 + 1}$

9. $f(x) = (x^2 + 1)^3$

10. $f(x) = (x^2 + 1)^{3/2}$

11. $f(x) = (x^2 - 1)^3$

12. $f(x) = (x^2 - 1)^{3/2}$

13. $f(x) = (x^2 + x)^2$

14. $f(x) = \dfrac{1}{x + 1}$

15. $f(x) = \dfrac{x^2 + 2x}{x + 1}$

16. $f(x) = \dfrac{x - 1}{x + 1}$

REVIEW SECTION

The following definitions are stated here for convenience.

(a) A function f is *increasing (decreasing)* on $[a, b]$ if $a \leq x_1 < x_2 \leq b$ implies $f(x_1) < f(x_2) (f(x_1) > f(x_2))$.
(b) A function f has a *maximum (minimum)* at $x = a$ if $f(a) \geq f(x) (f(a) \leq f(x))$ for all x where f is defined.
(c) A function f has a *relative maximum (relative minimum)* at $x = c$ if there exists an interval $[c - h, c + h]$ such that $f(c) \geq f(x) (f(c) \leq f(x))$ for all x in $[c - h, c + h]$ where f is defined.
(d) The number c is a *critical point* of f if $f'(c) = 0$, $f'(c)$ does not exist, or c is an endpoint $c = a$ or $c = b$.
(e) The graph of f is *concave upward (concave downward)* on $[a, b]$ if the graph of f lies above (below) each of its tangent lines.
(f) A function f has an *inflection point* at $x = c$ if f' has an extreme value at $x = c$.
(g) A function f is *even (odd)* if $f(-x) = f(x)(f(-x) = -f(x))$ for all x.
(h) A line $y = mx + b$ is *asymptotic* to the graph of f if $\lim\limits_{x \to +\infty} (mx + b - f(x)) = 0$ or $\lim\limits_{x \to -\infty} (mx + b - f(x)) = 0$.
(i) A line $x = a$ is a *vertical asymptote* to the graph of f if $\lim\limits_{x \to a} f(x) = \pm\infty$.
(j) A line $y = b$ is a *horizontal asymptote* if $\lim\limits_{x \to +\infty} f(x) = b$ or $\lim\limits_{x \to -\infty} f(x) = b$.

The following facts and theorems are useful in sketching the graph of a function.

(a) If $f' > 0$ $(f' < 0)$ on $[a, b]$, then f is increasing (decreasing) on $[a, b]$.

(b) If f is defined only on $[a, b]$ and has an extreme value at $x = c$, then c is a critical point.
(c) **First Derivative Test.** Suppose that f is continuous at $x = c$. If there exists $h > 0$ such that $f' \geq 0$ on $[c - h, c]$ and $f' \leq 0$ on $[c, c + h]$, then f has a relative maximum at $x = c$. If there exists $h > 0$ such that $f' \leq 0$ on $[c - h, c]$ and $f' \geq 0$ on $[c, c + h]$, then f has a relative minimum at $x = c$.
(d) If f' is increasing (decreasing) on $[a, b]$, then f is concave upward (downward) on $[a, b]$.
(e) **Second Derivative Test.** If $f'(c) = 0$ and $f''(c) < 0$ ($f''(c) > 0$), then f has a relative maximum (minimum) at $x = c$.
(f) If f has an inflection point at $x = c$, then $f''(c) = 0$ or $f''(c)$ does not exist.
(g) The graph of each even function is symmetric with respect to the y-axis.
(h) The graph of each odd function is symmetric with respect to the origin.

REVIEW PROBLEMS

Sketch the graph of each of the following functions. Locate and label all extreme values, inflection points, all asymptotes, and all regions where the graph is concave upward. If the function is even or odd, state which.

1. $f(x) = x\sqrt{9 - x^2}$
2. $y = x^3 - 2x^2 + 2x - 1$
3. $y = \dfrac{x + 1}{x - 1}$
4. $x \to 3x^4 - 4x^3 - 12x^2 + 24$
5. $h(x) = \dfrac{3x^2 + 4}{x^2 + 1}$
6. $g(x) = \dfrac{2x^2 + 3x + 2}{x + 1}$
7. $f(x) = \sqrt{x(x - 1)^3}$
8. $y = \dfrac{\sqrt{x + 1}}{\sqrt{x}}$
9. $f(x) = x(x - 1)^3$
10. $x \to x^2(x - 2)^2$
11. $f(x) = \dfrac{x^3 + 4x^2 - 2}{2x^2}$
12. $y = x\sqrt{1 - x^2}$
13. $g(x) = \dfrac{1}{x} + \dfrac{1}{x - 1}$
14. $h(x) = \sqrt[3]{x^3 - 1}$
15. $f(x) = \dfrac{1}{x^2} + \dfrac{1}{(x - 1)^2}$
16. $f(x) = \sqrt{x + 1} - \sqrt{x}$
17. $f(x) = \dfrac{x}{x + 1}$
18. $f(x) = \sqrt{x^2 + 1} + 2$
19. $f(x) = x\sqrt{x^2 - 9}$
20. $x \to 2x^2 - \sqrt{x}$
21. $y = x - x^{3/2}$
22. $f(x) = x - \sqrt{x}$
23. $f(x) = \sqrt{x} - x\sqrt{x}$
24. $f(x) = \dfrac{1}{x} - \left(\dfrac{1}{x}\right)^2$

25. $f(x) = \dfrac{x-1}{x} + \dfrac{x}{x-1}$

26. $y = \dfrac{x^2+1}{2x+1}$

27. $y = \sqrt{1 - \sqrt{x}}$

28. $f(x) = \sqrt{x(x-1)^2}$

29. $f(x) = \sqrt{1 + \sqrt{x}}$

30. $f(x) = \dfrac{1}{\sqrt{1 - \sqrt{x}}}$

CHAPTER 5

APPLICATIONS

The derivative has already been used to help sketch the graph of a function. Other applications of the derivative are also of interest, such as the use of differentiation to solve max-min problems. In solving such problems, it is necessary to find the cheapest, the most profitable, the highest, the fastest, the biggest, the smallest, and so forth. In other words, we look for extreme values of some function, and here the methods developed in Chapter 4 will prove useful. The next collection of problems deals with motion. The derivative is used to study relationships between position, velocity, and acceleration. Finally, the derivative will be used to approximate roots of equations of the form $f(x) = 0$.

As applications of integration, we will solve differential equations and compute volumes of solids of revolution.

1. APPLIED MAXIMA AND MINIMA

We begin by looking at ways in which tests for extreme values can be used to solve what are called max-min problems.

EXAMPLE 1. Find a rectangle whose perimeter is 24 and whose area is as large as possible.

Solution. Let x be the width, so that $12 - x$ is the length of the rectangle. We wish to maximize the function $f(x) = x(12 - x) = 12x - x^2$. The derivative $f'(x) = 12 - 2x$ equals zero only when $x = 6$. Since $f''(x) = -2 < 0$, the function is concave downward; and by the Second Derivative Test, a maximum occurs at $x = 6$. The largest area is enclosed by a rectangle of dimensions 6 by 6; that is, the largest area is enclosed by a square.

EXAMPLE 2. A rectangular piece of cardboard 12 by 24 centimeters is to be made into a box with open top by cutting a square of dimensions x by x out of each corner and folding up the edges. How should x be chosen so that the volume of the box is maximal? (See Figure 5.1.)

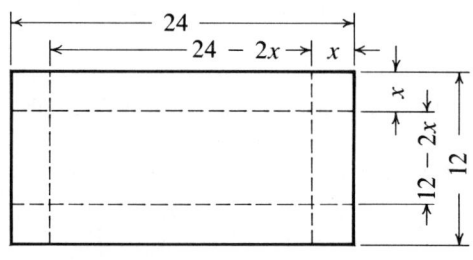

FIGURE 5.1

Solution. The volume of the box is given by
$$V(x) = x(12 - 2x)(24 - 2x)$$
$$= 4x^3 - 72x^2 + 288x,$$
so that
$$V'(x) = 12x^2 - 144x + 288$$
$$= 12(x^2 - 12x + 24).$$
By the quadratic formula, $V'(x) = 0$ if
$$x = \frac{12 \pm \sqrt{144 - 96}}{2}$$
$$= 6 \pm 2\sqrt{3}.$$
By Figure 5.1, it is clear that x must be between 0 and 6, so the maximum volume must occur when x is 0, $6 - 2\sqrt{3}$, or 6. Since $V''(x) = 24x - 144$, it follows that

$V''(6 - 2\sqrt{3}) = -48\sqrt{3} < 0$, and hence the maximum volume occurs when $x = 6 - 2\sqrt{3}$ by the Second Derivative Test.

EXAMPLE 3. During a 100-day growing season, a certain plant grows at a rate of approximately $0.2 + 0.01t - 0.0001t^2$ inch per day, where t is the number of days since the start of the growing season. When is the plant growing most rapidly?

Solution. We need to maximize the function $f(t) = 0.2 + 0.01t - 0.0001t^2$, $0 \leq t \leq 100$. The derivative $f'(t) = 0.01 - 0.0002t$ is zero at $t = 50$. Since $f''(t) = -0.0002$ for all t, the Second Derivative Test shows that the plant is growing most rapidly 50 days after the start of the growing season.

EXAMPLE 4. A gardener wants to use 100 feet of fencing materials to enclose two circular plots of ground. How should he distribute the materials between the two plots to maximize the enclosed area?

Solution. We need to find two circles such that the sum of the perimeters is 100 feet and the sum of the areas is as large as possible. If one perimeter is x, then the other perimeter is $100 - x$. The radius of the first circle is $x/2\pi$, and the radius of the second circle is $(100 - x)/2\pi$. The sum of the areas is given by

$$A = \pi \left(\frac{x}{2\pi}\right)^2 + \pi \left(\frac{100 - x}{2\pi}\right)^2$$

$$= \frac{1}{4\pi}(x^2 + 100^2 - 200x + x^2)$$

$$= \frac{1}{4\pi}(2x^2 - 200x + 100^2).$$

To maximize A, we need to check the derivative:

$$\frac{dA}{dx} = \frac{1}{4\pi}(4x - 200).$$

The derivative is zero when $x = 50$. The Second Derivative Test shows, however, that there is a relative minimum at $x = 50$. The function

$$x \to \pi\left(\frac{x}{2\pi}\right)^2 + \pi\left(\frac{100 - x}{2\pi}\right)^2$$

is defined only for x in the interval $[0, 100]$, so we need to consider possible endpoint extreme values. The maximum must occur at an endpoint, but, in fact, $A(0) = A(100)$, so the maximum occurs at both endpoints. If the gardener is actually required to enclose *two* plots, then one should be made as small as possible.

EXAMPLE 5. In probability theory, we find the formula

$$P_{n,r}(p) = \frac{n!}{r!(n-r)!} p^r (1-p)^{n-r},$$

where $P_{n,r}(p)$ is the probability of exactly r successes in n trials when p is the probability of success in each trial. Suppose that n and r are given fixed integers for which $0 < r < n$. Find the number p that maximizes $P_{n,r}(p)$.

Solution. Differentiating, we obtain

$$\frac{d}{dp}[P_{n,r}(p)] = \frac{n!}{r!(n-r)!}[rp^{r-1}(1-p)^{n-r} - (n-r)p^r(1-p)^{n-r-1}]$$

$$= \frac{n!}{r!(n-r)!} p^{r-1}(1-p)^{n-r-1}[r(1-p) - (n-r)p],$$

and hence the derivative equals zero only when $p = 0$, $p = 1$, or when $r(1-p) - (n-r)p = 0$. Solving for p, we obtain $p(r - n - r) = -r$, so that $p = r/n$. We notice that $P_{n,r}(0) = 0$, $P_{n,r}(1) = 0$, and $P_{n,r}(x) > 0$ for each x such that $0 < x < 1$, so there must be a maximum at $p = r/n$. This answer is not surprising. It says, in particular, that a tossed coin is most likely to come up heads exactly 50 times out of 100 if the probability of heads on a single toss is $\frac{1}{2}$. (But, by the way, it is not *very* likely that such will occur because

$$P_{100,50}\left(\frac{1}{2}\right) = \frac{100!}{50!50!}\left(\frac{1}{2}\right)^{50}\left(\frac{1}{2}\right)^{50} \approx 0.1.$$

It is just *more* likely than any other split between heads and tails.)

PROBLEMS

1. Find the points on the graph of $y = x^2$ that are closest to the point $(0, 3/2)$.

2. A metal can is to contain $4000/\pi$ cubic centimeters. Find the dimensions of the least-expensive can if the material for the cylindrical surface costs \$0.0001 per square centimeter and the material for the ends costs \$0.0002 per square centimeter. What is the cost of such a can?

3. An alert executive observes that one secretary will work 30 hours per week but that each additional secretary hired reduces the effectiveness of all the secretaries by entering into conversations, asking questions, and so forth. On further examination, the executive finds that if there are x secretaries, x not exceeding 30, then each one will work effectively only

$$30 - \frac{x^2}{30}$$

hours per week. Find the number of secretaries that will produce the most hours of work per week.

160   CHAPTER 5: APPLICATIONS

4. The specific weight S of water at a temperature t degrees centigrade is given by the formula

$$S = 1 + at + bt^2 + ct^3,$$

where $0° \leq t° \leq 100°$, $a = 5.3 \times 10^{-5}$, $b = -6.53 \times 10^{-6}$, and $c = 1.4 \times 10^{-8}$. At what temperature does water have a maximum specific weight? [*Hint:* Show that if x_1 is a root of $Ax^2 + Bx + C = 0$, then the other root is $x_2 = C/(Ax_1)$.] Can you explain why most lakes never freeze solid, even in very cold climates?

5. A brewery wants to be ecological and use as little aluminum as possible in its returnable beer cans. What should be the ratio between the height and the diameter of the top for such an all-aluminum can?

6. A college student spends a summer working in a lighthouse. Early in the summer he meets a beautiful girl who lives in a house on the beach. The lighthouse is on an island one mile from the nearest point on the straight shoreline, and the girl lives a mile away from this point. If he can walk twice as fast as he can row a boat, what path should he take to the girl's house in order to spend as much of his leisure time there as possible? (See Figure 5.2.)

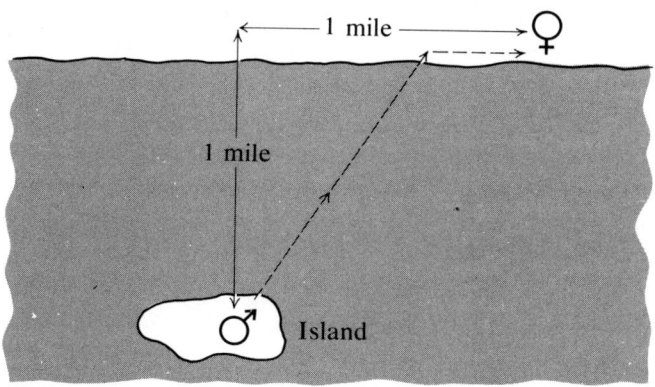

FIGURE 5.2

7. A rectangular piece of cardboard of dimensions a by b is to be made into a box by cutting squares out of the corners, as in Example 2. Maximize the volume of such a box.

8. A rectangular beam is to be cut from a log of diameter a. What are the dimensions of the strongest beam that can be cut out of this log, assuming that the strength of a beam is proportional to its width and proportional to the square of its depth? (See Figure 5.3.)

9. What number exceeds its square by the largest amount?

10. Find two positive numbers x and y whose sum is 10 and such that the product of the first with the square of the second is as large as possible.

11. Find the maximum area of a rectangle inscribed inside a semicircle of radius 1.

12. A concrete ditch is to have 2-foot straight sides and a 2-foot flat bottom. How wide should the top of the ditch be for maximum capacity?

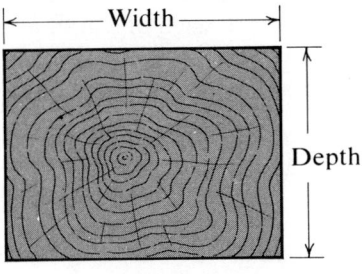

FIGURE 5.3

13. At 1 P.M., a bear is one mile due east of a hiker. If the hiker then walks north at 2 miles per hour and the bear walks west at 3 miles per hour, when are they closest together, and how close do they get?
14. Find the maximum volume of a right circular cylinder that can be inscribed inside a right circular cone.
15. Find the maximum volume of a right circular cone that can be inscribed inside a sphere.
16. A grain storage bin is to be made in the form of a right circular cylinder. If materials for a roof cost twice as much per square meter as materials for the walls and floor, what are the dimensions of the most economical storage bin of volume x cubic meters?
17. A quarter-mile (440-yard) track is constructed by placing two semicircles on the opposite ends of a rectangle. What should the dimensions of the rectangle be in order to maximize the total area enclosed by the track? What should the dimensions be if the area of the rectangle is to be maximized?
18. A local building code requires that a rectangular-shaped manufacturing plant be bordered by 100 feet of landscaping on the front and rear and by 80 feet of landscaping on each side. If 800,000 square feet of land are required for parking lots, buildings, and so forth, what is the minimum number of square feet that must be purchased for the manufacturing plant?
19. The cost of producing x units is $C(x) = 16 + 6x + 0.2x^{3/2}$ for an item that sells for \$24. How many units should be produced to maximize profit? What is the maximum profit?

2. VELOCITY AND ACCELERATION

A rock is dropped into a deep well and it takes 5 seconds to hit bottom. How deep is the well? From physics, the velocity in feet per second of the rock after t seconds (neglecting air friction) is given by $v(t) = 32t$. Again from elementary physics,

(1) $\qquad$ Distance = rate × time

or

(2) $$d = rt$$

for short. We will use Formula (2) to estimate the distance traveled by subdividing the interval $[0, 5]$ into n subintervals, each of length $5/n$. If n is large, then during the kth subinterval the rate (or velocity) is nearly constant, and hence the distance traveled by the rock during this subinterval of time is approximately

$$v\left(\frac{5k}{n}\right)\left(\frac{5}{n}\right) = 32\frac{5k}{n} \cdot \frac{5}{n}.$$

Note that in the formula $d = rt$ we have replaced r by $v(5k/n)$, the approximate velocity during the kth subinterval, and t by $5/n$, the length of time during the kth subinterval. Summing over all the subintervals, we see that the total distance traveled by the rock is approximately

$$\sum_{k=1}^{n} 32\frac{5k}{n} \cdot \frac{5}{n}.$$

This is precisely a Riemann sum for the integral

$$\int_0^5 32t \, dt = 16t^2 \Big|_0^5$$

$$= 400 \text{ feet.}$$

It follows that the well is 400 feet deep.

If $v(t) \geq 0$ is the velocity of an object and $s(t)$ is the distance traveled by the object at time t, then as in the preceding example,

$$s(t) = \int_0^t v(x) \, dx,$$

and hence

$$s'(t) = v(t).$$

We will generalize this result slightly to allow negative velocities. Suppose that a dot is traveling along the x-axis and that the position of the dot at time t is $x(t)$. If the dot is traveling to the right, we say that the velocity $v(t)$ is positive; and if the dot is traveling to the left, the velocity is negative. What is the relationship between $x(t)$ and $v(t)$? If $h \neq 0$, then the average velocity between t and $t + h$ is

$$\frac{x(t+h) - x(t)}{h},$$

and the exact velocity is the limit

$$v(t) = \lim_{h \to 0} \frac{x(t+h) - x(t)}{h}$$

$$= x'(t).$$

Here, as before, velocity is the derivative of position.

EXAMPLE 6. The speedometer of a car reads approximately $20\sqrt{t+4}$ miles per hour for t between 0 and 5 seconds. How many feet does the car travel during those 5 seconds?

Solution. We first convert $v(t)$ to feet per second by multiplying by the factor 22/15 to obtain

$$v(t) = \frac{88}{3}\sqrt{t+4}$$

feet per second. The position $s(t)$ is an antiderivative of $v(t)$, and hence the distance traveled is

$$\int_0^5 \frac{88}{3}\sqrt{t+4}\,dt = \frac{88}{3}\cdot\frac{2}{3}(t+4)^{3/2}\Big|_0^5$$

$$= 371.6 \text{ feet.}$$

Derivatives can be interpreted as rates of change. In particular, the slope of a tangent line measures how fast the graph is rising or falling, and velocity is the rate of change of the position of an object. Velocity can also change. The rate of change of velocity is called *acceleration*. Thus $s''(t) = v'(t) = a(t)$, the acceleration at time t.

EXAMPLE 7. An object starts from rest and has an acceleration given by $a(t) = 6t$. What is the velocity at $t = 7$? How far has the object traveled at $t = 7$?

Solution. Since $v(t)$ is an antiderivative of $a(t)$, it follows that $v(t) = 3t^2 + C$ for some constant C. But the object starts at rest, and hence $v(0) = 0 = C$. At $t = 7$, the velocity is given by $v(7) = 147$. The distance is

$$\int_0^7 3t^2\,dt = t^3\Big|_0^7$$

$$= 343.$$

Notice that in the preceding example we did not say anything about units of distance and time. Had we said $a(t) = 6t$ feet/sec² then we would have $v(7) = 147$ feet/sec and $s(7) = 343$ feet. The solutions would be $v(7) = 147$ kilometers/hour and $s(7) = 343$ kilometers if we had started with $a(t) = 6t$ kilometers/hour². Since the numerical answers will be the same, we will often work problems without specific mention of the units involved.

EXAMPLE 8. A ball is thrown straight up from the surface of the Earth with an initial velocity of 64 feet/sec. How high does the ball go before it begins to fall? What is the velocity of the ball just before it hits the surface?

Solution. On the Earth, $a(t) = -32$ feet/sec². Hence $v(t) = -32t + C$. But $v(0) = 64 = -32 \cdot 0 + C$, so $C = 64$. The ball reaches its highest point when $v(t) = 0$—that is, when $-32t + 64 = 0$, which implies $t = 2$. Now $s(t) = -16t^2 + 64t + k$ for some constant k. But $s(0) = 0$ implies $k = 0$. The highest point is thus $s(2) = 64$ feet. The ball hits the ground when $s(t) = 0$. Solving, we get $-16t^2 + 64t = 0$ when $t = 0$ and also when $t = 4$. The ball is thrown at $t = 0$, so it must return at $t = 4$. Since $v(4) = -64$ feet per second, it follows that the speed (absolute value of velocity) when the ball returns is the same as the initial speed.

EXAMPLE 9. If a feather is dropped from a point 2 meters above the surface of the moon, how fast is the feather traveling when it hits the surface of the moon, and how long does it take to fall?

Solution. On Earth, $a(t) = -9.8$ meters/sec² and the force due to gravity on the moon is roughly one-sixth as great; that is, on the moon

$$a(t) \approx -1.6 \text{ meters/sec}^2.$$

It follows that $v(t) \approx -1.6t + C$, where $C = 0$ because the initial velocity is zero. Also, $s(t) \approx -0.8t^2 + k$, where $k = 2$ because $s(0) = 2$. The feather lands on the surface of the moon when $s(t) = 0$, which is true when $0.8t^2 = 2$—that is, when $t = \sqrt{2.5} \approx 1.58$ seconds. At that same instant, $v(1.58) \approx -2.5$ meters/sec.

PROBLEMS

A particle is traveling along the x-axis and the position of the particle at time t is $x(t)$. Find the velocity $v(t)$ and the acceleration $a(t)$ at the given time in each of the following problems.

1. $x(t) = t^2 + 3t - 1$, $t = 5$

2. $x(t) = \dfrac{1}{t+1}$, $t = 2$

3. $x(t) = \sqrt{t}$, $t = 4$

4. $x(t) = \sqrt{t^2 + 1}$, $t = 0$

5. $x(t) = \dfrac{t^2 + 3t}{t + 1}$, $t = 3$

A particle is traveling along the x-axis with acceleration $a(t)$. Find $v(t)$ and $s(t)$ at time t.

6. $a(t) = t - \sqrt{t}$, $v(0) = 0$, $s(0) = 1$
7. $a(t) = \sqrt{t+1} - \sqrt{t}$, $v(0) = 1$, $s(0) = 2$
8. $a(t) = (t+1)^3 - t^3$, $v(1) = 0$, $s(1) = 3$

9. $a(t) = \dfrac{1}{\sqrt{t+1}}$, $v(0) = 1$, $s(0) = 2$

10. $a(t) = -12t^2 + 6t - 4$, $v(0) = 0$, $s(0) = 1$

11. A careless hunter's gun accidentally discharges while it is ponting straight up. How long does the hunter have to seek the protection of a nearby cave if the muzzle velocity of the gun is 1000 feet per second?

12. The velocity of a particle is given by the equation

$$v(t) = 5 + \dfrac{t}{t+1}.$$

How far does the particle travel from $t = 0$ to $t = 3$?

3. RELATED RATES

A change in one object is often affected by or is related to one or more other objects. For example, a change in the volume of air inside a balloon affects the surface area of the balloon, and the rate of change in the distance between two automobiles is related to the velocity of each of the two vehicles. Problems that concern relationships between rates of change of two or more objects are called **related rates** problems. Since a derivative can be interpreted as a rate of change, we are going to study relationships between derivatives of functions that are related in some manner. This section is devoted to related rates problems.

EXAMPLE 10. In a football game, a receiver runs north at a rate of 30 feet per second, while the quarterback runs east at a rate of 20 feet per second.

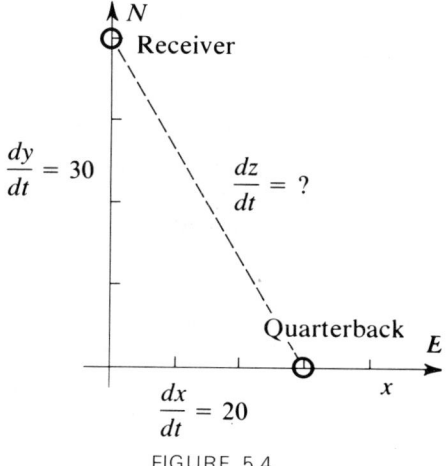

FIGURE 5.4

How fast are the two players moving away from each other at an instant when the receiver is 80 feet downfield and the quarterback is 60 feet east of center, as in Figure 5.4?

Solution. We know that

$$\frac{dy}{dt} = 30 \quad \text{and} \quad \frac{dx}{dt} = 20.$$

By the Pythagorean Theorem, $x^2 + y^2 = z^2$. Differentiating term by term, we get

$$2x\frac{dx}{dt} + 2y\frac{dy}{dt} = 2z\frac{dz}{dt}.$$

When $x = 60$ and $y = 80$, it follows that $z = \sqrt{60^2 + 80^2} = 100$. At this same instant,

$$\frac{dz}{dt} = \frac{60 \cdot 20 + 80 \cdot 30}{100} = 36 \text{ feet per second}.$$

Hence at this instant the distance between the two players is increasing at the rate of 36 feet per second.

Several steps are necessary in a solution of a related rates problem.
1. A rough sketch should be made and all the vital parts labeled.
2. Write down all the information that is given or that can be derived from the picture. (In the preceding example it was evident that $dx/dt = 20$, $dy/dt = 30$, and $x = 60$ when $y = 80$. From the picture, we can see that $x^2 + y^2 = z^2$ is the desired relationship among x, y, and z.)
3. An equation must be differentiated, and the given information must be used to solve for the unknown rate of change.

EXAMPLE 11. A glass, which has the shape of a cone of height 20 centimeters and base radius 4 centimeters, is being filled from a tap at the rate of 25 cubic centimeters per second. How fast is the level of beer rising at the instant when the height of beer in the glass is 10 centimeters, as in Figure 5.5?

Solution. The volume V of a cone of height h and radius r is equal to $\pi r^2 h/3$. We are given that $dV/dt = 25$. On the other hand, by looking at similar triangles, we see that

$$\frac{h}{r} = \frac{20}{4} = 5,$$

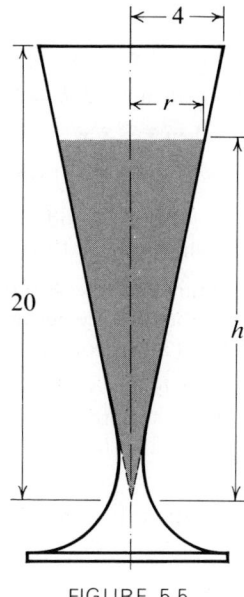

FIGURE 5.5

so that $r = h/5$. Therefore,

$$V = \frac{\pi r^2 h}{3} = \frac{\pi h^3}{75}.$$

Differentiating, we get

$$\frac{dV}{dt} = 25 = \left(\frac{\pi 3 h^2}{75}\right)\frac{dh}{dt}.$$

Evaluating at $h = 10$ gives

$$\frac{dh}{dt} = 25 \cdot 75 \frac{1}{300\pi} = \frac{25}{4\pi} \approx 1.989,$$

and hence the level of beer is rising at the rate of approximately 2 centimeters per second.

PROBLEMS

1. The volume of a balloon is $V = \frac{4}{3}\pi r^3$. How fast is the radius of the balloon changing at an instant when the radius is 3 and the volume is changing at a rate of 10 cubic inches per second?

2. The gas law equation of chemistry states that $PV = kT$, where P = pressure, V = volume, T = temperature, and k is a constant. If volume is held constant, how fast is the pressure changing if $dT/dt = 5$?

3. If $x^2 = y^3$, determine dx/dt if $dy/dt = -1$ when $x = 8$ and $y = 4$.

4. Assume that $V = \frac{1}{3}\pi r^2 h$. Find dV/dt when $r = 3$, $h = 2$, $dr/dt = -2$, and $dh/dt = 4$.

5. If $A = xy$, find dy/dt at an instant when $x = 5$, $y = 5$, $dA/dt = 4$, and $dx/dt = -3$.

6. Suppose that $x^2 - y^2 = 1$. Find dy/dt at an instant when $x = 3$, $y = 2\sqrt{2}$, and $dx/dt = 1$.

7. Suppose that $m = m_0/\sqrt{1 - (v^2/c^2)}$, where m_0 and c are constants and v depends on t. Find dm/dt if $v = c/2$ at an instant when $dv/dt = 1000$.

8. A man is standing 80 feet from a railroad track. A train is traveling at a rate of 100 feet per second on this track. How fast is the distance between the train and the man changing at an instant when the train is 100 feet away from the man?

9. A car traveling south at 30 miles per hour collides with a second car traveling west at 40 miles per hour. How fast were the cars approaching each other at the instant of impact? [*Hint*: If z is the distance between the cars, show that $dz/dt = -50$ at any time t prior to the accident.]

10. Give physical or geometric interpretations to as many of Problems 3 to 7 as possible.

4. NEWTON'S METHOD

Let f be a function. Then c is a *root* of f if $f(c) = 0$. If $f(x) = ax + b$ with $a \neq 0$, then $c = -b/a$ is a root of f. The roots of the function $x \to Ax^2 + Bx + C$ ($A \neq 0$) are given by the quadratic formula

$$(3) \qquad x = \frac{-B \pm \sqrt{B^2 - 4AC}}{2A}.$$

There are more complicated formulas for finding roots of polynomial functions of degrees 3 and 4. In the early 1800s a mathematical proof was given that there is no such formula for finding roots of general polynomial functions of degree 5 or higher. Since finding roots is so basic (for example, in sketching the graph of f, it is helpful to find the roots of f' and to find where the graph crosses the coordinate axes), we need methods that will at least enable us to find approximations to roots if these roots cannot be computed exactly.

In order to find a root of f, we need to find a point where the graph of f crosses the x-axis. Pick a number x_0, preferably close to a root (obtained, say, from a rough sketch of the graph), but for now any x_0 will do. Consider the tangent line at the point $(x_0, f(x_0))$. Since the equation

$$y = f(x_0) + f'(x_0)(x - x_0)$$

is a polynomial function of degree 1, it is easy to find a root—namely,

$$x_1 = x_0 - \frac{f(x_0)}{f'(x_0)},$$

if $f'(x_0) \neq 0$. If we are lucky, x_1 will be closer to a root than was x_0, as in Figure 5.6. We now take x_1 to be our approximation to a root and then compute x_2 by

$$x_2 = x_1 - \frac{f(x_1)}{f'(x_1)}.$$

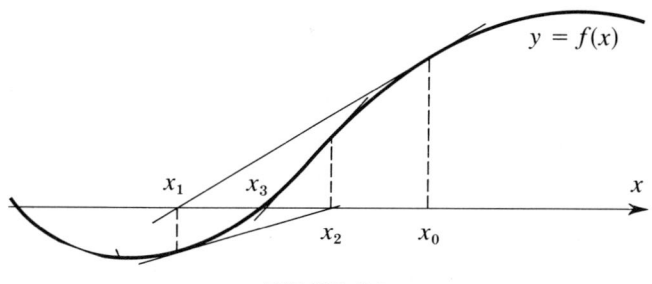

FIGURE 5.6

In general, if x_n is the nth approximation, then we define x_{n+1} by the equation

(4) $$x_{n+1} = x_n - \frac{f(x_n)}{f'(x_n)}.$$

If $f'(x_n) \neq 0$ for each n, then this defines a sequence $\{x_n\}_{n=1}^{\infty}$ that may or may not converge. Fortunately, this sequence usually converges, and when it does, it converges rapidly. To converge rapidly means roughly that x_n is very close to the limit for relatively small values of n. This method for approximating roots is known as **Newton's method** and is illustrated in Figure 5.6.

From the theory of equations, if p/q is a rational root of

$$a_n x^n + a_{n-1} x^{n-1} + \cdots + a_1 x + a_0,$$

where the a_i's are integers, then q must divide a_n exactly and p must divide a_0 exactly; that is, a_n/q and a_0/p must be integers. This is known appropriately as the **Rational Root Theorem** (see Problem 9). It is useful for testing polynomials for rational roots.

EXAMPLE 12. Find a root of $x^3 + x + 1$.

Solution. From the Rational Root Theorem, if p/q is a root, then $p|1$ and $q|1$. Hence $p = \pm 1$ and $q = \pm 1$, which means that $p/q = \pm 1$. However, $f(1) = 3 \neq 0$ and $f(-1) = -1 \neq 0$, which means that f can have no rational roots.

Since $f(-1) = -1$ and $f(0) = 1$, there must be some real root between -1 and 0. Let $x_0 = -0.5$ be our initial guess. Since $f'(x) = 3x^2 + 1$, we get

$$x_1 = -0.5 - \frac{-0.125 - 0.5 + 1}{0.75 + 1}$$

$$= -0.714.$$

Repeating this process, we get

$$x_2 = -0.685,$$
$$x_3 = -0.6825,$$

and
$$x_4 = -0.682375.$$

EXAMPLE 13. Use Newton's method to approximate a root to $f(x) = \sqrt[3]{x}$. (For the purpose of the example, pretend not to notice that $x = 0$ is a root.)

Solution. We have $f'(x) = 1/(3x^{2/3})$. Let $x_0 = 1$. Then

$$x_{n+1} = x_n - \frac{f(x_n)}{f'(x_n)}$$

$$= x_n - (x_n)^{1/3} 3(x_n)^{2/3}$$

$$= -2x_n.$$

We get the following sequence.

$$x_0 = 1, \quad x_1 = -2, \quad x_2 = 4, \quad x_3 = -8, \quad x_4 = 16, \quad x_5 = -32,$$
$$x_6 = 64, \quad \ldots, \quad x_n = (-1)^n 2^n, \quad \ldots$$

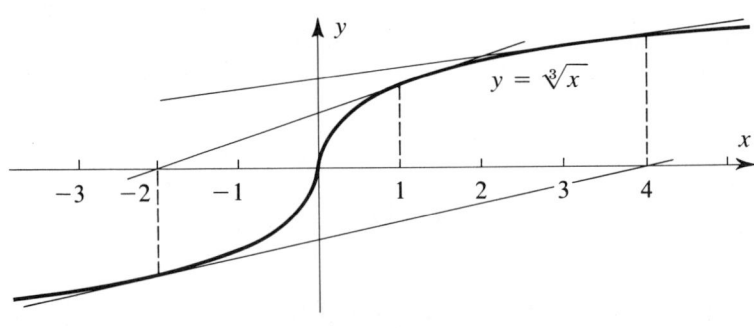

FIGURE 5.7

By looking at Figure 5.7, we can get some idea as to why this sequence fails so miserably. The following example shows another way in which Newton's method can fail.

EXAMPLE 14. Let $f(x) = x^3 + x^2 - 2x$, so that $f'(x) = 3x^2 + 2x - 2$. Notice that 0 and 1 are roots of f. The number $\frac{1}{2}$ is between 0 and 1, so we might expect Newton's method to yield a sequence that would converge to either 0 or 1 if we start with $x_0 = \frac{1}{2}$. However, Newton's method gives

$$x_1 = x_0 - \frac{f(x_0)}{f'(x_0)}$$

$$= 0.5 - \frac{0.125 + 0.25 - 1}{0.75 + 1 - 2}$$

$$= -2,$$

$$x_2 = -2 - 0$$

$$= -2,$$

and from there on, $x_2 = x_3 = x_4 = \cdots = -2$. Since $f(-2) = 0$, Newton's method succeeds in finding a root, but it fails to find a root that is closest to the original guess. By picking $x_0 = 0.1$, however, we get

$$x_0 = 0.1,$$

$$x_1 = -0.0067796,$$

$$x_2 = -0.0000225,$$

and the resulting sequence appears to converge to 0.

Thus there are times when Newton's method fails. It is fortunate that it is much easier to find examples where Newton's method works than to find examples where the method fails. Also, when Newton's method works, it works very well indeed. For example, $\sqrt{5}$ can be computed by using Newton's method to find a root of $f(x) = x^2 - 5$ starting with $x_0 = 1$. After only eight iterations, $x_8 = 2.236\ldots$ approximates $\sqrt{5}$ with error less than 10^{-25}.

PROBLEMS

Use Newton's method to solve the following problems. Repeat the iteration until $|x_n - x_{n-1}| < 10^{-3}$.

1. Compute $\sqrt{2}$. [*Hint*: Find a root of $f(x) = x^2 - 2$, starting with $x_0 = 1.5$.]
2. Compute $\sqrt[3]{2}$.
3. Compute $\sqrt{15}$.

4. Compute $\sqrt[3]{26}$.
5. Find a root of $x^2 - x - 15$ between 4 and 5.
6. Find a root of $x^3 - x^2 + x - 3$ between 1 and 2.
7. Find a root of $x^4 + x - 1$ between 0 and 1.
8. Find a root of $x^3 + 2x - 10$ between 1 and 2.
9. Show that if p/q is a rational root of the polynomial $a_n x^n + \cdots + a_1 x + a_0$, where the a_i's are integers, then q divides a_n and p divides a_0. [*Hint:* Assume that $a_n(p/q)^n + \cdots + a_0 = 0$ and multiply both sides by q^n.]
10. Use Problem 9 to show that $x^3 + 2x - 10$ has no rational roots.
11. Use Problem 9 to find a root of $2x^3 - 3x^2 + 2x - 3$.

SOME SPECIAL ALGORITHMS (OPTIONAL)

Newton's method gives a useful technique for computing square roots on an electronic hand calculator in the event that a square root function is not available. Suppose that we want to compute $\sqrt{a}$. This problem is equivalent to finding a positive root of $f(x) = x^2 - a$. If x is an initial guess, then the next estimate is given by

$$x - \frac{f(x)}{f'(x)} = x - \frac{x^2 - a}{2x}$$

$$= \frac{2x^2 - x^2 + a}{2x}$$

$$= \frac{x^2 + a}{2x}.$$

The expression $(x^2 + a)/2/x$ is easy to compute, even on calculators without memory. An algorithm for computing $\sqrt{a}$ is given in Figure 5.8.

The flow chart in Figure 5.8 should be interpreted for use on an electronic calculator as follows: to compute $\sqrt{a}$,
    1. Make a guess x, and write it down.
    2. Square x.
    3. Add a.
    4. Divide by 2.
    5. Divide by the original guess.
    6. Take the result in part (5) to be the new guess.
    7. Return to part (1) and repeat the whole process until x^2 is close to a.

EXAMPLE 15. Estimate $\sqrt{5}$.

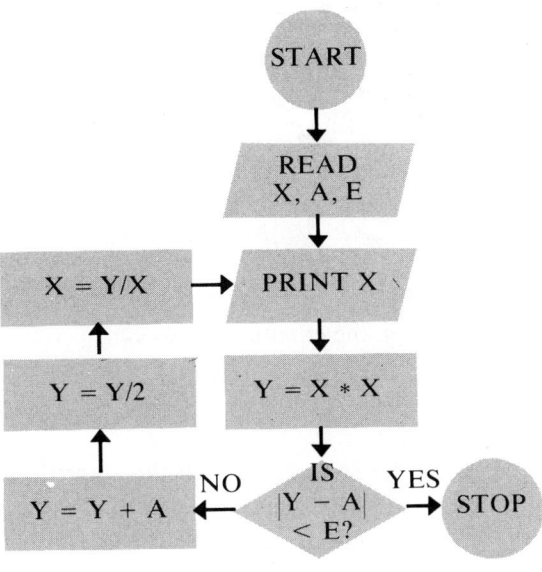

FIGURE 5.8

Solution. Let $x = 2$ be the initial guess. We square x to get 4, add 5 to get 9, divide by 2 to obtain 4.5, and then divide by the initial guess 2 to get 2.25. We repeat the whole process: $(2.25)^2 = 5.0625$, $5.0625 + 5 = 10.0625$, $10.0625 \div 2 = 5.03125$, $5.03125 \div 2.25 = 2.2361111$. From here on, the successive estimates tend to get very close to $\sqrt{5}$. The next estimate is 2.2360677, followed by 2.2360679. We stop at this point because $(2.2360679)^2 = 4.9999996$, which is indeed close to 5. Unless a calculator has more than eight significant digits, there is no reason to continue.

Other examples of computations of square roots using the algorithm in Figure 5.8 are tabulated in Table 5.1.

Table 5.1

a	2	3	5	7	10
x_0	1.5	1.5	2	3	3
x_1	1.4166666	1.75	2.25	2.6666666	3.1666666
x_2	1.4142156	1.7321428	2.2361111	2.6458331	3.1622804
x_3	1.4142135	1.7320507	2.2360677	2.6457511	3.1622774
x_4	1.4142135	1.7320507	2.2360679	2.6457511	3.1622776
x_5	1.4142135	1.7320507	2.2360679	2.6457511	3.1622775

CHAPTER 5: APPLICATIONS

An algorithm for computing nth roots can also be found by using Newton's method. We again start by guessing a possible positive root x of $f(x) = x^n - a$. Then the next estimate is given by

$$x - \frac{f(x)}{f'(x)} = x - \frac{x^n - a}{nx^{n-1}}$$

$$= \frac{nx^n - x^n + a}{nx^{n-1}}$$

$$= \frac{(n-1)x^n + a}{nx^{n-1}}.$$

An algorithm based on the equation

$$\text{Next } x = \frac{(n-1)x^n + a}{nx^{n-1}}$$

is given in Figure 5.9. Table 5.2 shows how well the algorithm estimates $\sqrt[n]{a}$ for various choices of n and a and for some selected initial guesses. Notice that in the computation of $\sqrt[10]{10}$, the initial guess $x = 1.2$ leads to a solution much faster than the initial guess $x = 1.0$.

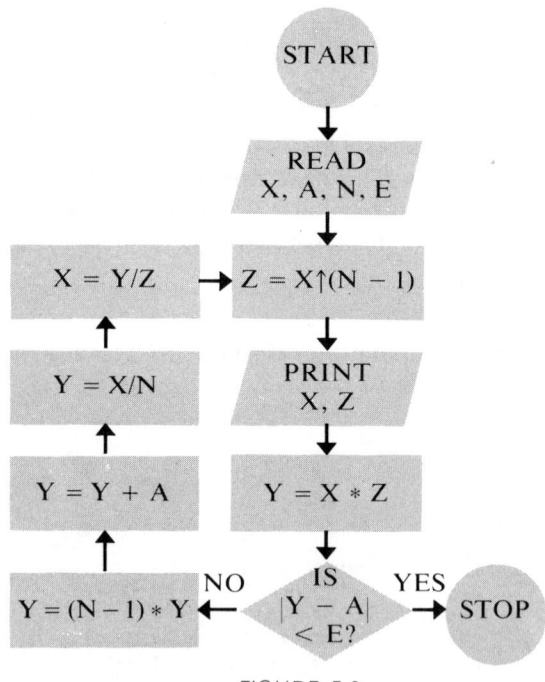

FIGURE 5.9

Table 5.2

a	5	3	1	10	10	10
n	3	5	3	10	10	10
x_0	2	1.2	2	1	1.1	1.2
x_1	1.75	1.2493517	1.4166666	1.9	1.4140975	1.2738066
x_2	1.7108843	1.2457516	1.1105342	1.7130985	1.3169145	1.2596837
x_3	1.7099762	1.2457306	1.0106366	1.5496578	1.2691654	1.2589273
x_4	1.7099758	1.2457307	1.0001114	1.4140954	1.2592892	1.2589253
x_5	1.7099758	1.2457307	0.9999999	1.3169132	1.2589257	1.2589254
x_6				1.2691650	1.2589251	1.2589254
x_7				1.2592891	1.2589254	
x_8				1.2589257	1.2589254	
x_9				1.2589251		
x_{10}				1.2589254		
x_{11}				1.2589254		

PROBLEMS

Estimate each of the following to six decimal places.

1. $\sqrt{1.5}$
2. $\sqrt{0.5}$
3. $\sqrt{11}$
4. $\sqrt{50}$
5. $\sqrt[3]{25}$
6. $\sqrt[3]{2}$
7. $\sqrt[4]{3}$
8. $\sqrt[3]{1.5}$
9. $\sqrt[10]{5}$
10. $\sqrt[5]{7}$
11. $\sqrt[5]{201}$
12. $\sqrt[6]{0.23}$

5. VOLUMES OF REVOLUTION

Let f be a function and suppose that the graph of f is rotated about the x-axis. What is the volume of the generated solid between $x = a$ and $x = b$? The volume can be approximated by subdividing the interval $[a, b]$ into n equal parts, as in Figure 5.10.

The slice of the solid between x_{i-1} and x_i has roughly the same volume as the right circular cylinder of height $(b - a)/n$, which is given in Figure 5.10 (see Appendix). The volume of the cylinder is

$$\pi f(x_i)^2 \frac{b-a}{n}.$$

Hence the volume of the solid should be approximately

$$\pi \sum_{i=1}^{n} f(x_i)^2 \frac{b-a}{n}.$$

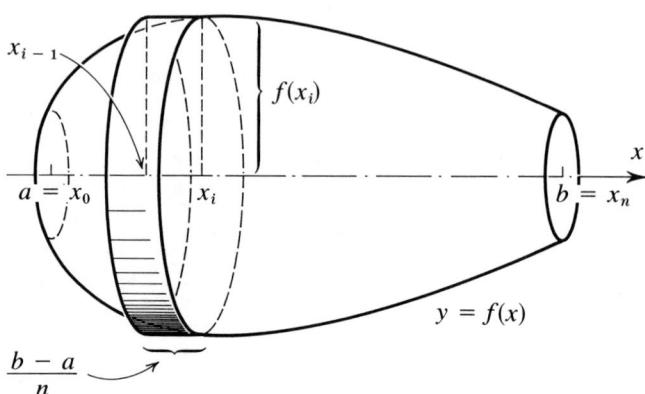

FIGURE 5.10

But we recognize this as being a Riemann sum, and hence the limit as n tends to infinity is the integral

(5) $$\pi \int_a^b f(x)^2 \, dx.$$

We *define* the volume of the solid of revolution generated by rotating the graph of $y = f(x)$ about the x-axis, $a \leq x \leq b$, to be the integral (5), if this integral exists.

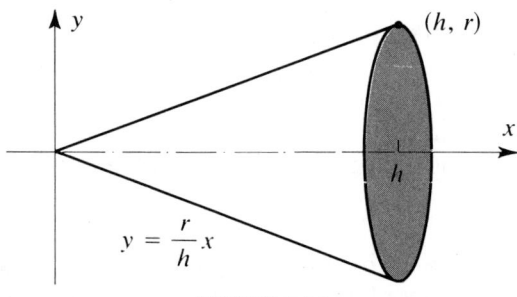

FIGURE 5.11

EXAMPLE 16. Compute the volume of a right circular cone whose height is h and whose base has radius r.

Solution. By Figure 5.11, this problem is equivalent to finding the volume of the cone generated by rotating the line $y = (r/h)x$ about the x-axis $0 \leq x \leq h$. Thus the volume is given by

$$\pi \int_0^h \left(\frac{r}{h}x\right)^2 dx = \frac{\pi r^2}{h^2} \int_0^h x^2\, dx$$

$$= \frac{\pi r^2}{3h^2} x^3 \Big|_0^h$$

$$= \frac{\pi r^2 h^3}{3h^2}$$

$$= \frac{1}{3}\pi r^2 h.$$

That is, the volume of a right circular cone is one-third the area of the base times the height.

EXAMPLE 17. Compute the volume of a sphere of radius r.

Solution. Such a sphere can be generated by rotating $y = \sqrt{r^2 - x^2}$ about the x-axis, $-r \le x \le r$. The volume is then given by

$$\pi \int_{-r}^{r} (\sqrt{r^2 - x^2})^2\, dx = \pi \int_{-r}^{r} (r^2 - x^2)\, dx$$

$$= \pi r^2 x - \frac{1}{3}\pi x^3 \Big|_{-r}^{r}$$

$$= \pi r^2 [r - (-r)] - \frac{1}{3}\pi [r^3 - (-r)^3]$$

$$= \pi \left(2r^3 - \frac{2r^3}{3}\right)$$

$$= \frac{4\pi r^3}{3}.$$

Suppose that the x-axis passes through a solid that starts at $x = a$ and ends at $x = b$. If the plane perpendicular to the x-axis intersects the solid in a region of area equal to $A(x)$ for each x, what is the volume of the solid? As usual, subdivide the interval $[a, b]$ into n equal subintervals. Then by Figure 5.12, the kth slab has volume approximately equal to

$$A(x_k)\frac{b - a}{n}.$$

178 CHAPTER 5: APPLICATIONS

The total volume is thus approximately

$$\sum_{k=1}^{n} A(x_k) \frac{b-a}{n}.$$

Taking the limit as n tends to infinity, we get

(6) $$\lim_{n \to \infty} \sum_{k=1}^{n} A(x_k) \frac{b-a}{n} = \int_{a}^{b} A(x)\,dx,$$

which, by definition, is the volume of the solid (if the integral exists).

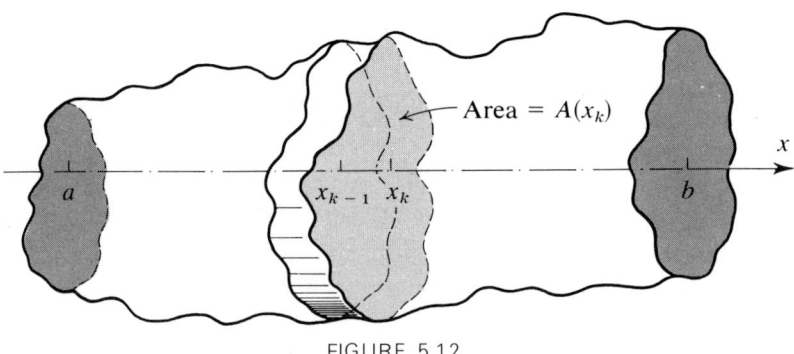

FIGURE 5.12

EXAMPLE 18. Find the volume of a pyramid with height 1 and base a unit square.

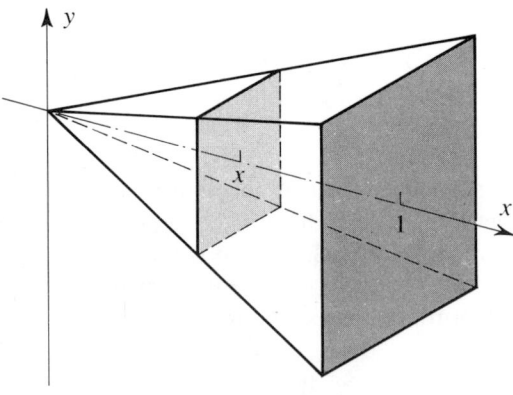

FIGURE 5.13

SECTION 5: VOLUMES OF REVOLUTION

Solution. By Figure 5.13, $A(x) = x^2$. Thus the volume of the pyramid is

$$\int_0^1 x^2 \, dx = \frac{1}{3} x^3 \Big|_0^1$$

$$= \frac{1}{3},$$

which is one-third the area of the base times the height. (Compare this result with the volume of the cone found in Example 16.)

PROBLEMS

Find the volume of each solid of revolution formed by rotating the graph of the given function about the x-axis from $x = a$ to $x = b$.

1. $y = \sqrt{x^2 + 1}$, $a = 0$, $b = 10$
2. $y = x + 1$, $a = 0$, $b = 1$
3. $y = \dfrac{1}{\sqrt[3]{x}}$, $a = 1$, $b = 8$
4. $y = \sqrt{x(x^2 + 1)^5}$, $a = 0$, $b = 1$
5. $y = x^2$, $a = 0$, $b = 5$
6. $y = \sqrt{x}$, $a = 0$, $b = 1$
7. $y = \dfrac{\sqrt{x}}{x^2 + 1}$, $a = 0$, $b = 4$
8. $y = \sqrt{1 + \sqrt{x}}$, $a = 0$, $b = 1$
9. $y = \dfrac{\sqrt{x^2 + 1}}{x}$, $a = 1$, $b = 2$
10. $y = \dfrac{1}{x}$, $a = 1$, $b = 2$
11. $y = x^{-3/2}$, $a = 1$, $b = 2$
12. $y = x^{3/2}$, $a = 0$, $b = 1$

13. Find the volume of the solid of revolution formed by rotating the region bounded by $y = x^2$, $x = 5$, and $y = 0$ about the y-axis. [*Hint:* Subdivide the interval $[0, 25]$ on the y-axis into n parts and draw a picture of the ith slab.]

14. Find the volume of the solid of revolution formed by rotating the region bounded by $y = \sqrt{x}$, $x = 1$, and $y = 0$ about the y-axis.

15. Find the volume of the solid of revolution formed by rotating the region bounded by $y = x^2$, $x = 5$, and $y = 0$ about the line $y = -1$.

16. Find the volume of the solid of revolution formed by rotating the region bounded by $y = \sqrt{x}$, $x = 1$, and $y = 0$ about the line $x = 3$.

17. A solid lies on the x-axis, and a plane through the point $(x, 0)$ perpendicular to the x-axis intersects the solid in a region of area $A(x) = \sqrt{x + 4}$, $a = 0$, $b = 5$. What is the volume of the solid between $x = 0$ and $x = 5$?

18. What is the volume of a tetrahedron if the length of each edge is a? (A *tetrahedron* is a regular solid with four faces, each of which is an equilateral triangle.)

6. DIFFERENTIAL EQUATIONS

The acceleration of an object is given by $a = t/v^2$, where v is the velocity of the object. If it is known that $v = 1$ when $t = 0$, what is an expression for the velocity of the object at time t? Since

$$a = \frac{dv}{dt},$$

the equation $a = t/v^2$ is a **differential equation**; that is, an equation that includes one or more derivatives. Many physical problems can be translated into differential equations, and the original problems can be solved by solving the corresponding differential equations. The translation of physical problems into mathematical problems is called **mathematical modeling.** This process usually requires some background in a certain area, such as physics, sociology, or biology, plus a plentiful supply of ingenuity and insight. We will restrict ourselves primarily to the solution of the corresponding mathematical problems.

Replacing a by dv/dt in the equation $a = t/v^2$, we get

$$\frac{dv}{dt} = \frac{t}{v^2}.$$

If we treat dv and dt as differentials, we can multiply both sides of the equation by $v^2 \, dt$ to get

$$v^2 \, dv = t \, dt.$$

Notice that the left-hand side is an expression in v and the right-hand side is an expression in t. Thus we can now compute the indefinite integral on both sides.

$$\int v^2 \, dv = \int t \, dt$$

implies

$$\frac{1}{3} v^3 = \frac{1}{2} t^2 + C$$

for some constant C. But $v = 1$ when $t = 0$, and hence $C = \frac{1}{3}$. Simplifying, we have

$$2v^3 = 3t^2 + 2$$

or

$$v = \sqrt[3]{\frac{3t^2 + 2}{2}}.$$

The foregoing example was solved by a technique known as **separation of variables.** When this technique works, it reduces the solution of a differential equation to the computation of indefinite integrals.

EXAMPLE 19. Solve $y' + yy' = x + 2$ if $y = 0$ when $x = 0$.

Solution. The equation is equivalent to $y'(y + 1) = x + 2$. Multiplying both sides by dx, we get
$$(y + 1)\, dy = (x + 2)\, dx.$$
Computing the indefinite integral on both sides gives
$$\frac{1}{2}y^2 + y = \frac{1}{2}x^2 + 2x + C,$$
where $C = 0$, since $y = 0$ when $x = 0$. The quadratic formula, together with $y = 0$ when $x = 0$, now leads to
$$y = -1 + \sqrt{x^2 + 4x + 1}.$$

EXAMPLE 20. Solve $x^2 yy' = 1$ if $y = 1$ when $x = 1$.

Solution. Multiply by dx/x^2 to get
$$y\, dy = \frac{dx}{x^2}$$
and hence
$$\frac{1}{2}y^2 = -\frac{1}{x} + C.$$
Since $y = 1$ when $x = 1$, it follows that $C = 3/2$ and
$$y = \sqrt{3 - \frac{2}{x}}$$
$$= \frac{\sqrt{3x - 2}}{\sqrt{x}}.$$

EXAMPLE 21. Find $s(t)$ if
$$\frac{d^2 s}{dt^2} = -32, \qquad \frac{ds}{dt} = 0$$
when $t = 0$, and $s = 5$ when $t = 0$.

Solution. Let $v = ds/dt$. Then
$$\frac{dv}{dt} = \frac{d^2 s}{dt^2} = -32,$$

and hence $dv = -32\,dt$; that is, $v = -32t + C$, where $C = 0$ because $v = 0$ when $t = 0$. Now

$$\frac{ds}{dt} = -32t$$

is equivalent to $ds = -32t\,dt$, and hence $s = -16t^2 + K$, where $K = 5$ because $s = 5$ when $t = 0$. It follows that $s(t) = -16t^2 + 5$.

Often differential equations are given without **initial conditions**—that is, without information of the form $y = 5$ when $x = 0$. Solutions to such differential equations must include arbitrary constants, as in the following example.

EXAMPLE 22. Solve the differential equation $\sqrt{xy}\,dy = dx$.

Solution. We divided both sides by $\sqrt{x}$ to get

$$\sqrt{y}\,dy = \frac{dx}{\sqrt{x}}.$$

Integrating both sides gives

$$\frac{2}{3}y^{3/2} = 2\sqrt{x} + C.$$

PROBLEMS

Solve the following differential equations.

1. $x^2\,dy + \dfrac{x^2 + 1}{y}\,dx = 0$

2. $x\sqrt{y^2 + 2}\,dx + y\sqrt{3 - x^2}\,dy = 0$

3. $x^2 y\,\dfrac{dy}{dx} = 1 + x^2 + x^3$

4. $\dfrac{dy}{dx} = \dfrac{x\sqrt{x^2 + 1}}{y^2\sqrt{y^3 + 1}}$

5. $s''(t) = \dfrac{1}{\sqrt{t + 1}}$, $s(0) = 1$, $s'(0) = 2$

6. $s(t)^2 = v(t)$, where $v(t) = s'(t)$ and $s(0) = 1$

7. $x\sqrt{y^3 + 1}\,dx + y^2\sqrt{x^2 + 1}\,dy = 0$

8. $\dfrac{d^2s}{dt^2} + t\left(\dfrac{ds}{dt}\right)^2 = 0$, where $\dfrac{ds}{dt}\bigg|_{t=1} = 2$

REVIEW PROBLEMS

1. The **profit** P is the revenue R minus the cost C. If the total revenue from selling n bicycles is $50n - 0.1n^2$ and the cost of manufacturing n bicycles is $30n + 500 + 0.03n^2$, what output will maximize profit? What is the maximum profit?

2. A rumor is spreading among the students at a state university. If the rate at which the rumor spreads is given by $r(t) = kp(t)[1 - p(t)]$, where $p(t)$ is the ratio of students who have heard the rumor and k is some constant, for what value of $p(t)$ does the rumor spread the fastest?

3. A group of students can charter a flight to England if at least 40 students agree to go. The fare is $200 per person for 40 people, and all the fares decrease by $1 for each additional person. How many people will it take to maximize the revenue for the airline company? What is the maximum revenue? Should the airline permit these fares on a 240-passenger plane? What is the probable capacity of the plane?

4. A policeman on horseback sees a car run a red light and wishes to record the license number of the car. The car is traveling east at a constant 70 feet per second. At the instant the car is halfway through the intersection, the policeman is 100 feet north of the intersection. If the horse can run only 50 feet per second, and a city park makes it possible for the horse to take a shortcut, how close can the policeman get to the car? (See Figure 5.14.) *Hint:*
   a. Assume that a solution can be found and consider the final location of the car and the policeman.
   b. Argue that the policeman should travel in a straight line.
   c. Argue that the final location of the policeman should lie on the hypotenuse of a right triangle with vertices at $(0, 100)$, $(0, 0)$, and $(70t, 0)$, as in Figure 5.14.
   d. Use the Pythagorean Theorem to find a relationship between s and t.
   e. Use implicit differentiation and find t such that $ds/dt = 0$.

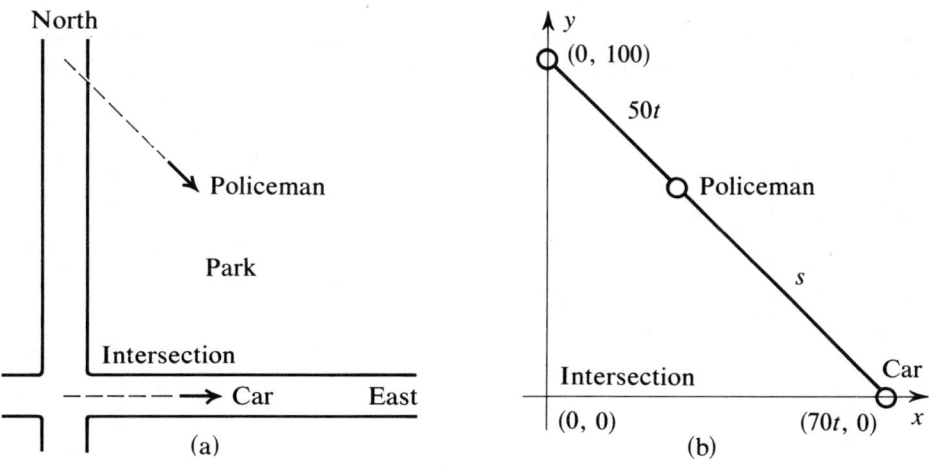

FIGURE 5.14

5. A car starts from rest and accelerates at a rate of 4 feet/sec^2 for 10 seconds, travels at a constant speed for 30 seconds, and then decelerates at a rate of 5 feet/sec^2 until the car comes to rest.
   a. Construct a graph of acceleration, below it a graph of velocity, and below that a graph of the position of the car.
   b. For how many seconds is the car in motion?
   c. How far does the car travel?
   d. What is the maximum velocity?

6. A motor boat requires $v^3/5400$ gallons of fuel per hour when the boat is traveling at v miles per hour in still water. What is the minimal amount of fuel required to travel 100 miles upstream in a river that is flowing downstream at 20 miles per hour? How fast should the boat travel to minimize fuel consumption? How much fuel would be used if the boat traveled at 40 miles per hour?

7. Assume that the amount of energy required by a fish to swim is kv^3, where k is some constant and v is the velocity of the fish. How fast should a fish swim upstream against a current v_0 in order to conserve energy so that the fish can migrate upstream as far as possible to spawn?

8. The height x of a building is computed experimentally by 10 different people. Due to slight imperfections in equipment and people, the measurements are $x_1 = 101.2$, $x_2 = 102.3$, $x_3 = 100.4$, $x_4 = 99.8$, $x_5 = 102.4$, $x_6 = 100.8$, $x_7 = 101.7$, $x_8 = 103.0$, $x_9 = 100.5$, and $x_{10} = 101.5$ meters. Find a value of x that minimizes the sum of the squares $(x - x_1)^2 + (x - x_2)^2 + (x - x_3)^2 + \cdots + (x - x_{10})^2$.

9. The cost of making plastic Christmas trees increases as the number of trees increases, but it does not necessarily cost twice as much to produce twice as many trees. There are fixed costs, such as property taxes, rent, and certain utilities. The cost per pound of material depends on the quantities ordered. Wages are higher for overtime. Suppose that the *average cost* (total cost divided by the number of trees) is given by $c(t) = 0.001t + 0.5 + 1000/t$ dollars, where t is the number of trees produced per month. How many trees should be produced per month to minimize the average cost per tree? What is the minimum average cost?

Use Newton's method in Problems 10 through 15 to find a root between a and b. Continue applying Newton's method until x_{n-1} and x_n differ by less than 10^{-3} (10^{-6} if you have access to a calculator).

10. $x^2 + x - 1$, $a = 0$, $b = 1$
11. $x^3 + x - 1$, $a = 0$, $b = 1$
12. $x^4 - 5x^3 + 2x^2 + 34$, $a = 3$, $b = 4$
13. $x^3 - 3x^2 - 1$, $a = 3$, $b = 4$
14. $x^5 + x^4 + x^3 + x^2 + x - 7$, $a \doteq 1$, $b = 2$
15. $x^4 - x^2 + 0.1$, $a = 0$, $b = 0.5$

CHAPTER 6

LOGARITHMS AND EXPONENTIALS

So far we have restricted our attention to what are called *algebraic functions*—that is, functions that can be expressed in terms of polynomial functions, quotients of polynomial functions, nth roots of such functions, and so forth. In the sciences, there are many functions that occur naturally but that are not algebraic. We shall now begin to introduce and study these functions, the so-called *elementary transcendental functions*.

1. THE NATURAL LOGARITHM

In Chapter 2 it was noted that

$$\frac{d}{dx} x^n = nx^{n-1}$$

for all real numbers n. It follows that

$$\int x^n \, dx = \frac{n^{n+1}}{n+1} + C$$

186   CHAPTER 6: LOGARITHMS AND EXPONENTIALS

for all real numbers n *except* $n = -1$. If $0 < a < b$, then $f(x) = 1/x$ is continuous on $[a, b]$, and hence

$$\int_a^b \frac{1}{x}\, dx$$

does exist. The only trouble is that $1/x$ does not have an antiderivative that is an algebraic function. It is convenient to give a name to a certain antiderivative of $1/x$.

For each positive real number x, define the function $\ln(x)$ by

(1)
$$\ln(x) = \int_1^x \frac{1}{t}\, dt.$$

We saw in Chapter 2, Theorem 6, that if f is continuous,

$$\frac{d}{dx} \int_a^x f(t)\, dt = f(x),$$

so, in particular,

(2)
$$\frac{d}{dx} \ln x = \frac{1}{x},$$

and hence $\ln x$ is indeed an antiderivative of $1/x$. Since every differentiable function is continuous, it follows that $\ln x$ is continuous. The function $\ln x$ is called the **natural logarithm** of x.

If t is positive, $1/t$ is also positive; and if $x > 1$, then $\ln x$ is the area of the region in Figure 6.1. In particular, if $x > 1$, then $\ln x$ is positive. On the other hand, if $0 < x < 1$, then

$$\ln x = \int_1^x \frac{1}{t}\, dt = -\int_x^1 \frac{1}{t}\, dt,$$

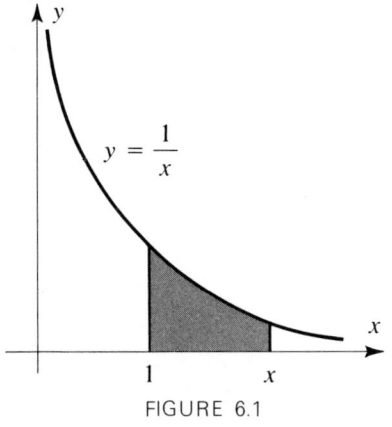

FIGURE 6.1

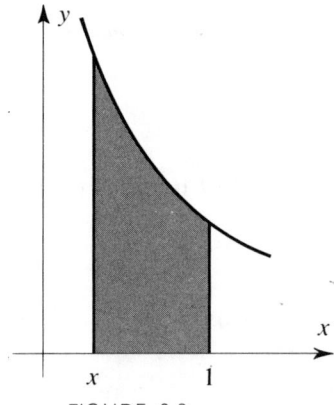

FIGURE 6.2

and hence ln x is the negative of the area of the region in Figure 6.2. Thus ln x < 0 if 0 < x < 1. At x = 1,

$$\ln 1 = \int_1^1 \frac{1}{t} dt = 0.$$

The function ln x is not defined for $x \leq 0$. Summarizing, we have

(3) $$\ln x \text{ is} \begin{cases} \text{positive} & \text{if } x > 1 \\ \text{zero} & \text{if } x = 1 \\ \text{negative} & \text{if } 0 < x < 1 \\ \text{undefined} & \text{if } x \leq 0 \end{cases}$$

What does the graph of $y = \ln x$ look like? In order to sketch the graph, we refer back to the techniques of sketching that were developed in Chapter 4. Differentiating, $y' = 1/x$ and $y'' = -1/x^2$. Since $x > 0$, it follows that the graph of $y = \ln x$ is always increasing and concave downward. There are no extreme values. Also, $\ln (1) = 0$ and $y'(1) = 1$. Thus the graph must look something like Figure 6.3.

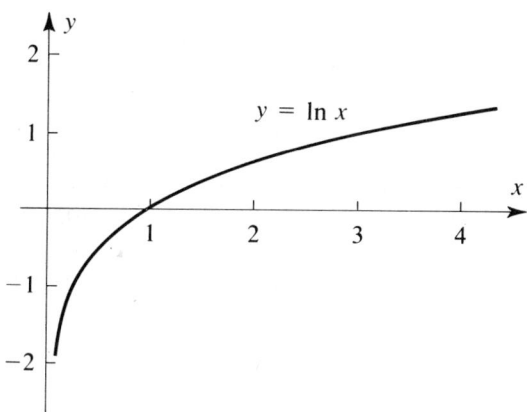

FIGURE 6.3

A short table of approximate values of ln x is given in Table 6.1. To compute ln 5.2, locate 5 in the left-hand column, then go to the column labeled 2 and read the entry 1.649. Interpret this to mean that ln 5.2 ≈ 1.649. We will occasionally relax the notation a bit and write ln 5.2 = 1.649 instead of ln 5.2 ≈ 1.649. Keep in mind, however, that the logarithms in Table 6.1 are merely approximations and are not exact.

Table 6.1. Natural Logarithms

x	0	1	2	3	4	5	6	7	8	9
1	0.000	0.095	0.182	0.262	0.336	0.405	0.470	0.531	0.588	0.642
2	0.693	0.742	0.788	0.833	0.875	0.916	0.956	0.993	1.030	1.065
3	1.099	1.131	1.163	1.194	1.224	1.253	1.281	1.308	1.335	1.361
4	1.386	1.411	1.435	1.459	1.482	1.504	1.526	1.548	1.569	1.589
5	1.609	1.629	1.649	1.668	1.686	1.705	1.723	1.740	1.758	1.775
6	1.792	1.808	1.825	1.841	1.856	1.872	1.887	1.902	1.917	1.932
7	1.946	1.960	1.974	1.988	2.001	2.015	2.028	2.041	2.054	2.067
8	2.079	2.092	2.104	2.116	2.128	2.140	2.152	2.163	2.175	2.186
9	2.197	2.208	2.219	2.230	2.241	2.251	2.262	2.272	2.282	2.293

What we have done may seem artificial. We were unable to find a legitimate antiderivative, so we just made one up. This step can be justified by showing that this antiderivative is very useful. Indeed the function ln x, together with Table 6.1, can be used to give numerical approximations to such integrals as

$$\int_2^4 \frac{1}{x}\,dx \quad \text{and} \quad \int_1^5 \frac{x\,dx}{x^2+1}.$$

For example,

$$\int_2^4 \frac{1}{x}\,dx = \ln x \Big|_2^4$$

$$= \ln 4 - \ln 2$$

$$= 1.386 - 0.693$$

$$= 0.693.$$

In order to deal with more complicated expressions, we must look at properties of the function $x \to \ln x$ more closely. First, suppose that a is a positive real number and define a new function f by $f(x) = \ln(ax)$, $x > 0$. Then using the chain rule for differentiating a composition of functions, we see that

$$f'(x) = \frac{1}{ax} \cdot a$$

$$= \frac{1}{x}$$

$$= \frac{d}{dx}(\ln x).$$

Thus $f(x) = \ln(x) + C$ for some constant C. To find C, we evaluate both sides of the equation at $x = 1$ and recall that $\ln(1) = 0$:

$$f(1) = \ln(1) + C$$

$$= C.$$

But $f(1) = \ln(a \cdot 1) = \ln(a)$, so $C = \ln(a)$. It follows that
$$\ln(ax) = \ln(a) + \ln(x),$$
since each side of the equation is equal to $f(x)$. In particular, if a and b are any positive real numbers, then

(4) $$\ln(ab) = \ln(a) + \ln(b).$$

Therefore the natural logarithm of a product of two positive numbers equals the sum of the natural logarithms of the numbers taken separately.

Next, we define a function g by $g(x) = \ln(x^a)$, where a is any fixed rational number and $x > 0$. Then again using the differentiation trick, we have

$$g'(x) = \frac{1}{x^a} \cdot ax^{a-1}$$

$$= a \cdot \frac{1}{x}$$

$$= a \frac{d}{dx}[\ln(x)]$$

$$= \frac{d}{dx}[a \ln(x)],$$

and hence
$$g(x) = a \ln(x) + C$$

for some constant C. By again evaluating both sides at $x = 1$, it follows that $C = 0$. Thus if a is any rational number and b is any positive real number, then

(5) $$\ln(b^a) = a \ln b.$$

Hence the natural logarithm of a power equals the exponent times the natural logarithm of the base.

Finally, if a and b are positive real numbers, then

$$\ln\left(\frac{a}{b}\right) = \ln(ab^{-1})$$
$$= \ln(a) + \ln(b^{-1})$$
$$= \ln a - \ln b;$$

that is,

(6) $$\ln\left(\frac{a}{b}\right) = \ln a - \ln b.$$

In words, the natural logarithm of a quotient equals the natural logarithm of the numerator minus the natural logarithm of the denominator.

It is known that ln x is an antiderivative of $1/x$ for x positive. Moreover, for $x \neq 0$, we have

$$\frac{d}{dx}(\ln|x|) = \frac{d}{dx}[\ln(\sqrt{x^2})]$$

$$= \frac{d}{dx}[\ln([x^2]^{1/2})]$$

$$= \frac{1}{2}\frac{d}{dx}[\ln(x^2)]$$

$$= \frac{1}{2} \cdot \frac{1}{x^2} \cdot 2x$$

$$= \frac{1}{x}.$$

It follows that

(7) $$\int \frac{1}{x} dx = \ln|x| + C.$$

We now look at some examples that illustrate how to integrate and differentiate functions involving the natural logarithm.

EXAMPLE 1. Differentiate $y = \ln(x^2 + 1)$.

Solution. Using the rule for differentiating a composition of functions, we see that

$$\frac{d}{dx}[\ln(x^2 + 1)] = \frac{1}{x^2 + 1} 2x$$

$$= \frac{2x}{x^2 + 1}.$$

EXAMPLE 2. Compute $\int_0^1 \frac{x}{x^2 + 1} dx$.

Solution. By Example 1, we see that

$$\int_0^1 \frac{x\, dx}{x^2 + 1} = \frac{1}{2}\int_0^1 \frac{2x\, dx}{x^2 + 1}$$

$$= \frac{1}{2}\ln(x^2 + 1)\Big|_0^1$$

$$= \frac{1}{2}[\ln 2 - \ln 1]$$

$$= 0.346,$$

using Table 6.1.

EXAMPLE 3. Find an antiderivative of $\dfrac{\ln x}{x}$.

Solution. Using the method of substitution with $u = \ln x$, we get $du = (1/x)\,dx$, and hence

$$\int \frac{\ln x}{x}\,dx = \int u\,du$$

$$= \frac{1}{2}u^2 + C$$

$$= \frac{1}{2}(\ln x)^2 + C.$$

EXAMPLE 4. Differentiate $y = \ln\sqrt{x}$.

Solution.

$$\frac{dy}{dx} = \frac{d}{dx}(\ln\sqrt{x}) = \frac{1}{\sqrt{x}} \cdot \frac{d}{dx}(\sqrt{x}) = \frac{1}{\sqrt{x}} \cdot \frac{1}{2\sqrt{x}} = \frac{1}{2x}.$$

Alternate Solution.

$$\frac{dy}{dx} = \frac{d}{dx}(\ln\sqrt{x}) = \frac{d}{dx}[\ln(x^{1/2})] = \frac{1}{2}\frac{d}{dx}(\ln x) = \frac{1}{2}\cdot\frac{1}{x} = \frac{1}{2x}.$$

EXAMPLE 5. Compute the derivative of $f(x) = \dfrac{\sqrt{x}}{(x+1)^2}$.

Solution. This step can be done directly, using the quotient rule for derivatives. It can also be differentiated by using a method known as **logarithmic differentiation**. Taking the logarithm of both sides of the equation, we obtain

$$\ln[f(x)] = \ln(\sqrt{x}) - \ln[(x+1)^2]$$

$$= \frac{1}{2}\ln(x) - 2\ln(x+1).$$

Differentiating, we see that
$$\frac{f'(x)}{f(x)} = \frac{1}{2x} - \frac{2}{x+1},$$
and hence
$$f'(x) = f(x)\left(\frac{1}{2x} - \frac{2}{x+1}\right)$$
$$= \frac{\sqrt{x}}{(x+1)^2}\left(\frac{1}{2x} - \frac{2}{x+1}\right).$$

This computation can be done for any function f where $f(x) \neq 0$. Since
$$\frac{d}{dx}[\ln|f(x)|] = \frac{f'(x)}{f(x)},$$
it follows that
$$f'(x) = f(x)\frac{d}{dx}[\ln|f(x)|].$$

EXAMPLE 6. Differentiate $f(x) = \dfrac{x^3(x+1)^2\sqrt{x-1}}{(2x+1)^2(x-2)^3}$.

Solution. Taking the logarithm of the absolute value of both sides, we obtain
$$\ln|f(x)| = \ln\left|\frac{x^3(x+1)^2\sqrt{x-1}}{(2x+1)^2(x-2)^3}\right|$$
$$= 3\ln|x| + 2\ln|x+1| + \frac{1}{2}\ln|x-1| - 2\ln|2x+1| - 3\ln|x-2|.$$

Differentiating both sides gives
$$\frac{f'(x)}{f(x)} = \frac{3}{x} + \frac{2}{x+1} + \frac{1}{2(x-1)} - \frac{4}{2x+1} - \frac{3}{x-2},$$
and hence
$$f'(x) = f(x)\left[\frac{3}{x} + \frac{2}{x+1} + \frac{1}{2(x-1)} - \frac{4}{2x+1} - \frac{3}{x-2}\right].$$

PROBLEMS

Differentiate each of the following functions.

1. $y = \ln x^3$

2. $y = \ln x + \ln \dfrac{1}{x}$

3. $y = \ln x^2(x+1)^3$

4. $y = x \ln x$

5. $y = \ln (\ln x)$

6. $f(x) = (\ln x)^2$

7. $f(x) = \ln 5x^2$

8. $y = \ln [\sqrt{x}(x+1)^2]$

Sketch the graph of each of the following functions.

9. $y = \ln \dfrac{x}{x+1}$

10. $y = x^2 - 2 \ln x$

11. $y = x \ln x$

12. $y = (\ln x)^2$

Use logarithmic differentiation to compute $f'(x)$.

13. $f(x) = \dfrac{x^{10}\sqrt{x^3}}{(x^2+1)^5}$

14. $f(x) = \dfrac{x(x+1)(x+2)}{(x-1)(x-2)}$

15. $f(x) = (x^2+1)^9(x^3+2x+3)^5(x^4+1)^3$

16. $f(x) = \dfrac{x^2\sqrt{x^2+1}}{(x+1)^3(x-1)^3}$

Compute the following integrals.

17. $\displaystyle\int_0^2 \dfrac{x\,dx}{x^2+1}$

18. $\displaystyle\int_1^4 \dfrac{\ln\sqrt{x}}{x}\,dx$

19. $\displaystyle\int \dfrac{dx}{x \ln x}$

20. $\displaystyle\int_1^3 \dfrac{(\ln x)^2}{x}\,dx$

21. $\displaystyle\int \dfrac{3x^2+2}{x^3+2x+4}\,dx$

22. $\displaystyle\int_1^3 \dfrac{x^2+x+1}{x^2+1}\,dx$

23. Find the average value of the function $y = 1/x$ on the interval $[5, 8]$. (Recall that the average value of f on $[a, b]$ is given by $f_{\text{av}} = [1/(b-a)] \int_a^b f(x)\,dx$. See Section 5 of Chapter 2.)

24. Find the area of the region bounded between the graph of $y = 1/[\sqrt{x}(\sqrt{x}+1)]$ and the graph of $y = -3$, $1 \le x \le 3$. [*Hint*: Let $u = \sqrt{x}+1$.]

2. THE EXPONENTIAL FUNCTION

From the definition of $\ln(x)$ we know that the derivative of $\ln(x)$ is $1/x$. Suppose, however, that we compute this derivative by using the definition of derivative and known properties of the natural logarithm. Then

$$\dfrac{d}{dx}(\ln x) = \lim_{h \to 0} \dfrac{\ln(x+h) - \ln(x)}{h}$$

194 CHAPTER 6: LOGARITHMS AND EXPONENTIALS

$$= \lim_{h \to 0} \frac{1}{h} \ln\left(\frac{x+h}{x}\right)$$

$$= \lim_{h \to 0} \frac{1}{x} \frac{x}{h} \ln\left(1 + \frac{h}{x}\right)$$

$$= \lim_{h \to 0} \frac{1}{x} \ln\left(1 + \frac{h}{x}\right)^{x/h}$$

$$= \frac{1}{x} \ln\left[\lim_{h \to 0}\left(1 + \frac{h}{x}\right)^{x/h}\right]$$

because the natural logarithm is a continuous function. [Recall from Theorem 2 of Chapter 2 that if $\lim_{x \to a} g(x) = L$ and if f is continuous at L, then $\lim_{x \to a} f(g(x)) = f(\lim_{x \to a} g(x)) = f(L)$.] If we let $z = x/h$, then z tends to infinity as h tends to zero, and hence

$$\frac{d}{dx}(\ln x) = \frac{1}{x} \ln\left[\lim_{z \to \infty}\left(1 + \frac{1}{z}\right)^z\right].$$

Since.

$$\frac{d}{dx}(\ln x) = \frac{1}{x},$$

it follows that the limit on the right-hand side of the preceding equation must exist, say

(8) $$\lim_{z \to \infty}\left(1 + \frac{1}{z}\right)^z = e.$$

Notice that e must be a number such that $\ln e = 1$; that is, such that $\int_1^e (1/t)\, dt = 1$, as in Figure 6.4. We **define** e to be the number whose natural logarithm is 1. From Table 6.1 we see that $\ln 2.7 = 0.993$ and $\ln 2.8 = 1.030$. Since $\ln x$ is an increasing function, it follows that if $\ln e = 1$, then e must be some number between 2.7 and 2.8. In fact,

$$e = 2.71828182845904523536028\ldots$$

is an irrational number. (Recall that an irrational number is a real number that cannot be written as a quotient of integers.)

The number e arose naturally in the computation of the derivative of the natural logarithm. It should not be surprising that the functions $x \to \ln x$ and $x \to e^x$ are closely related.

We need to make sense out of the expression e^x. The **exponential function** $x \to e^x$ is defined by

(9) $$e^x = y \quad \text{if and only if} \quad x = \ln y.$$

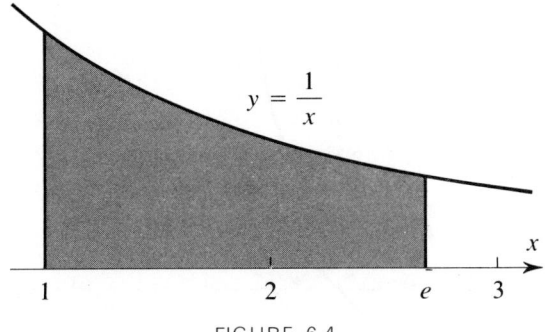

FIGURE 6.4

We can now compute dy/dx by differentiating $x = \ln y$ implicitly:

$$1 = \frac{1}{y}\frac{dy}{dx},$$

or solving for dy/dx, we get

$$\frac{dy}{dx} = y = e^x;$$

that is,

(10) $$\frac{d}{dx} e^x = e^x.$$

The exponential function $x \to e^x$ thus has a rather interesting property—namely, the derivative of the exponential function is the exponential function back again.

What does the graph of $y = e^x$ look like? Since $y' = e^x$, $y'' = e^x$, and $e^x > 0$ for every x, it follows that the graph of $y = e^x$ is always increasing and always concave upward. There are no extreme values, and $e^0 = 1$. The slope at $(0, 1)$ is equal to 1, and hence the graph must look something like Figure 6.5. Notice that the graph of $y = e^x$ is the mirror image of the graph of $y = \ln x$, reflected about the line $y = x$.

Suppose that u is any real number and let $x = e^u$. Then $\ln x = u$; that is,

(11) $$\ln(e^u) = u.$$

In particular,

$$\ln x = \ln(e^{\ln x}).$$

But $y = \ln x$ is an increasing function (see Figure 6.3), and hence if $\ln a = \ln b$, then necessarily $a = b$. We must conclude that

(12) $$x = e^{\ln x}.$$

196   CHAPTER 6: LOGARITHMS AND EXPONENTIALS

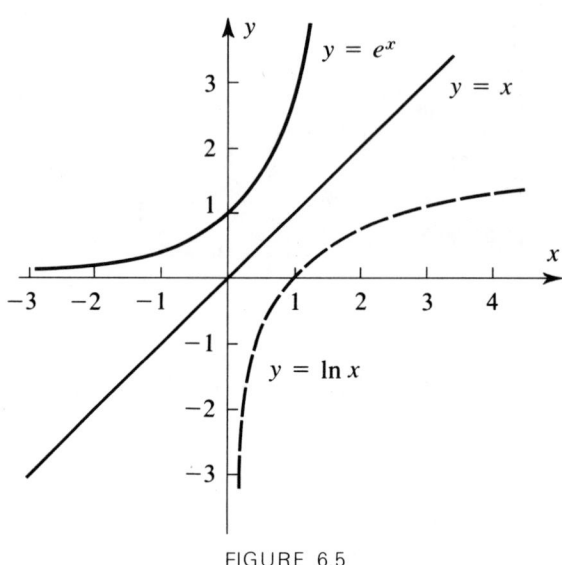

FIGURE 6.5

Let $f(u) = e^u$ and $g(x) = \ln x$. Equations (11) and (12) say that $f(g(x)) = x$ and $g(f(u)) = u$ for all allowable choices of x and u. Such pairs of functions are called **inverse functions.**

EXAMPLE 7. Show that the functions $f(x) = x^3$ and $g(x) = \sqrt[3]{x}$ are inverse functions.

Solution. $g(f(x)) = g(x^3) = \sqrt[3]{x^3} = x$ and $f(g(x)) = f(\sqrt[3]{x}) = (\sqrt[3]{x})^3 = x$ for all x. Thus f and g are inverse functions. The graphs of f and g are given in Figure 6.6. Notice that the graph of f is the mirror image of the graph of g, reflected about the line $y = x$.

EXAMPLE 8. Find the inverse function for $f(x) = 3x + 4$.

Solution. We solve the equation $y = 3x + 4$ for x in terms of y. Thus $3x = y - 4$, and hence $x = (y - 4)/3$. Let $g(y) = (y - 4)/3$. Then for each real number y,

$$f(g(y)) = f\left(\frac{y-4}{3}\right)$$

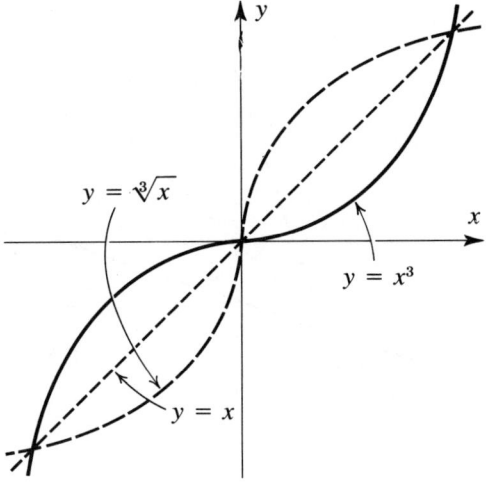

FIGURE 6.6

$$= 3\left(\frac{y-4}{3}\right) + 4$$
$$= (y-4) + 4$$
$$= y.$$

Similarly,

$$g(f(x)) = g(3x+4)$$
$$= \frac{(3x+4) - 4}{3}$$
$$= x$$

for each real number x. Therefore g is an inverse function for f (see Figure 6.7).

In all the foregoing examples of inverse functions f and g, the graph of f is the **reflection** of the graph of g about the line $y = x$. Another way to say this is

(13) $\qquad b = f(a) \quad \text{if and only if} \quad a = g(b);$

that is, the point (a, b) is on the graph of f if and only if the point (b, a) is on the graph of g. In order for a function f to have an inverse, the reflection of the graph of f about the line $y = x$ must also be the graph of some function g, and so each vertical line must intersect the graph of g once at most. Since the

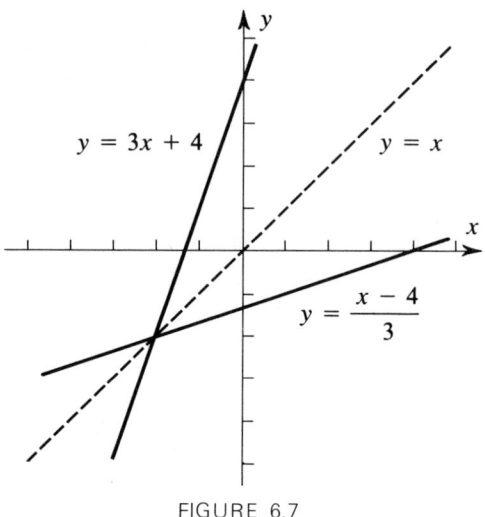

FIGURE 6.7

reflection of a horizontal line about $y = x$ is a vertical line, it follows that each horizontal line must intersect the graph of f once at most. (Such functions are called **one-to-one functions**.) The following theorem summarizes these remarks.

THEOREM 1. A function f has an inverse if and only if every horizontal line intersects the graph of f at most once. In this case, the inverse g is given by $g(x) = y$ if and only if $x = f(y)$.

We will find other examples of pairs of inverse functions in Chapter 7.

The exponential functions occur naturally in many problems related to population growth, radioactive decay, growth of bacteria in a culture, growth of a cancerous tumor, and so forth. Consider, for example, the unrestricted growth of bacteria under ideal conditions. If $f(t)$ is the number of bacteria at time t, then the rate of growth at time t is proportional to the number of bacteria present at time t; that is, $f'(t) = kf(t)$. Hence the function $y = f(t)$ satisfies the differential equation

$$\frac{dy}{dt} = ky.$$

Separating variables, this leads to

$$\frac{dy}{y} = k\, dt.$$

SECTION 2: THE EXPONENTIAL FUNCTION

Taking the indefinite integral on both sides, we see that
$$\ln y = kt + C$$
for some constant C. But $\ln y = x$ is equivalent to $y = e^x$, and hence
$$y = e^{kt+C}$$
$$= e^{kt} e^C$$
$$= N e^{kt},$$
where $N = e^C$. Hence the number of bacteria at time t is given by the exponential function
$$f(t) = N e^{kt}.$$

We will treat such functions in more detail in the next section.

There are other more complicated types of exponential functions. We can use previously developed methods to compute derivatives and integrals for many such functions.

EXAMPLE 9. Differentiate the function $f(x) = e^{x^2 + 2x}$.

Solution. By the composition rule with $g(x) = x^2 + 2x$ and $h(x) = e^x$, we obtain
$$f'(x) = h'(g(x))g'(x)$$
$$= e^{x^2 + 2x}(2x + 2)$$
$$= 2(x + 1)e^{x^2 + 2x}.$$

In general, if $y = e^{u(x)}$, then by the chain rule,

(14) $$\frac{dy}{dx} = e^{u(x)} \frac{du}{dx}.$$

EXAMPLE 10. Compute $\int (x + 1)e^{x^2 + 2x} dx$.

Solution. Use substitution with $u = x^2 + 2x$. Then $du = (2x + 2) dx$, and hence
$$\int (x + 1)e^{x^2 + 2x} dx = \frac{1}{2} \int (2x + 2)e^{x^2 + 2x} dx$$
$$= \frac{1}{2} \int e^u du$$

$$= \frac{1}{2}e^u + C$$

$$= \frac{1}{2}e^{x^2+2x} + C.$$

For $a > 0$ and any real number x, we **define** a^x by

(15) $$a^x = e^{x \ln a}.$$

Hence $\ln a^x = \ln e^{x \ln a} = x \ln a$; that is,

(16) $$\ln a^x = x \ln a.$$

(We had this formula only for rational x in Section 1.)

Suppose that $g(x) > 0$ for all x. Then $[g(x)]^{f(x)} = e^{\ln [g(x)]^{f(x)}} = e^{f(x) \ln g(x)}$; that is,

(17) $$[g(x)]^{f(x)} = e^{f(x) \ln g(x)}.$$

This gives us a method of differentiating functions of the form $[g(x)]^{f(x)}$. This method is illustrated in the following examples.

EXAMPLE 11. Compute the derivative of x^x, $x > 0$.

Solution. We note that $x^x = e^{x \ln x}$, and hence

$$\frac{d}{dx}(x^x) = \frac{d}{dx}(e^{x \ln x})$$

$$= e^{x \ln x} \frac{d}{dx}(x \ln x)$$

$$= x^x \left(x \frac{1}{x} + \ln x \right)$$

$$= x^x(1 + \ln x).$$

EXAMPLE 12. Derive the formula

$$\frac{d}{dx}(x^a) = ax^{a-1} \qquad (x > 0),$$

where a is any fixed real number.

SECTION 2: THE EXPONENTIAL FUNCTION

Solution. Since $x^a = e^{a \ln x}$, we see that

$$\frac{d}{dx}(x^a) = \frac{d}{dx}(e^{a \ln x})$$

$$= e^{a \ln x} \frac{d}{dx}(a \ln x)$$

$$= x^a \frac{a}{x}$$

$$= ax^{a-1}.$$

PROBLEMS

Differentiate the following functions.

1. $f(x) = e^{2x}$
2. $f(x) = e^x + e^{-x}$
3. $f(t) = e^{t^2}$
4. $g(t) = e^{(e^t)}$
5. $f(x) = e^{\ln x}$
6. $h(x) = \ln(e^x + x)$
7. $f(x) = \sqrt{e^x + 1}$
8. $f(x) = (e^x + e^{-x})^2$

Compute the following integrals.

9. $\int_0^1 2xe^{x^2} dx$
10. $\int e^{-x} dx$
11. $\int x^3 e^{x^4} dx$
12. $\int_1^2 (2x+1)e^{3x^2+3x} dx$
13. $\int e^x \sqrt{e^x + 1}\, dx$
14. $\int (e^x + e^{-x})^5 (e^x - e^{-x})\, dx$
15. $\int \frac{e^x}{e^x + 1} dx$
16. $\int e^{\ln x}\, dx$

Differentiate the following functions.

17. $y = (x^2 + 1)^{2x}$
18. $y = 2^x$
19. $f(x) = \left(x + \frac{1}{x}\right)^x$
20. $g(t) = t^2 + 2^t$
21. $y = 2^{\ln x}$
22. $y = x^{\ln x}$
23. $y = x^{x^x} (= x^{(x^x)})$
24. $f(x) = x^{x^2} (= x^{(x^2)})$

25. The number I of ions in a solution satisfies $(A + I)/(A - I) = e^{2At}$, where A is a constant and t is time. Use implicit differentiation to compute dI/dt in terms of A and I.

26. The relationship between the pressure p and the temperature T of a certain vapor is given by $p = Ce^{T/(K+T)}$, where C and K are constants. Compute dp/dT.

27. According to a model of E. Heinz [*Biochem. Zeitschrift*, **319** (1949), 482], the concentration y of a drug in the bloodstream at time t is given by

$$y = \frac{c}{b-a}(e^{-at} - e^{-bt}),$$

where a, b, and c are positive constants with $a < b$ and $t \geq 0$.
a. Find the time of maximum concentration.
b. Find the maximum concentration.

3. LOGARITHMS BASE a

Let a be a positive real number distinct from 1. The function $\log_a(x)$ is defined for $x > 0$ by

(18) $\qquad \log_a(x) = y \quad \text{if and only if} \quad a^y = x.$

Thus the function $x \to \log_a(x)$ is the inverse of the function $x \to a^x$. Notice that $\log_a(x)$ is the exponent of the power to which a must be raised to equal x. For example, $10^2 = 100$, and hence $\log_{10}(100) = 2$.

If we pick $a = e$, what can be said about the function $y = \log_e(x)$? From the definition of $\log_e(x)$, we know that $\log_e(x) = y$ if and only if $x = e^y$, which in turn is true if and only if $\ln x = y$. This statement implies that

$$\log_e(x) = \ln x$$

for all $x > 0$. It follows that the natural logarithm $\ln x$, which was defined in terms of an integral, is a logarithm in the usual sense; that is, the natural logarithm is the exponent of the power to which e must be raised to equal x.

Let a be any positive real number distinct from 1 and let $x = a^y = e^{y \ln a}$. Then $\log_a(x) = y$ and $\ln x = y \ln a$. Hence

$$y = \frac{\ln x}{\ln a} = \log_a(x);$$

that is,

(19) $\qquad \log_a(x) = \dfrac{\ln x}{\ln a}.$

If b is another positive real number distinct from 1, then

$$\log_a(b) \log_b(x) = \frac{\ln b \ln x}{\ln a \ln b}$$

$$= \frac{\ln x}{\ln a}$$

$$= \log_a(x).$$

SECTION 3: LOGARITHMS BASE a

This result verifies the following *change-of-basis formula*:

(20) $$\log_a(b) \log_b(x) = \log_a(x).$$

The change-of-basis formula allows all logarithmic computations to be carried out in your favorite base. If base 7 is your thing, then $\log_a(x) = (\log_7 x)/(\log_7 a)$, and all the logarithms on the right-hand side are in base 7.

The usual rules for the logarithm base b can be derived from the corresponding rules for the natural logarithm developed in Section 1. For example,

$$\log_b(cd) = \frac{\ln cd}{\ln b}$$
$$= \frac{\ln c + \ln d}{\ln b}$$
$$= \frac{\ln c}{\ln b} + \frac{\ln d}{\ln b}$$
$$= \log_b(c) + \log_b(d).$$

It is left as an exercise to derive the rules

$$\log_b(c^d) = d \log_b(c),$$

and

$$\log_b\left(\frac{c}{d}\right) = \log_b(c) - \log_b(d).$$

The function $y = \log_a(x)$ can be differentiated by using Formula (12) as follows:

$$\frac{d}{dx}[\log_a(x)] = \frac{d}{dx}\frac{\ln x}{\ln a} = \frac{1}{\ln a} \cdot \frac{1}{x} = \frac{1}{x \ln a}.$$

Thus

(21) $$\frac{d}{dx}[\log_a(x)] = \frac{1}{x \ln a}.$$

Notice that

$$\frac{d}{dx}[\log_a(x)]$$

is positive if $a > 1$ and negative if $0 < a < 1$. The function $y = \log_a(x)$ is not defined if $a = 1$. Using the techniques developed in Chapter 4, it is now possible to sketch the graph of $y = \log_a(x)$, as in Figure 6.8.

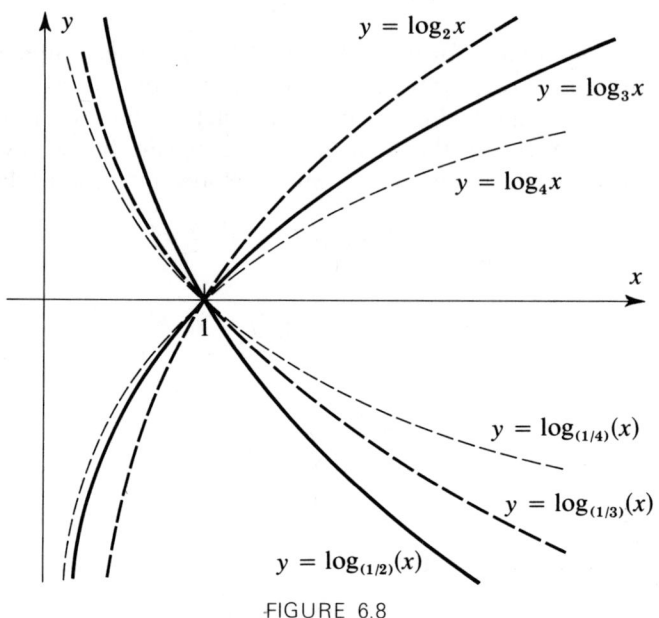
FIGURE 6.8

EXAMPLE 13. Differentiate $y = \log_{10}(2x^2 + 5)$.

Solution.

$$\frac{dy}{dx} = \frac{1}{(2x^2 + 5)\ln 10} \frac{d}{dx}(2x^2 + 5)$$

$$= \frac{4x}{(2x^2 + 5)\ln 10}$$

$$= \frac{4x}{2x^2 + 5}\log_{10} e$$

$$\approx \frac{1.737x}{2x^2 + 5}.$$

EXAMPLE 14. Differentiate $f(x) = \log_2(2x)$.

Solution.

$$f'(x) = \frac{d}{dx} \log_2 (2x)$$

$$= \frac{d}{dx} \frac{\ln 2x}{\ln 2}$$

$$= \frac{1}{x \ln 2}.$$

EXAMPLE 15. Differentiate $y = \log_x e$.

Solution. Since

$$y = \log_x e = \frac{\ln e}{\ln x} = \frac{1}{\ln x},$$

it follows that

$$\frac{dy}{dx} = -\frac{1}{(\ln x)^2} \frac{1}{x}$$

$$= -\frac{1}{x} (\log_x e)^2.$$

Because we usually work in base 10 arithmetic, it is convenient to work with *common logarithms*—that is, logarithms base 10. We will write $\log x$ instead of $\log_{10}(x)$. In computing with common logarithms, it is convenient to first express numbers in *scientific notation,* which requires that we write $x = A \times 10^k$, where $1 \le A < 10$ and k is an integer. For example, $543 = 5.43 \times 10^2$ and $0.00000543 = 5.43 \times 10^{-6}$. Since

$$\log (A \times 10^k) = \log A + \log 10^k$$

$$= \log A + k \log 10$$

$$= k + \log A,$$

it is possible to compute the common logarithm of any positive number if we know common logarithms of the numbers between 1 and 10. A short table of logarithms is given in Table 6.2. To compute $\log 5.43$, locate 54 under the column labeled N, locate the column labeled 3, then find the entry 7348 in the row opposite 54 and in the column labeled 3. The entry 7348 should be interpreted as 0.7348; thus $\log 5.43 = 0.7348$.

Table 6.2. Common Logarithms

N	0	1	2	3	4	5	6	7	8	9
10	0000	0043	0086	0128	0170	0212	0253	0294	0334	0374
11	0414	0453	0492	0531	0569	0607	0645	0682	0719	0755
12	0792	0828	0864	0899	0934	0969	1004	1038	1072	1106
13	1139	1173	1206	1239	1271	1303	1335	1367	1399	1430
14	1461	1492	1523	1553	1584	1614	1644	1673	1703	1732
15	1761	1790	1818	1847	1875	1903	1931	1959	1987	2014
16	2041	2068	2095	2122	2148	2175	2201	2227	2253	2279
17	2304	2330	2355	2380	2405	2430	2455	2480	2504	2529
18	2553	2577	2601	2625	2648	2672	2695	2718	2742	2765
19	2788	2810	2833	2856	2878	2900	2923	2945	2967	2989
20	3010	3032	3054	3075	3096	3118	3139	3160	3181	3201
21	3222	3243	3263	3284	3304	3324	3345	3365	3385	3404
22	3424	3444	3464	3483	3502	3522	3541	3560	3579	3598
23	3617	3636	3655	3674	3692	3711	3729	3747	3766	3784
24	3802	3820	3838	3856	3874	3892	3909	3927	3945	3962
25	3979	3997	4014	4031	4048	4065	4082	4099	4116	4133
26	4150	4166	4183	4200	4216	4232	4249	4265	4281	4298
27	4314	4330	4346	4362	4378	4393	4409	4425	4440	4456
28	4472	4487	4502	4518	4533	4548	4564	4579	4594	4609
29	4624	4639	4654	4669	4683	4698	4713	4728	4742	4757
30	4771	4786	4800	4814	4829	4843	4857	4871	4886	4900
31	4914	4928	4942	4955	4969	4983	4997	5011	5024	5038
32	5051	5065	5079	5092	5105	5119	5132	5145	5159	5172
33	5185	5198	5211	5224	5237	5250	5263	5276	5289	5302
34	5315	5328	5340	5353	5366	5378	5391	5403	5416	5428
35	5441	5453	5465	5478	5490	5502	5514	5527	5539	5551
36	5563	5575	5587	5599	5611	5623	5635	5647	5658	5670
37	5682	5694	5705	5717	5729	5740	5752	5763	5775	5786
38	5798	5809	5821	5832	5843	5855	5866	5877	5888	5899
39	5911	5922	5933	5944	5955	5966	5977	5988	5999	6010
40	6021	6031	6042	6053	6064	6075	6085	6096	6107	6117
41	6128	6138	6149	6160	6170	6180	6191	6201	6212	6222
42	6232	6243	6253	6263	6274	6284	6294	6304	6314	6325
43	6335	6345	6355	6365	6375	6385	6395	6405	6415	6425
44	6435	6444	6454	6464	6474	6484	6493	6503	6513	6522
45	6532	6542	6551	6561	6571	6580	6590	6599	6609	6618
46	6628	6637	6646	6656	6665	6675	6684	6693	6702	6712
47	6721	6730	6739	6749	6758	6767	6776	6785	6794	6803
48	6812	6821	6830	6839	6848	6857	6866	6875	6884	6893
49	6902	6911	6920	6928	6937	6946	6955	6964	6972	6981
50	6990	6998	7007	7016	7024	7033	7042	7050	7059	7067
51	7076	7084	7093	7101	7110	7118	7126	7135	7143	7152
52	7160	7168	7177	7185	7193	7202	7210	7218	7226	7235
53	7243	7251	7259	7267	7275	7284	7292	7300	7308	7316
54	7324	7332	7340	7348	7356	7364	7372	7380	7388	7396
N	0	1	2	3	4	5	6	7	8	9

Table 6.2 (Continued)

N	0	1	2	3	4	5	6	7	8	9
55	7404	7412	7419	7427	7435	7443	7451	7459	7466	7474
56	7482	7490	7497	7505	7513	7520	7528	7536	7543	7551
57	7559	7566	7574	7582	7589	7597	7604	7612	7619	7627
58	7634	7642	7649	7657	7664	7672	7679	7686	7694	7701
59	7709	7716	7723	7731	7738	7745	7752	7760	7767	7774
60	7782	7789	7796	7803	7810	7818	7825	7832	7839	7846
61	7853	7860	7868	7875	7882	7889	7896	7903	7910	7917
62	7924	7931	7938	7945	7952	7959	7966	7973	7980	7987
63	7993	8000	8007	8014	8021	8028	8035	8041	8048	8055
64	8062	8069	8075	8082	8089	8096	8102	8109	8116	8122
65	8129	8136	8142	8149	8156	8162	8169	8176	8182	8189
66	8195	8202	8209	8215	8222	8228	8235	8241	8248	8254
67	8261	8267	8274	8280	8287	8293	8299	8306	8312	8319
68	8325	8331	8338	8344	8351	8357	8363	8370	8376	8382
69	8388	8395	8401	8407	8414	8420	8426	8432	8439	8445
70	8451	8457	8463	8470	8476	8482	8488	8494	8500	8506
71	8513	8519	8525	8531	8537	8543	8549	8555	8561	8567
72	8573	8579	8585	8591	8597	8603	8609	8615	8621	8627
73	8633	8639	8645	8651	8657	8663	8669	8675	8681	8686
74	8692	8698	8704	8710	8716	8722	8727	8733	8739	8745
75	8751	8756	8762	8768	8774	8779	8785	8791	8797	8802
76	8808	8814	8820	8825	8831	8837	8842	8848	8854	8859
77	8865	8871	8876	8882	8887	8893	8899	8904	8910	8915
78	8921	8927	8932	8938	8943	8949	8954	8960	8965	8971
79	8976	8982	8987	8993	8998	9004	9009	9015	9020	9025
80	9031	9036	9042	9047	9053	9058	9063	9069	9074	9079
81	9085	9090	9096	9101	9106	9112	9117	9122	9128	9133
82	9138	9143	9149	9154	9159	9165	9170	9175	9180	9186
83	9191	9196	9201	9206	9212	9217	9222	9227	9232	9238
84	9243	9248	9253	9258	9263	9269	9274	9279	9284	9289
85	9294	9299	9304	9309	9315	9320	9325	9330	9335	9340
86	9345	9350	9355	9360	9365	9370	9375	9380	9385	9390
87	9395	9400	9405	9410	9415	9420	9425	9430	9435	9440
88	9445	9450	9455	9460	9465	9469	9474	9479	9484	9489
89	9494	9499	9504	9509	9513	9518	9523	9528	9533	9538
90	9542	9547	9552	9557	9562	9566	9571	9576	9581	9586
91	9590	9595	9600	9605	9609	9614	9619	9624	9628	9633
92	9638	9643	9647	9652	9657	9661	9666	9671	9675	9680
93	9685	9689	9694	9699	9703	9708	9713	9717	9722	9727
94	9731	9736	9741	9745	9750	9754	9759	9763	9768	9773
95	9777	9782	9786	9791	9795	9800	9805	9809	9814	9818
96	9823	9827	9832	9836	9841	9845	9850	9854	9859	9863
97	9868	9872	9877	9881	9886	9890	9894	9899	9903	9908
98	9912	9917	9921	9926	9930	9934	9939	9943	9948	9952
99	9956	9961	9965	9969	9974	9978	9983	9987	9991	9996
N	0	1	2	3	4	5	6	7	8	9

EXAMPLE 16. Compute log (0.00000543) and log (54300).

Solution. Since $0.00000543 = 5.43 \times 10^{-6}$, it follows that

$$\log (0.00000543) = -6 + \log 5.43$$
$$= -6 + 0.7348$$
$$= -5.2652.$$

Also,

$$\log 54300 = 4 + \log 5.43$$
$$= 4 + 0.7348$$
$$= 4.7348.$$

The *characteristic* of a number $x = A \times 10^k$ in scientific notation is the integer k, and the *mantissa* is log A. Hence the characteristic of 9.43 is 0, the characteristic of 0.00943 is -3, and the characteristic of 94300 is 4. The mantissa of a number is independent of the location of the decimal point. Thus the mantissa of 0.00943 is the same as the mantissa of 94300—namely, log 9.43 $= 0.9745$.

EXAMPLE 17. Compute $\sqrt{\dfrac{550 \times 9.1}{(2.3)^4 \times 0.012}}$.

Solution. We compute the given number by first computing the logarithm of the given number.

$$\log \sqrt{\frac{550 \times 9.1}{(2.3)^4 \times 0.012}} = \log \left[\frac{550 \times 9.1}{(2.3)^4 \times 0.012}\right]^{1/2}$$

$$= \frac{1}{2} \log \left[\frac{550 \times 9.1}{(2.3)^4 \times 0.012}\right]$$

$$= \frac{1}{2} (\log 550 + \log 9.1 - 4 \log 2.3 - \log 0.012)$$

$$= \frac{1}{2} [2 + \log 5.5 + \log 9.1 - 4 \log 2.3 - (-2 + \log 1.2)]$$

$$= 2.086.$$

Now log $1.22 = 0.086$, and hence log $122 = 2.086$. It follows that

$$\sqrt{\frac{550 \times 9.1}{(2.3)^4 \times 0.012}} = 122.$$

SECTION 3: LOGARITHMS BASE a

The following example illustrates how logarithms can be used in exponential growth and decay problems.

EXAMPLE 18. A certain radioactive material produced in a laboratory has a *half-life* of 3 minutes; that is, one-half of the remaining material decays every 3 minutes. After 6 minutes, only one-fourth of the material remains, and so forth. How long does it take for 99% of the material to decay?

Solution. The amount of radioactive material left after t minutes is given by $R(t) = C \times 10^{kt}$ for some constants C and k. If R_0 is the amount of radioactive material at time $t = 0$, then $R_0 = C \times 10^0 = C$. One-half the material decays after 3 minutes, and hence $R_0/2 = R_0 \times 10^{3k}$, which simplifies to $0.5 = 10^{3k}$. We can solve for k by first taking the common logarithm of both sides:

$$3k = \log_{10}(0.5)$$
$$= -\log_{10} 2$$
$$= -0.301,$$

or
$$k = -0.1003.$$

It follows that
$$R(t) = R_0 \times 10^{-0.1003t}.$$

We need to find t so that $R(t) = 0.01 R_0 = R_0 \times 10^{-0.1003t}$. Taking the common logarithm of both sides gives

$$-0.1003t = \log(0.01) = -2,$$

and hence $t \approx 20$ minutes.

EXAMPLE 19. The population of the United States is growing at a rate that is roughly proportional to the population itself, and hence $P(t) \approx P_0 \times 10^{kt}$ for some constants P_0 and k. If we take t to be the number of years since 1900, then $P(0) = P_0 \times 10^{0k} = 76{,}000{,}000$, and hence $P_0 = 76{,}000{,}000$. In 1970, the population was 204 million, which implies

$$204{,}000{,}000 = 76{,}000{,}000 \times 10^{70k}.$$

Taking the logarithm of both sides, we get

$$\log 204{,}000{,}000 = 70k + \log 76{,}000{,}000,$$

and hence
$$70k = 8 + \log 2.04 - (7 + \log 7.6)$$
$$= 1 + 0.3096 - 0.8808$$
$$= 0.4288,$$

210 CHAPTER 6: LOGARITHMS AND EXPONENTIALS

which leads to

$$k = 0.006126.$$

Thus the population of the United States t years after 1900 should be approximately

$$P(t) = 76{,}000{,}000 \times 10^{0.006126t}.$$

In particular, in 1910

$$P(10) = 76{,}000{,}000 \times 10^{(0.006126) \times 10},$$

which implies

$$\log P(10) = 7.8808 + 0.06126$$
$$= 7.9421,$$

and hence

$$P(10) = 87{,}500{,}000.$$

The population in 1910 was actually closer to 92 million. A comparison between $P(t)$ and the population during census years is given in Table 6.3.

Table 6.3. United States Population (in millions)

Year	$P(t)$	Official Census
1900	76.0	76
1910	87.5	92
1920	100.8	106
1930	116.0	123
1940	133.6	132
1950	153.8	151
1960	177.2	178
1970	204.0	204

EXAMPLE 20. If $1000 is deposited in a savings account at 6% interest compounded yearly and no additional money is deposited or withdrawn, how much is the account worth after 12 years?

Solution. After one year, the account is worth $1000(1.06)$, after 2 years the account is worth $1000(1.06)(1.06) = \$1000(1.06)^2$, and after 12 years the account is worth $1000(1.06)^{12}$. We can compute $(1.06)^{12}$ using logarithms:

$$\log (1.06)^{12} = 12 \log 1.06$$
$$= 12(0.0253)$$
$$= 0.3036$$
$$= \log 2.01,$$

and hence $\$1000(1.06)^{12} = \$2010$, the amount of money in the account after 12 years.

PROBLEMS

Differentiate the following functions.

1. $y = \log_a (x) \log_b (x)$
2. $F(x) = \log_3 (\sqrt{x})$
3. $f(x) = \log_a (a^x)$
4. $y = x \log_2 (x)$
5. $y = \log_{10} (x) \log_x (x^2 + 1)$
6. $y = \log_5 \dfrac{x\sqrt{x+1}}{(x^2+1)^3}$
7. Derive the formula $\log_b (c^d) = d \log_b (c)$.
8. Derive the formula $\log_b \left(\dfrac{c}{d}\right) = \log_b (c) - \log_b (d)$.

Simplify the following expressions.

9. $\log_2 (16)$
10. $\log_3 \left(\dfrac{1}{27}\right)$
11. $\log_3 (25) \log_5 (3)$
12. $\log_2 \left(\dfrac{2}{3}\right) - \log_2 \left(\dfrac{1}{3}\right)$
13. $\log_{10} (100^x)$
14. $\log_9 (3)$
15. $\log_{10} (e) \ln (10^{-3})$
16. $\log_3 \left(\dfrac{2}{9}\right) + 2 \log_3 \left(\dfrac{1}{\sqrt{2}}\right)$

17. The population of Dullsville was 75 at the time of the 1900 census and 59 at the time of the 1910 census. Assuming exponential "growth" (decay might be a more appropriate term), what was the population in 1924? 1885?

18. A man invested $1000 in a savings account that advertised 6% interest compounded continuously. How much will the account be worth after 20 years if no deposits or withdrawals are made?

19. Assuming the population of the United States is increasing at a rate of 1% per year, how many years will it take for the population to double? How many to quadruple?

20. If a pound of radioactive material has a half-life of one year, how much will remain after 10 years?

21. A certain type of bacteria doubles every 3 hours under ideal conditions, such as in a culture. Twenty hours after exposure, it is estimated that there are 40 million bacteria in the culture. How many bacteria were placed in the culture during exposure?

22. A new variety of larva-eating insect is developed and allowed to multiply under ideal conditions. After 7 days, 50 insects multiply to 2800. What is the doubling time? How long will it take to produce 50 million of these insects?

23. Sketch the graph of $y = \log_x (x^2 + 1)$.
24. Sketch the graph of $y = e^{(1 - \ln x)}$.

CHAPTER 6: LOGARITHMS AND EXPONENTIALS

FOR HAND-HELD CALCULATORS ONLY (OPTIONAL)

In 1614 John Napier wrote *A Description of the Wonderful Law of Logarithms*. Within a few years 14-place tables of common logarithms were published, and these tables had a major impact on science by greatly reducing the time required to perform difficult computations.

Today, however, the use of computers and hand-held calculators has almost made the use of mathematical tables obsolete. It is much easier to design a computer that can compute $\log_{10} 2$ to 14 places directly than it would be to develop enough high-speed memory to store a complete 14-place table of common logarithms. Even using a hand-held calculator, it is not difficult to compute approximate values of logarithms (but to fewer than 14 decimal places).

The following formulas may be used on any hand-held calculator having the four basic operations $(+, -, *, \div)$. These formulas are easier to use, however, if at least one bank of memory is available.

(22) $\quad \ln x \approx ((B*B*3/5 + 1)*B*B/3 + 1)*B*2 \quad (0.5 \leq x \leq 1.5),$

where $\quad B = \dfrac{x-1}{x+1}.$

If memory is available, store B. Otherwise compute B and write it down.

Formula (22) should be read from left to right. The following steps should be performed in order.

Instruction	New Result
1. Compute B.	$B = \dfrac{x-1}{x+1}$
2. Multiply by B.	B^2
3. Multiply the result by 3.	$3B^2$
4. Divide by 5.	$\dfrac{3B^2}{5}$
5. Add 1.	$\dfrac{3B^2}{5} + 1$
6. Multiply by B.	$\left(\dfrac{3B^2}{5} + 1\right)B$
7. Multiply by B.	$\left(\dfrac{3B^2}{5} + 1\right)B^2$
8. Divide by 3.	$\left(\dfrac{3B^2}{5} + 1\right)\dfrac{B^2}{3}$

9. Add 1. $\qquad\left(\dfrac{3B^2}{5}+1\right)\dfrac{B^2}{3}+1$

10. Multiply by B. $\qquad\left(\left(\dfrac{3B^2}{5}+1\right)\dfrac{B^2}{3}+1\right)B$

11. Multiply by 2. $\qquad\left(\left(\dfrac{3B^2}{5}+1\right)\dfrac{B^2}{3}+1\right)2B$

Formula (22) gives the best results for x close to 1. If x is not between 0.5 and 1.5, then multiply or divide by successive powers of

$$e \approx 2.718281828$$

until the result is between 0.5 and 1.5. For example, in order to compute $\ln 10$, note that $10/e^2 \approx 1.3533528$, which implies $10 \approx 1.3533528e^2$, and hence $\ln 10 \approx \ln(1.3533528e^2) = 2 + \ln(1.3533528)$. It is therefore sufficient to compute $\ln(1.3533528)$.

To compute common logs, it is sufficient to compute the natural logarithm $\ln x$ and then multiply by

$$\log_{10} e \approx 0.4342944819$$

because $\log_{10} x = \log_{10} e \ln x$.

The exponential function can be computed by using the following formula.

(23) $\quad e^x \approx ((((x/5 + 1)x/4 + 1)x/3 + 1)x/2 + 1)x + 1 \qquad \left(-\dfrac{1}{2} \le x \le \dfrac{1}{2}\right).$

Formula (23) works best for x close to 0. To compute $e^{3.64}$, divide by e^4 and compute $e^{-0.36}$; then multiply the result by e^4, using $e \approx 2.718281828$.

The following formulas are more convenient in case a reciprocal function $1/x$ is available.

(24) $\qquad \ln x \approx ((-3B^2/5 + 1)^{-1} * 5 + 4) * 2B/9 \qquad (0.5 \le x \le 1.5),$

where

$$B = \dfrac{x-1}{x+1} = (x+1)^{-1}(-2) + 1.$$

(25) $\qquad e^x \approx (((x^2/60 + 1)^{-1} * (-5) + 6)/x - 0.5)^{-1} + 1,$

where $0 < x \le 1$.

PROBLEMS

1. Estimate $\ln 3$ by using
   a. the midpoint rule on $\ln 3 = \int_1^3 (1/x)\,dx$ with $n = 4$;
   b. Formula (22) or (24). [Remember to compute $\ln(3/e)$ first.]

2. Estimate ln 10, using Formula (22) or (24).
3. Estimate $\log_{10} e$ by computing ln 10 and using the fact that $\log_{10} e = 1/\ln 10$.
4. Estimate $e^{-5.847}$, using Formula (23) or (25).
5. Estimate $e^{2.5}$, using Formula (23) or (25).
6. Make a table of values of e^x for $x = 0, 0.1, 0.2, \ldots, 0.9, 1.0$.
7. Make a table of values of ln x, $x = 0.5, 0.6, \ldots, 1.4, 1.5$.
8. Compute $(5.4)^{7.27}$. [*Hint:* Use $b^a = e^{a \ln b}$, together with Formulas (22) through (25).]

REVIEW SECTION

The following definitions are repeated for convenience.

(a) $\ln x = \int_1^x \frac{1}{t} dt \quad (x > 0)$

(b) $e = \lim_{z \to \infty} \left(1 + \frac{1}{z}\right)^z$

(c) Functions f and g are *inverse* if $f(g(x)) = x$ and $g(f(u)) = u$ for all allowable choices of x and u.

(d) Let $a > 0$, $a \neq 1$. The function $\log_a(x)$ is defined for $x > 0$ by $\log_a(x) = y$ if and only if $x = a^y$.

(e) $x = N \times 10^k$ is expressed in *scientific notation* if $1 \leq N < 10$ and k is an integer.

(f) The *characteristic* of $x = N \times 10^k$ (written in scientific notation) is the integer k.

(g) The *mantissa* of $x = N \times 10^k$ (written in scientific notation) is log N.

The following facts were developed in this chapter.

(a) $\frac{d}{dx} \ln x = \frac{1}{x}$

(b) $\ln(ab) = \ln a + \ln b$

(c) $\ln b^a = a \ln b$

(d) $\ln\left(\frac{a}{b}\right) = \ln a - \ln b$

(e) $\frac{d}{dx} e^u = e^u \frac{du}{dx}$

(f) $\ln(e^u) = u$

(g) $x = e^{\ln x}$

(h) If the rate of change of a function $y = f(t)$ is proportional to the function itself, then $f(t) = Ne^{kt} = C10^{at}$ for some constants N, k, C, and a.

(i) $\log_a x = \dfrac{\ln x}{\ln a}$

(j) $\log_a (b) \log_b (x) = \log_a (x)$

(k) $\dfrac{d}{dx} \log_a x = \dfrac{1}{x \ln a}$

REVIEW PROBLEMS

Sketch the graph of each of the following functions.

1. $y = e^x - e^{-x}$
2. $y = e^{x^2}$
3. $y = xe^x$
4. $y = e^{2\ln x}$
5. $y = x^x$
6. $y = x^2 e^x$
7. $y = x^{x^2+1}$
8. $y = \ln(xe^x)$

Find the volume of revolution generated by rotating the graph of $y = f(x)$ about the x-axis.

9. $f(x) = e^x, 0 \le x \le 1$
10. $f(x) = \dfrac{1}{\sqrt{x}}, 1 \le x \le 3$
11. $f(x) = \sqrt{\dfrac{\ln x}{x}}, 1 \le x \le 2$
12. $f(x) = \sqrt{x e^{x^2}}, 0 \le x \le 1$

Solve for x:

13. $\log_2 x^2 = 3 \log_2 x + \log_2 8 - 4 \log_2 16$
14. $\log_3 x^2 = 3 \log_3 x + \log_3 8 - 4 \log_3 16$
15. $3^{2x-1} = 12$
16. $2 \log_3 \left(\dfrac{1}{2}\right) \log_{(1/2)} (x) = \log_3 4$
17. $3^{x-1} = 2^{2x-3}$
18. $e^{2x} + e^x - 2 = 0$

Use logarithms to compute the following numbers.

19. $\sqrt[2]{15}$
20. $\sqrt[3]{7.9}$
21. 2^{64}
22. $3^{1.37}$
23. $\dfrac{(45)^2 \times 986 \times 0.45}{1.3 \times (2.6)^3 \times 5.2}$
24. $\left(\dfrac{58 \times 7.2 \times 6.9}{(3.7)^3 \times (4.1)^2}\right)^{1/3}$

25. A merchandising expert claims that the amount of Brand X soap sold depends logarithmically on the advertising budget for the soap; that is, the amount $X(A)$ of Brand X sold is equal to $K + C \log_{10}(A)$ for some constants K and C, where A is the amount spent for advertising. If \$10,000 worth of advertising will cause \$100,000 worth of Brand X to be sold, and \$20,000 worth of advertising will sell \$160,000 of Brand X, how much Brand X should the expert predict that \$50,000 of advertising would sell?

26. Money is invested in a savings account and interest is compounded continuously. What is the annual interest rate if doubling time is 20 years? If doubling time is 14 years?

27. Which is larger, 100^{99} or 99^{100}?

28. If $a = \left(1 + \dfrac{1}{n}\right)^n$ and $b = \left(1 + \dfrac{1}{n}\right)^{n+1}$, show that $a^b = b^a$. (In particular, $n = 1$ yields the equation $2^4 = 4^2$.)

29. Compute $\lim_{n\to\infty}\left(\dfrac{1}{n+1}+\dfrac{1}{n+2}+\dfrac{1}{n+3}+\cdots+\dfrac{1}{2n}\right)$. [*Hint:* Recognize the sum inside the brackets as being a Riemann sum of an integral of the form $\int_1^2 f(x)\,dx$ and compute the integral.]

30. The amount of candy consumed by a child on Halloween is given approximately by $C(t)=K(1-e^{-at})$, where K and a are constants determined by the capacity and aggressiveness of the child. Find the rate of consumption at time $t=t_1$.

31. Atmospheric pressure is related to altitude above sea level by the equation $dp/dh=kp$. At sea level, the pressure p is 1013.25 millibars; that is, the atmosphere will support a column of mercury 1013.25 millimeters high. At an altitude of 1000 meters, the pressure is 898.74 millibars. Find a formula for p in terms of the altitude h. What is the atmospheric pressure at the top of Pikes Peak (altitude 4300 meters)?

32. Fechner's law states that $dS/d\sigma=k/\sigma$, where S is the sensation, σ is the stimulus, and k is a constant. Find S in terms of σ if $S=0$ when $\sigma=\sigma_0$.

33. Carbon 14, denoted C^{14}, is used in archaeology to date artifacts because the amount of C^{14} in living organisms is known, and C^{14} decays exponentially after death, with a half-life of 5570 years. How old is a piece of wood out of a cliff dwelling if the concentration of C^{14} is 0.8 of the original concentration?

34. The number of coyotes and rabbits in a certain region is determined by the predator-prey relationship $dA/dP=cA/P$, where P is the number of coyotes, A is the frequency with which a coyote attacks a rabbit, and c is a constant. Find P in terms of A.

CHAPTER 7

THE CIRCULAR FUNCTIONS

A function f is called *periodic* if for some $a > 0$, $f(x + a) = f(x)$ for all x. Examples of periodic functions are found by investigating sound waves, alternating electrical current, yearly business cycles, and the pendulum of a clock. Many of these functions can be written in terms of the so-called circular functions, which are additional examples of elementary transcendental functions (we have already seen the logarithmic and exponential functions).

1. THE CIRCULAR FUNCTIONS

Consider a circle of radius 1 with center at the origin. Given a real number α, let (x, y) be the terminal point of an arc of length $|\alpha|$ with initial point $(1, 0)$ on the circle. Arc length is measured in a counterclockwise direction if α is positive and in a clockwise direction if α is negative. We define the functions *sine* and *cosine* by the equations

(1) $$\sin \alpha = y$$

and

(2) $\cos \alpha = x.$

The circumference of the unit circle is 2π, where π is the irrational number 3.14159265358979323846.... From Figure 7.1, we can compile Table 7.1.

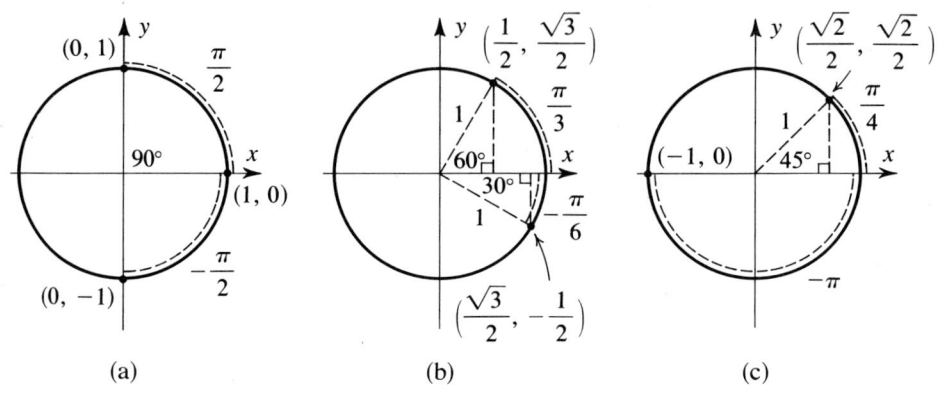

FIGURE 7.1

Table 7.1

α	$\sin \alpha$	$\cos \alpha$
0	0.	1
$\dfrac{\pi}{6}$	$\dfrac{1}{2}$	$\dfrac{\sqrt{3}}{2}$
$\dfrac{\pi}{4}$	$\dfrac{\sqrt{2}}{2}$	$\dfrac{\sqrt{2}}{2}$
$\dfrac{\pi}{3}$	$\dfrac{\sqrt{3}}{2}$	$\dfrac{1}{2}$
$\dfrac{\pi}{2}$	1	0
$\dfrac{2\pi}{3}$	$\dfrac{\sqrt{3}}{2}$	$-\dfrac{1}{2}$
$\dfrac{3\pi}{4}$	$\dfrac{\sqrt{2}}{2}$	$-\dfrac{\sqrt{2}}{2}$
$\dfrac{5\pi}{6}$	$\dfrac{1}{2}$	$-\dfrac{\sqrt{3}}{2}$
π	0	-1
$-\dfrac{\pi}{2}$	-1	0

We can also think of α as the angle (in **radians**) that the line generated by the terminal point and the origin makes with the positive x-axis. The correspondence between radians and degrees is given by

$$\pi \text{ radians} = 180 \text{ degrees,}$$

or
$$1 \text{ radian} = \frac{180}{\pi} \text{ degrees}$$

$$\approx 57.29578 \text{ degrees.}$$

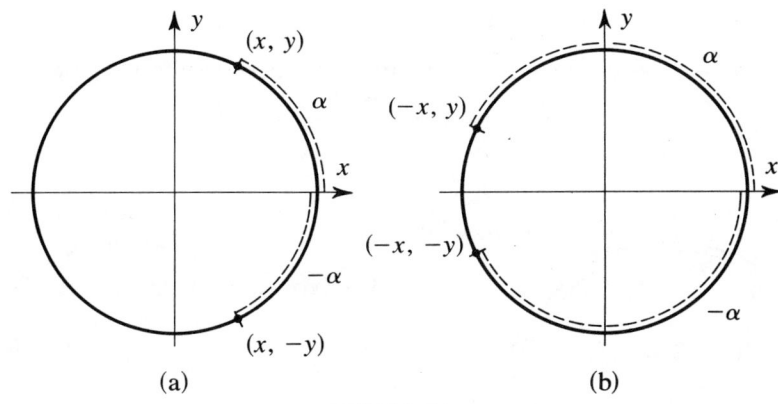

FIGURE 7.2

From Figure 7.2, we see that

(3) $$\sin(-\alpha) = -\sin \alpha$$

and

(4) $$\cos(-\alpha) = \cos \alpha;$$

that is, sine is an odd function and cosine is an even function.

Let $\sin \alpha = y$ and $\cos \alpha = x$. Since the point (x, y) lies on the unit circle, it follows that $x^2 + y^2 = 1$; in other words, $(\cos \alpha)^2 + (\sin \alpha)^2 = 1$. This equation is usually rewritten in the form

(5) $$\sin^2 \alpha + \cos^2 \alpha = 1.$$

We can also derive an identity for $\cos(\alpha - \beta)$. Let $\overline{AB}$ be the distance from A to B in Figure 7.3. By the distance formula,

$$(\overline{AB})^2 = (\cos\alpha - \cos\beta)^2 + (\sin\alpha - \sin\beta)^2$$
$$= \cos^2\alpha - 2\cos\alpha\cos\beta + \cos^2\beta + \sin^2\alpha$$
$$\quad - 2\sin\alpha\sin\beta + \sin^2\beta.$$

But $\cos^2\alpha + \sin^2\alpha = \cos^2\beta + \sin^2\beta = 1$, and hence

$$(\overline{AB})^2 = 2 - 2(\cos\alpha\cos\beta + \sin\alpha\sin\beta).$$

On the other hand, if the circle is rotated as in Figure 7.4, we see that

$$(\overline{AB})^2 = [\cos(\alpha - \beta) - 1]^2 + [\sin(\alpha - \beta)]^2$$
$$= \cos^2(\alpha - \beta) - 2\cos(\alpha - \beta) + 1 + \sin^2(\alpha - \beta)$$
$$= 2 - 2\cos(\alpha - \beta),$$

since $\cos^2(\alpha - \beta) + \sin^2(\alpha - \beta) = 1$. Comparing these results, we note

$$\cos(\alpha - \beta) = \cos\alpha\cos\beta + \sin\alpha\sin\beta.$$

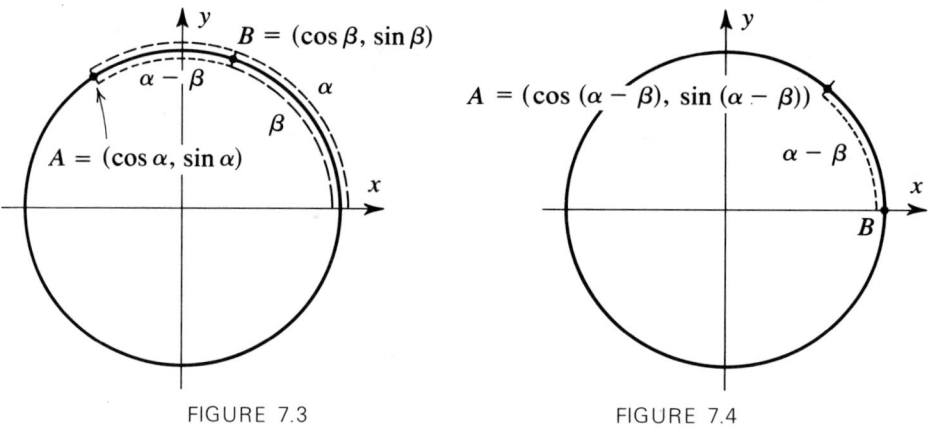

FIGURE 7.3   FIGURE 7.4

Replacing β by $-\beta$, we see that

$$\cos(\alpha + \beta) = \cos[\alpha - (-\beta)]$$
$$= \cos\alpha\cos(-\beta) + \sin\alpha\sin(-\beta)$$
$$= \cos\alpha\cos\beta - \sin\alpha\sin\beta,$$

since cosine is an even function and sine is an odd function. Thus

(6) $$\cos(\alpha \pm \beta) = \cos\alpha\cos\beta \mp \sin\alpha\sin\beta.$$

We leave it as an exercise to show that

(7) $$\cos \alpha = \sin\left(\alpha + \frac{\pi}{2}\right),$$

and hence that

$$\sin \alpha = \cos\left(\alpha - \frac{\pi}{2}\right) = \cos\left(\frac{\pi}{2} - \alpha\right).$$

It follows from Equations (6) and (7) that

$$\sin(\alpha + \beta) = \cos\left[\frac{\pi}{2} - (\alpha + \beta)\right]$$

$$= \cos\left[\left(\frac{\pi}{2} - \alpha\right) - \beta\right]$$

$$= \cos\left(\frac{\pi}{2} - \alpha\right)\cos \beta + \sin\left(\frac{\pi}{2} - \alpha\right)\sin \beta$$

$$= \sin \alpha \cos \beta + \cos \alpha \sin \beta.$$

Replacing β by $-\beta$, we see that

$$\sin(\alpha - \beta) = \sin \alpha \cos(-\beta) + \cos \alpha \sin(-\beta)$$

$$= \sin \alpha \cos \beta - \cos \alpha \sin \beta.$$

Thus

(8) $$\sin(\alpha \pm \beta) = \sin \alpha \cos \beta \pm \cos \alpha \sin \beta.$$

The remainder of the **circular functions (tangent, cotangent, secant,** and **cosecant)** are defined as follows for all x such that the denominator is not zero:

(9) $$\tan x = \frac{\sin x}{\cos x}$$

(10) $$\cot x = \frac{\cos x}{\sin x}$$

(11) $$\sec x = \frac{1}{\cos x}$$

(12) $$\csc x = \frac{1}{\sin x}$$

[The symbol x is used instead of α or β because we wish to retain the notation $y = f(x)$ for functions. Do not confuse it with the x in the equation $x = \cos \alpha$.]

Starting with the identity $\sin^2 x + \cos^2 x = 1$, we can divide through by $\cos^2 x$ to get

$$\frac{\sin^2 x}{\cos^2 x} + \frac{\cos^2 x}{\cos^2 x} = \frac{1}{\cos^2 x},$$

which simplifies to

(13) $$\tan^2 x + 1 = \sec^2 x.$$

Table 7.2

x	$\sin x$	$\tan x$	$\cot x$	$\cos x$	Degrees
0.00	0.0000	0.0000	--	1.0000	0°
0.05	0.0500	0.0500	19.9800	0.9988	2.9
0.10	0.0998	0.1003	9.9670	0.9950	5.7
0.15	0.1494	0.1511	6.6170	0.9888	8.6
0.20	0.1987	0.2027	4.9330	0.9801	11.5
0.25	0.2474	0.2553	3.9160	0.9689	14.3
0.30	0.2955	0.3093	3.2330	0.9553	17.2
0.35	0.3429	0.3650	2.7400	0.9394	20.1
0.40	0.3894	0.4228	2.3650	0.9211	22.9
0.45	0.4350	0.4831	2.0700	0.9004	25.8
0.50	0.4794	0.5463	1.8300	0.8776	28.6
0.55	0.5227	0.6131	1.6310	0.8525	31.5
0.60	0.5646	0.6841	1.4620	0.8253	34.4
0.65	0.6052	0.7602	1.3150	0.7961	37.2
0.70	0.6442	0.8423	1.1870	0.7648	40.1
0.75	0.6816	0.9316	1.0730	0.7317	43.0
0.80	0.7174	1.0300	0.9712	0.6967	45.8
0.85	0.7513	1.1380	0.8785	0.6600	48.7
0.90	0.7833	1.2600	0.7936	0.6216	51.6
0.95	0.8134	1.3980	0.7151	0.5817	54.4
1.00	0.8415	1.5570	0.6421	0.5403	57.3
1.05	0.8674	1.7430	0.5736	0.4976	60.2
1.10	0.8912	1.9650	0.5090	0.4536	63.0
1.15	0.9128	2.2340	0.4475	0.4085	65.9
1.20	0.9320	2.5720	0.3888	0.3624	68.8
1.25	0.9490	3.0100	0.3323	0.3153	71.6
1.30	0.9636	3.6020	0.2776	0.2675	74.5
1.35	0.9757	4.4550	0.2245	0.2190	77.3
1.40	0.9854	5.7980	0.1725	0.1700	80.2
1.45	0.9927	8.2380	0.1214	0.1205	83.1
1.50	0.9975	14.1000	0.0709	0.0707	85.9
1.55	0.9998	48.0800	0.0208	0.0208	88.8
1.60	0.9996	−34.2300	−0.0292	−0.0292	91.7
1.65	0.9969	−12.6000	−0.0794	−0.0791	94.5
1.70	0.9917	−7.6970	−0.1299	−0.1288	97.4
1.75	0.9840	−5.5200	−0.1811	−0.1782	100.3
1.80	0.9738	−4.2860	−0.2333	−0.2272	103.1
1.85	0.9613	−3.4880	−0.2867	−0.2756	106.0
1.90	0.9463	−2.9270	−0.3416	−0.3233	108.9
1.95	0.9290	−2.5090	−0.3985	−0.3702	111.7
2.00	0.9093	−2.1850	−0.4577	−0.4161	114.6

SECTION 2: DERIVATIVES OF THE CIRCULAR FUNCTIONS

Other identities are contained in the exercises.

In Section 4 we will consider sine as a function of an angle in degrees and show, for example, that $\sin 30° = \frac{1}{2}$. For now, keep in mind that the x in $\sin x$ is just a number and is not measured in degrees.

A short table of values of $\sin x$, $\cos x$, $\tan x$, and $\cot x$ is given in Table 7.2.

PROBLEMS

Derive the following identities.

1. $\tan x = \dfrac{1}{\cot x}$
2. $\cos x = \sin\left(x + \dfrac{\pi}{2}\right)$
3. $\cos x = \sin\left(\dfrac{\pi}{2} - x\right)$
4. $1 + \cot^2 x = \csc^2 x$
5. $\sin 2x = 2 \sin x \cos x$
6. $\cos 2x = \cos^2 x - \sin^2 x$
7. $2 \cos^2 x = 1 + \cos 2x$
8. $\cos^2\left(\dfrac{x}{2}\right) = \dfrac{1}{2}(1 + \cos x)$
9. $\tan(x + \pi) = \tan x$
10. $\sin(x + \pi) = -\sin x$
11. $\sin^2 x = \dfrac{1}{2}(1 - \cos 2x)$
12. $\tan 2x = \dfrac{2 \tan x}{1 - \tan^2 x}$
13. $\sin x \csc x = 1$
14. $2 \csc 2x = \csc x \sec x$
15. $\tan(x + y) = \dfrac{\tan x + \tan y}{1 - \tan x \tan y}$
16. $\tan^2 x = \dfrac{\tan 2x - 2 \tan x}{\tan 2x}$

Evaluate the following expressions.

17. $\sin \dfrac{5\pi}{6}$
18. $\cos\left(-\dfrac{3\pi}{4}\right)$
19. $\tan \dfrac{\pi}{4}$
20. $\sec \dfrac{5\pi}{4}$
21. $\sin^2 \dfrac{\pi}{4} + \cos^2 \dfrac{\pi}{4}$
22. $\sin \dfrac{\pi}{6} \cos \dfrac{\pi}{3}$
23. $\tan \dfrac{\pi}{3} \sec \dfrac{\pi}{4}$
24. $\sin \dfrac{\pi}{3} \csc \dfrac{\pi}{3}$

2. DERIVATIVES OF THE CIRCULAR FUNCTIONS

Suppose that we attempt to compute the derivative of the sine function:

$$\lim_{h \to 0} \frac{\sin(x + h) - \sin x}{h} = \lim_{h \to 0} \frac{\sin x \cos h + \cos x \sin h - \sin x}{h}$$

$$= \lim_{h \to 0} \cos x \, \frac{\sin h}{h} + \lim_{h \to 0} \sin x \, \frac{\cos h - 1}{h}$$

$$= \cos x \lim_{h \to 0} \frac{\sin h}{h} + \sin x \lim_{h \to 0} \frac{\cos h - 1}{h}$$

if all of the limits on the last line exist. We first show that

(14) $$\lim_{h \to 0} \frac{\sin h}{h} = 1.$$

Assume that $0 < h < \pi/2$. From Figure 7.5, we see that $\sin h < h$, and hence $(\sin h)/h < 1$.

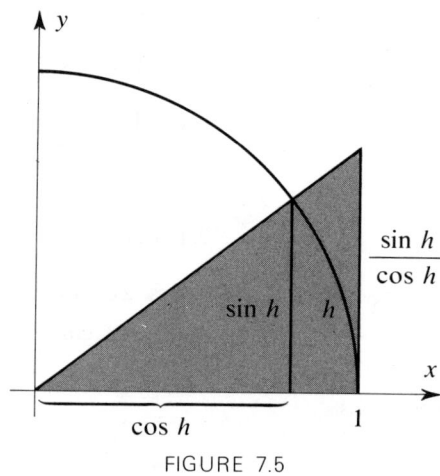

FIGURE 7.5

In a circle of radius r, the area of a sector of the circle is $hr/2$, where h is the arc length. Since $r = 1$ in Figure 7.5, it follows that the area of the sector is $h/2$.

By looking at similar triangles, we see that the height of the large triangle is $\sin h/\cos h$. Since the area of the large triangle is greater than the area of the sector of the circle, it follows that

$$\frac{1}{2} h < \frac{1}{2} \frac{\sin h}{\cos h}.$$

Hence $\cos h < (\sin h)/h$. Thus $(\sin h)/h$ is between $\cos h$ and 1 if $0 < h < \pi/2$. It is left as an exercise to show that

$$\cos h < \frac{\sin h}{h} < 1 \quad \text{if} \quad -\frac{\pi}{2} < h < 0.$$

SECTION 2: DERIVATIVES OF THE CIRCULAR FUNCTIONS

But $\lim_{h \to 0} \cos h = 1$. Since $(\sin h)/h$ is trapped in between $\cos h$ and 1, it follows that

$$\lim_{h \to 0} \frac{\sin h}{h} = 1.$$

It remains to compute

$$\lim_{h \to 0} \frac{\cos h - 1}{h}.$$

Multiplying and dividing by $\cos h + 1$, we obtain

$$\lim_{h \to 0} \frac{\cos h - 1}{h} = \lim_{h \to 0} \frac{\cos h - 1}{h} \cdot \frac{\cos h + 1}{\cos h + 1}$$

$$= \lim_{h \to 0} \frac{\cos^2 h - 1}{h(\cos h + 1)}$$

$$= \lim_{h \to 0} \frac{-\sin^2 h}{h(\cos h + 1)}$$

$$= \lim_{h \to 0} \frac{\sin h}{h} \lim_{h \to 0} \frac{-\sin h}{\cos h + 1}$$

$$= 1 \cdot \frac{-\lim_{h \to 0} (\sin h)}{\lim_{h \to 0} (1 + \cos h)}$$

$$= 0.$$

Thus

$$\frac{d}{dx}(\sin x) = \cos x \lim_{h \to 0} \frac{\sin h}{h} + \sin x \lim_{h \to 0} \frac{\cos h - 1}{h}$$

$$= \cos x \cdot 1 + \sin x \cdot 0$$

$$= \cos x.$$

Hence

(15)
$$\frac{d}{dx}(\sin x) = \cos x.$$

In other words, the derivative of the sine is the cosine.

To compute the derivative of cosine, we first note that

$$\cos x = \sin\left(\frac{\pi}{2} - x\right),$$

and then compute

$$\frac{d}{dx}(\cos x) = \frac{d}{dx}\sin\left(\frac{\pi}{2} - x\right)$$

$$= -\cos\left(\frac{\pi}{2} - x\right)$$

$$= -\sin x.$$

It follows that

(16) $$\frac{d}{dx}(\cos x) = -\sin x.$$

Thus the derivative of the cosine is the negative of the sine.

The following examples illustrate ways to compute derivatives of more complicated functions involving the sine and cosine.

EXAMPLE 1. Compute the derivative of $y = \sin(x^2 + 2)$.

Solution. Let $u = x^2 + 2$. Using the composition rule, we see that

$$\frac{dy}{dx} = \frac{dy}{du}\frac{du}{dx}$$

$$= (\cos u)2x$$

$$= 2x\cos(x^2 + 2).$$

EXAMPLE 2. Compute $\frac{d}{dx}[\sin x \cos(x^2)]$.

Solution. Using the product rule and the composition rule, we get

$$\frac{d}{dx}[\sin x \cos(x^2)] = \sin x[-\sin(x^2)]2x + \cos(x^2)\cos x$$

$$= \cos x \cos(x^2) - 2x \sin x \sin(x^2).$$

EXAMPLE 3. Find $f'(x)$ if $f(x) = \sin(e^x)$.

Solution.

$$f'(x) = \cos(e^x)\frac{d}{dx}(e^x)$$

$$= e^x \cos(e^x).$$

SECTION 2: DERIVATIVES OF THE CIRCULAR FUNCTIONS

The tangent function can be differentiated by using the quotient rule:

$$\frac{d}{dx} \tan x = \frac{d}{dx} \frac{\sin x}{\cos x}$$

$$= \frac{\cos x \cos x - \sin x(-\sin x)}{\cos^2 x}$$

$$= \frac{\cos^2 x + \sin^2 x}{\cos^2 x}$$

$$= \frac{1}{\cos^2 x}$$

$$= \sec^2 x,$$

so that

(17) $$\frac{d}{dx}(\tan x) = \sec^2 x.$$

Therefore the derivative of the tangent equals the secant squared.

The remainder of the circular functions can be differentiated by first rewriting them in terms of the sine and cosine functions. For example, $\sec x = 1/(\cos x)$, and hence

$$\frac{d}{dx}(\sec x) = \frac{d}{dx}\left(\frac{1}{\cos x}\right)$$

$$= \frac{-(-\sin x)}{\cos^2 x}$$

$$= \frac{1}{\cos x} \cdot \frac{\sin x}{\cos x}$$

$$= \sec x \tan x,$$

which shows that

(18) $$\frac{d}{dx} \sec x = \sec x \tan x.$$

Consequently, the derivative of the secant equals the product of the secant times the tangent.

It is left as an exercise to derive the remaining two formulas:

(19) $$\frac{d}{dx} \cot x = -\csc^2 x$$

and

(20) $$\frac{d}{dx} \csc x = -\csc x \cot x.$$

228 CHAPTER 7: THE CIRCULAR FUNCTIONS

It is also left as an exercise to use the methods of Chapter 4 to obtain the graphs in Figures 7.6 to 7.9.

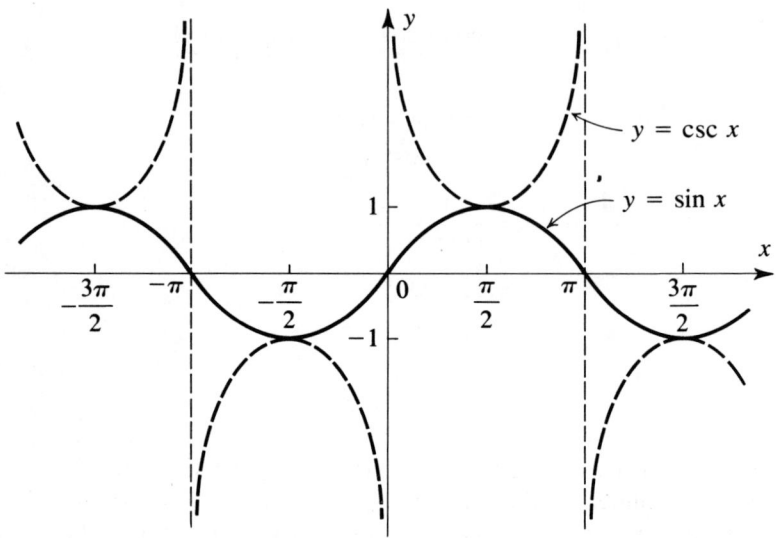

FIGURE 7.6

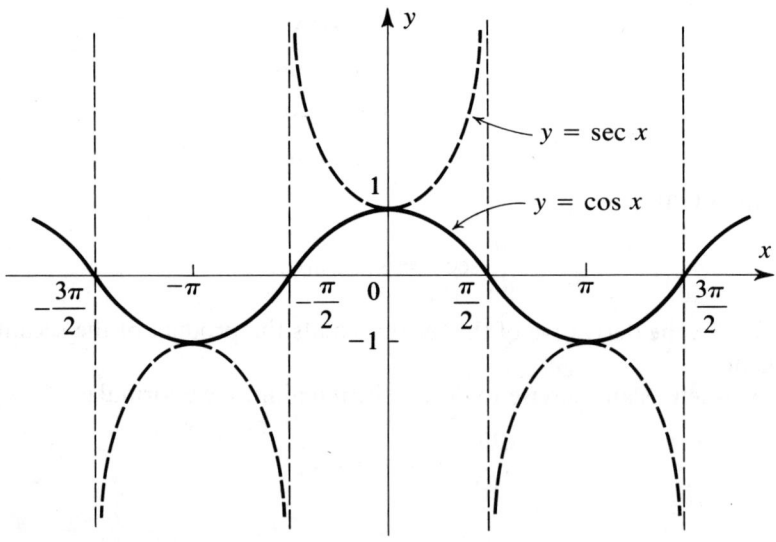

FIGURE 7.7

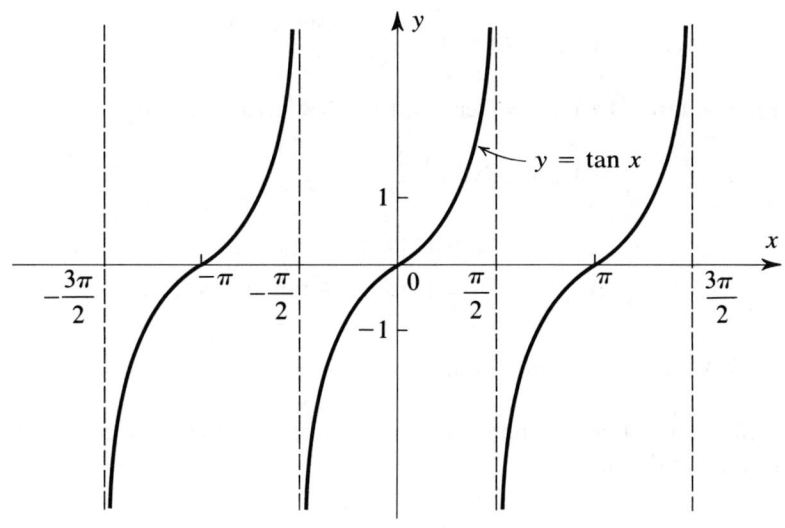

FIGURE 7.8

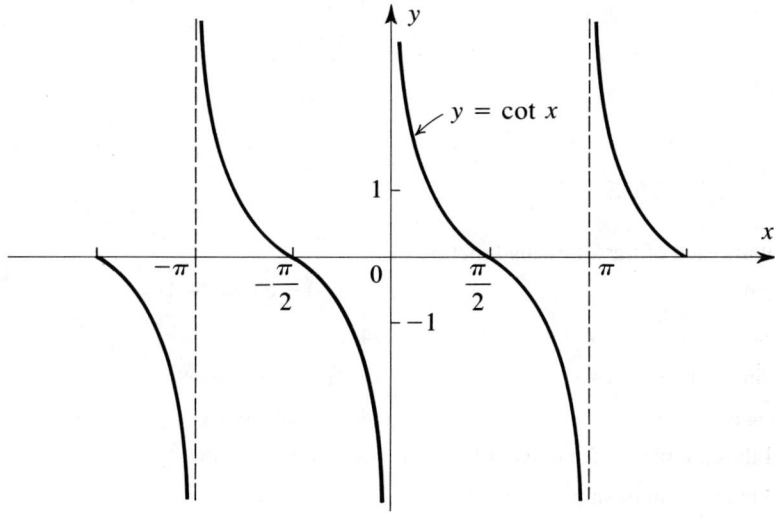

FIGURE 7.9

EXAMPLE 4. Find $\int \sec x \, dx$.

Solution. A trick is required. Multiplying and dividing by $\sec x + \tan x$, we have

$$\int \sec x \, dx = \int \frac{\sec^2 x + \sec x \tan x}{\sec x + \tan x} dx.$$

If $u = \sec x + \tan x$, then $du = (\sec x \tan x + \sec^2 x) \, dx$, and so

$$\int \sec x \, dx = \int \frac{du}{u}$$
$$= \ln |u| + C$$
$$= \ln |\sec x + \tan x| + C.$$

EXAMPLE 5. Find $\int \tan x \, dx$.

Solution. Recall that $\tan x = \sin x / \cos x$. Let $u = \cos x$. Then $du = -\sin x \, dx$, and hence

$$\int \tan x \, dx = \int \frac{\sin x}{\cos x} dx$$
$$= -\int \frac{-\sin x}{\cos x} dx$$
$$= -\int \frac{du}{u}$$
$$= -\ln |u| + C$$
$$= -\ln |\cos x| + C.$$

PROBLEMS

Differentiate each of the following functions.
1. $y = \sin 2x$
2. $f(x) = \cos(x^2 + 1)$
3. $f(t) = \tan \sqrt{t}$
4. $y = \sin x \cos x$
5. $y = \sin^2(x^2) + \cos^2(x^2)$
6. $f(x) = \sin(\cos x)$
7. $f(x) = e^{\sin x}$
8. $y = \sec x \tan x$
9. Find the equation of the tangent line to $y = \sin 2x$ at $x = \pi/6$.
10. Find the maximum value of $\sin x \cos x$.
11. Find the maximum value of $3 \sin(4x + 3)$.

12. Find the maximum value of $e^{-x^2} \sin x^2$.
13. Find the equation of the tangent line to $y = e^{\sin x}$ at $(0, 1)$.

Compute the following integrals.

14. $\displaystyle\int \sin x \cos x \, dx$

15. $\displaystyle\int \sin^2 x \, dx$

$\left[Hint: \sin^2 x = \dfrac{1}{2}(1 - \cos 2x). \right]$

16. $\displaystyle\int \sec x \tan x \, dx$

17. $\displaystyle\int \cot x \, dx$

18. $\displaystyle\int_0^{\pi/2} \sin^3 x \cos x \, dx$

19. $\displaystyle\int e^{3x} \cos(e^{3x}) \, dx$

Derive the following differentiation formulas.

20. $\dfrac{d}{dx} \cot x = -\csc^2 x$

21. $\dfrac{d}{dx} \csc x = -\csc x \cot x$

22. Use the definition of derivative to show that $(d/dx) \tan x = \sec^2 x$. $\Big[$ Hint: Use the identity
$\tan(A + B) = \dfrac{\tan A + \tan B}{1 - \tan A \tan B}$ and show that $\displaystyle\lim_{h \to 0} \dfrac{\tan h}{h} = 1.$ $\Big]$

23. Use the definition of derivative to show that $D_x \cos x = -\sin x$.

Use the methods of Chapter 4 to graph the following functions.

24. $y = \sin x$ 25. $y = \cos x$ 26. $y = \tan x$

27. $y = \cot x$ 28. $y = \sec x$ 29. $y = \csc x$

30. Find the volume of the solid generated by rotating the graph of $y = \sin x$ about the x-axis, $0 \le x \le \pi$. [Hint: To integrate $\sin^2 x$, show that $2 \sin^2 x = 1 - \cos 2x$.]

3. THE INVERSE CIRCULAR FUNCTIONS

One of our goals has been to find methods for computing various types of integrals. We will soon find antiderivatives for such functions as

$$\dfrac{1}{\sqrt{1 - x^2}}, \quad \dfrac{1}{x^2 + 1}, \quad \text{and} \quad \dfrac{1}{x\sqrt{x^2 - 1}}$$

in a most unlikely place; namely, we will find the antiderivatives by looking at functions that are closely related to the circular functions. The first of these functions is related to the sine function.

The equation $\sin x = \sqrt{3}/2$ has more than one solution. For example,

$$x = \dfrac{\pi}{3}, \quad x = \dfrac{2\pi}{3}, \quad x = -\dfrac{4\pi}{3}, \quad x = \dfrac{\pi}{3} + 8\pi, \quad \text{and} \quad x = \dfrac{2\pi}{3} + 10\pi$$

are all solutions. However, if the solutions are restricted to the interval $[-\pi/2, \pi/2]$, then by looking at Figure 7.10, we see that $x = \pi/3$ is the only allowable solution. In fact, if a is any real number between -1 and 1, the equation $\sin x = a$ has exactly one solution x such that $|x| \leq \pi/2$. This solution is denoted by $x = \arcsin a$. Thus

(21) $\qquad x = \arcsin a \quad \text{means} \quad \sin x = a,$

where

$$-\frac{\pi}{2} \leq x \leq \frac{\pi}{2} \quad \text{and} \quad -1 \leq a \leq 1.$$

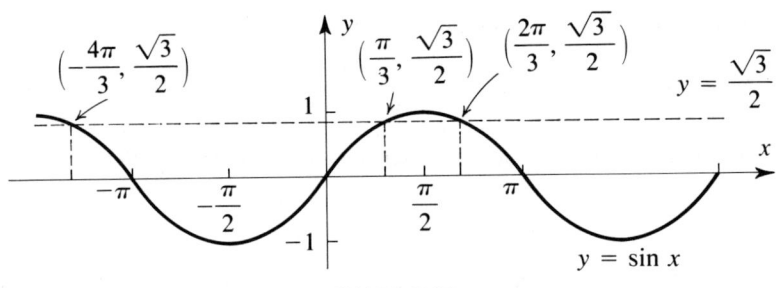

FIGURE 7.10

This definition makes **arcsine** into a function from $[-1, 1]$ to $[-\pi/2, \pi/2]$ that has the following properties:

1. If a is close to b, then $\arcsin a$ is close to $\arcsin b$.
2. The number $\arcsin a$ is relatively close to zero.

Notice that (1) implies that arcsine is a continuous function. Property (2) is merely an attempt to keep things as simple as possible. It would be possible (but highly impractical) to pick $(2\pi/3) + 990\pi$ as our "canonical" solution to $\sin x = \sqrt{3}/2$.

Notice that $\sin(\arcsin a) = \sin x = a$ and $\arcsin(\sin x) = \arcsin a = x$ if x is a solution to $\sin x = a$ such that $-\pi/2 \leq x \leq \pi/2$; that is,

$$\sin(\arcsin a) = a$$

for each a between -1 and 1, and

$$\arcsin(\sin x) = x$$

for each x between $-\pi/2$ and $\pi/2$. It follows that sine and arcsine are inverse functions if the domain of sine is restricted to the interval $[-\pi/2, \pi/2]$.

The graph of $y = \arcsin x$ is given in Figure 7.11. Notice that the graph is the reflection of the graph of $y = \sin x$ about the line $y = x$.

How can x be restricted so that $y = \cos x$ will have an inverse function? The equation $\cos x = \sqrt{3}/2$ has the two solutions $x = \pi/6$ and $x = -\pi/6$ even though x is restricted to $-\pi/2 \le x \le \pi/2$. On the other hand, the equation $\cos x = -\frac{1}{2}$ has no solutions between $-\pi/2$ and $\pi/2$. However, by looking at Figure 7.12, we can see that $\cos x = a$ has exactly one solution between 0 and π for each a between -1 and 1. Each horizontal line intersects the graph of $y = \cos x$ at most once if x is restricted to lie between 0 and π. We therefore define *arccosine* by

(22) $\qquad\qquad\qquad \arccos a = x \quad \text{if and only if} \quad \cos x = a,$

where $0 \le x \le \pi$ and $-1 \le a \le 1$. The graph of $y = \arccos x$ is given in Figure 7.12. Notice that arccosine is a function from $[-1, 1]$ to $[0, \pi]$, is continuous on $[-1, 1]$, and has values relatively close to zero.

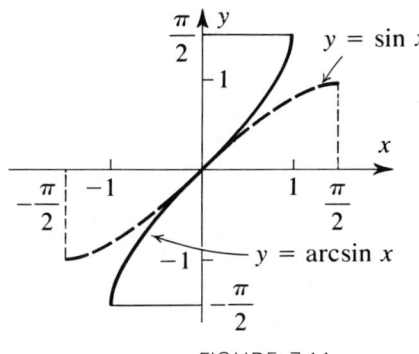

FIGURE 7.11

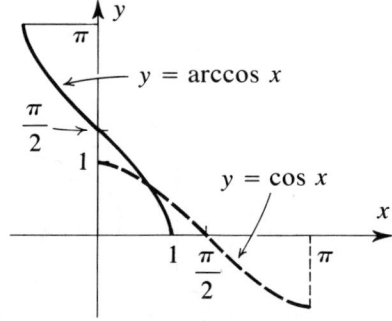

FIGURE 7.12

Using a similar line of reasoning, we define the remainder of the *inverse circular functions* as follows:

(23) $\qquad\qquad \arctan a = x \quad \text{means} \quad a = \tan x,$

where $-\pi/2 < x < \pi/2$;

(24) $\qquad\qquad \text{arccot } a = x \quad \text{means} \quad a = \cot x,$

where $0 < x < \pi$;

(25) $\qquad\qquad \text{arcsec } a = x \quad \text{means} \quad a = \sec x,$

where $-\pi \le x < -\pi/2$ or $0 \le x < \pi/2$ and $|a| \ge 1$; and

(26) $\qquad\qquad \text{arccsc } a = x \quad \text{means} \quad a = \csc x,$

where $-\pi < x \le -\pi/2$ or $0 < x \le \pi/2$ and $|a| \ge 1$.

Additional possible ways to restrict x in the solution of $\sec x = a$ also exist, and such restrictions are sometimes used in other books. However, the preceding restrictions on x will allow the derivative of arcsecant to be written in the simplest possible form.

The graphs of the remaining inverse circular functions are given in Figures 7.13 through 7.16.

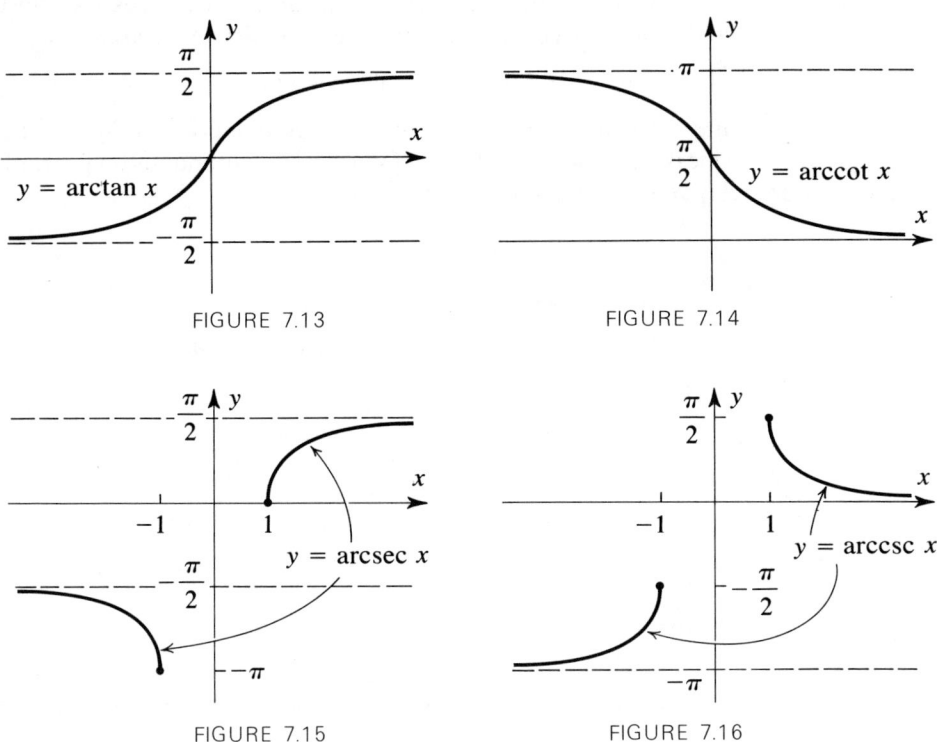

FIGURE 7.13

FIGURE 7.14

FIGURE 7.15

FIGURE 7.16

Another notation that is commonly used for the inverse sine function is $\sin^{-1} x$ instead of arcsine x. The symbolism $\sin^{-1} x$ should not be confused with $1/(\sin x)$, the reciprocal of $\sin x$. The same care should be taken with the other inverse circular functions in case the alternate notation $\cos^{-1} x$, $\tan^{-1} x$, $\cot^{-1} x$, $\sec^{-1} x$, and $\csc^{-1} x$ is used.

We will now differentiate the inverse circular functions so that we can, for example, find the slope of the tangent lines to the graphs of these functions.

Let $y = \arcsin x$. Then $x = \sin y$, where x is between -1 and 1 and y is between $-\pi/2$ and $\pi/2$. Differentiating both sides of $x = \sin y$ with respect to x, we obtain

$$1 = \cos y \frac{dy}{dx}.$$

SECTION 3: THE INVERSE CIRCULAR FUNCTIONS

Now $\cos^2 y + \sin^2 y = 1$, and hence
$$\cos y = \pm\sqrt{1 - \sin^2 y}$$
$$= \pm\sqrt{1 - x^2}.$$

But y is between $-\pi/2$ and $\pi/2$, and so $\cos y \geq 0$, whence $\cos y = \sqrt{1 - x^2}$. Solving for dy/dx, we get
$$\frac{dy}{dx} = \frac{1}{\cos y}$$
$$= \frac{1}{\sqrt{1 - x^2}},$$

which is an algebraic function! Rewritten, it becomes

(27) $$\frac{d}{dx} \arcsin x = \frac{1}{\sqrt{1 - x^2}} \qquad (-1 < x < 1).$$

It is left as an exercise for the reader to verify that

(28) $$\frac{d}{dx} \arccos x = \frac{-1}{\sqrt{1 - x^2}} \qquad (-1 < x < 1),$$

(29) $$\frac{d}{dx} \arctan x = \frac{1}{1 + x^2} \qquad (x \text{ any number}),$$

(30) $$\frac{d}{dx} \text{arccot } x = \frac{-1}{1 + x^2} \qquad (x \text{ any number}),$$

(31) $$\frac{d}{dx} \text{arcsec } x = \frac{1}{x\sqrt{x^2 - 1}} \qquad (|x| > 1),$$

and

(32) $$\frac{d}{dx} \text{arccsc } x = \frac{-1}{x\sqrt{x^2 - 1}} \qquad (|x| > 1).$$

The integration formulas
$$\int \frac{dx}{\sqrt{1 - x^2}} = \arcsin x + C,$$
$$\int \frac{dx}{1 + x^2} = \arctan x + C,$$

and
$$\int \frac{dx}{x\sqrt{x^2 - 1}} = \text{arcsec } x + C$$

are now immediate. (There is no need to state, for example, that

$$\int \frac{-dx}{\sqrt{1-x^2}} = \arccos x + C,$$

because it is just as simple to write this as

$$-\int \frac{dx}{\sqrt{1-x^2}} = -\arcsin x + C.\Big)$$

We have now significantly increased the number of functions that we can integrate.

EXAMPLE 6. Compute $\int_0^{1/2} \frac{dx}{\sqrt{1-x^2}}$.

Solution.

$$\int_0^{1/2} \frac{dx}{\sqrt{1-x^2}} = \arcsin x \Big|_0^{1/2}$$

$$= \arcsin\left(\frac{1}{2}\right) - \arcsin(0)$$

$$= \frac{\pi}{6}.$$

Notice that arcsin $(1/2) = \pi/6$ and *not* 30°. The preceding integral is equal to the real number $\pi/6 = 0.52359\ldots$, which can be interpreted as the area of a certain region in the plane. The expression 30° is not a real number but rather a measurement of an angle that is one-third the size of a right angle.

EXAMPLE 7. Find $\int \frac{x\,dx}{1+x^4}$.

Solution. This does not fit any of the formulas. However, if we make the substitution $u = x^2$, then $du = 2x\,dx$, and hence

$$\int \frac{x\,dx}{1+x^4} = \frac{1}{2}\int \frac{2x\,dx}{1+(x^2)^2}$$

$$= \frac{1}{2}\int \frac{du}{1+u^2}$$

SECTION 3: THE INVERSE CIRCULAR FUNCTIONS

$$= \frac{1}{2} \arctan u + C$$

$$= \frac{1}{2} \arctan (x^2) + C.$$

EXAMPLE 8. Compute $\displaystyle\int \frac{dx}{x\sqrt{x^2 - 9}}$.

Solution. This looks similar to the formula for inverse secant. Let $x = 3u$. Then $dx = 3\,du$, and hence

$$\int \frac{dx}{x\sqrt{x^2 - 9}} = \int \frac{3\,du}{3u\sqrt{(3u)^2 - 9}}$$

$$= \int \frac{du}{u\sqrt{9u^2 - 9}}$$

$$= \frac{1}{3} \int \frac{du}{u\sqrt{u^2 - 1}}$$

$$= \frac{1}{3} \operatorname{arcsec} u + C$$

$$= \frac{1}{3} \operatorname{arcsec} \left(\frac{x}{3}\right) + C.$$

PROBLEMS

Compute derivatives of the following functions.

1. $y = \arcsin (x^2)$
2. $f(x) = \sin (\arccos x)$
3. $y = (\arcsin x)^2$
4. $x \to \arctan (3x^2)$
5. $f(x) = \arcsin (e^x)$
6. $y = \arccos (\sin x)$
7. $y = \tan (\arccos x)$
8. $f(x) = \ln (\arctan x)$
9. $y = \arcsin (\sqrt{x})$
10. $y = \arctan (1 + x^2)$
11. $f(x) = e^{\arctan x}$
12. $f(x) = \sec (\arctan x)$
13. $y = x^{\arcsin x}$
14. $y = \ln (\sin x \arcsin x)$
15. $f(x) = (\operatorname{arcsec} x)(\arctan x)$
16. $y = (\arctan x)^x$

Compute the following integrals.

17. $\displaystyle\int \frac{dx}{x\sqrt{x^4 - 1}}$
18. $\displaystyle\int \frac{dx}{x\sqrt{x^2 - 4}}$

19. $\displaystyle\int \frac{x^2\,dx}{x^6+1}$

20. $\displaystyle\int \frac{dx}{4x^2+25}$

21. $\displaystyle\int \frac{x\,dx}{\sqrt{1-x^2}}$

22. $\displaystyle\int_0^{3\sqrt{2}/8} \frac{dx}{\sqrt{9-16x^2}}$

23. $\displaystyle\int \frac{x\,dx}{4+x^2}$

24. $\displaystyle\int \frac{4x\,dx}{\sqrt{1-4x^4}}$

25. $\displaystyle\int \frac{du}{2u\sqrt{u-1}}$

26. $\displaystyle\int \frac{dx}{2\sqrt{x}\sqrt{1-x}}$

27. Verify that $\dfrac{d}{dx}\arccos x = \dfrac{-1}{\sqrt{1-x^2}}$.

28. Verify that $\dfrac{d}{dx}\arctan x = \dfrac{1}{1+x^2}$.

29. Verify that $\dfrac{d}{dx}\operatorname{arcsec} x = \dfrac{1}{x\sqrt{x^2-1}}$.

30. Verify that $\dfrac{d}{dx}\operatorname{arccsc} x = \dfrac{-1}{x\sqrt{x^2-1}}$.

31. Find the average value of the function $f(x) = 5/(9x^2+4)$ on the interval $[0,1]$. (See Section 5 of Chapter 2.)

4. APPLICATIONS OF THE CIRCULAR FUNCTIONS

The circular functions were defined in terms of arc length. For many applications, however, it is more convenient to think of the circular functions as being trigonometric functions. That is, let A be an acute angle (in radians or in degrees) of a right triangle. (One radian equals $180/\pi$ degrees.) Then

(33) $$\sin A = \frac{\text{opposite side}}{\text{hypotenuse}},$$

(34) $$\cos A = \frac{\text{adjacent side}}{\text{hypotenuse}},$$

and

(35) $$\tan A = \frac{\text{opposite side}}{\text{adjacent side}},$$

as in Figure 7.17. These definitions will also be useful in Section 1 of Chapter 8.

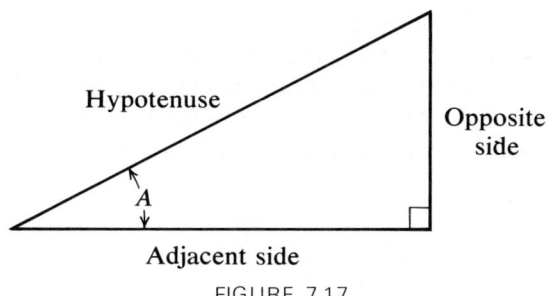

FIGURE 7.17

EXAMPLE 9. The top of a building makes an angle of 30° from one position on ground level and an angle of 20° from a position 100 feet farther away, as in Figure 7.18. How high is the building?

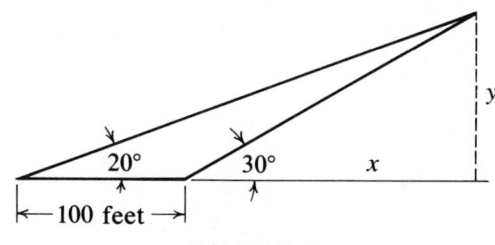

FIGURE 7.18

Solution. From Figure 7.18, we see that $\tan 30° = y/x$ and $\tan 20° = y/(x + 100)$. From a book of mathematical tables, $\tan 30° \approx 0.57735$ and $\tan 20° \approx 0.36397$, so that the two equations become

$$y = 0.57735x$$

and
$$y = (x + 100)(0.36397)$$
$$= 0.36397x + 36.397.$$

Subtracting the second equation from the first, we get

$$0 = 0.21338x - 36.397,$$

and hence

$$x = 170.57,$$

which implies

$$y = 98.48.$$

Thus the building is approximately 98.5 feet tall.

CHAPTER 7: THE CIRCULAR FUNCTIONS

EXAMPLE 10. A man is stranded on a circular island. Each day he must make a trip from a water hole near a beach to a clump of coconut trees near the opposite beach, as in Figure 7.19. He can walk around the island on the smooth beach at a rate of 2 miles per hour, he can walk through the rough interior at 1 mile per hour, or he can walk through the interior to some point on the beach and then follow the beach the rest of the way around the island. If the diameter of the island is 2 miles, which route will minimize the travel time?

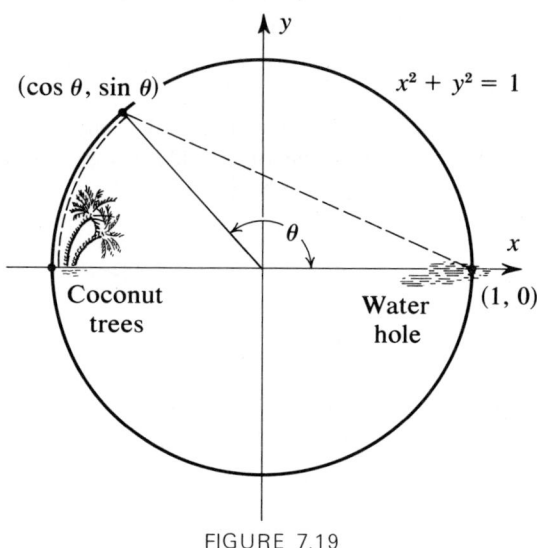

FIGURE 7.19

Solution. Suppose that he walks across the island to the point $(\cos\theta, \sin\theta)$ and then follows the beach the rest of the way. He walks a distance $\sqrt{(1-\cos\theta)^2 + \sin^2\theta}$ at a rate of 1 mile per hour; then he walks a distance $\pi - \theta$ at a rate of 2 miles per hour. The total travel time is given by

$$t = \sqrt{(1-\cos\theta)^2 + \sin^2\theta} + \frac{1}{2}(\pi - \theta)$$

$$= \sqrt{1 - 2\cos\theta + \cos^2\theta + \sin^2\theta} + \frac{1}{2}(\pi - \theta)$$

$$= \sqrt{2}\sqrt{1 - \cos\theta} + \frac{1}{2}(\pi - \theta).$$

To minimize t, it is necessary to compute $dt/d\theta$:

$$\frac{dt}{d\theta} = \frac{\sqrt{2}\sin\theta}{2\sqrt{1-\cos\theta}} - \frac{1}{2},$$

SECTION 4: APPLICATIONS OF THE CIRCULAR FUNCTIONS

which is zero when
$$\sqrt{2} \sin \theta = \sqrt{1 - \cos \theta}.$$
To solve for θ, we first square both sides to get
$$2 \sin^2 \theta = 1 - \cos \theta.$$
But $\sin^2 \theta = 1 - \cos^2 \theta = (1 - \cos \theta)(1 + \cos \theta)$, and hence
$$2(1 - \cos \theta)(1 + \cos \theta) = 1 - \cos \theta.$$
Thus
$$1 - \cos \theta = 0$$
or else
$$2(1 + \cos \theta) = 1.$$
In the first case, $\theta = 0$; and in the second case,
$$1 + \cos \theta = \frac{1}{2},$$
and hence
$$\cos \theta = -\frac{1}{2},$$
which is true when $\theta = 2\pi/3$. It follows that there is only one critical value between 0 and π. Notice that at $\theta = 0$,
$$t = \sqrt{2}\sqrt{1 - 1} + \frac{\pi}{2}$$
$$= \frac{\pi}{2}$$
$$\approx 1.571 \text{ hours}.$$
At $\theta = 2\pi/3$,
$$t = \sqrt{2}\sqrt{1 - \left(-\frac{1}{2}\right)} + \frac{1}{2}\left(\pi - \frac{2\pi}{3}\right)$$
$$= \sqrt{3} + \frac{\pi}{6}$$
$$\approx 2.256 \text{ hours},$$
and at $\theta = \pi$,
$$t = \sqrt{2}\sqrt{1 - (-1)} + \frac{1}{2}(0)$$
$$= 2 \text{ hours}.$$

242   CHAPTER 7: THE CIRCULAR FUNCTIONS

Since there is only one critical value between 0 and π, it follows that the minimum time occurs when $\theta = 0$—that is, when the man walks around on the beach. He can walk across the island in 120 minutes, and it takes 94 minutes to walk around the island on the beach. For the sake of variety, he may decide to walk across the island part of the time. In fact, for the sake of variety, he may eventually try almost anything.

EXAMPLE 11.   A man at a race track is following his favorite horse with a pair of binoculars. The horse is running south at a rate of 60 feet per second on a straight stretch of track. The man is 200 feet west of the finish line. At what rate must the man turn his head at an instant when the horse is 200 feet from the finish line, as in Figure 7.20?

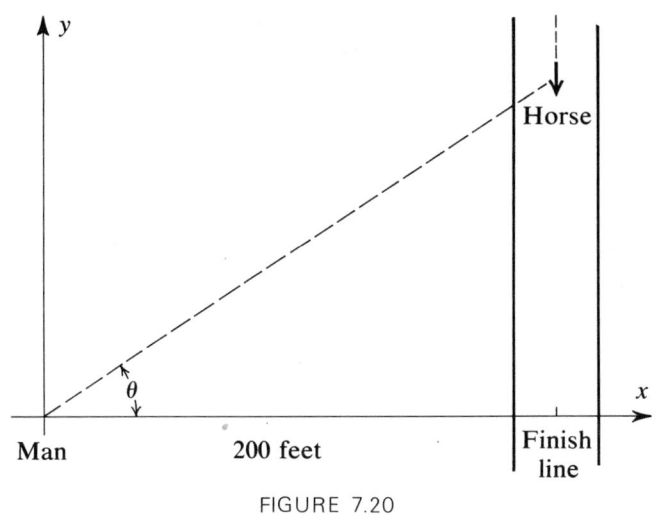

FIGURE 7.20

Solution. From Figure 7.20, we see that $dy/dt = -60$ and $y = 200 \tan \theta$. Thus

$$-60 = \frac{dy}{dt}$$

$$= 200 \sec^2 \theta \, \frac{d\theta}{dt}.$$

Notice that $\theta = \pi/4$ when $y = 200$, and hence at that instant

$$\frac{d\theta}{dt} = \frac{-60}{200 \sec^2 (\pi/4)}$$

SECTION 4: APPLICATIONS OF THE CIRCULAR FUNCTIONS

$$= \frac{-60}{200 \cdot 2}$$

$$= -\frac{3}{20},$$

since $\sec^2(\pi/4) = 2$. Thus, at the given instant, the man must turn his head at the rate of $-3/20$ radians per second, which is roughly -8.6 degrees per second. (A radian equals $180/\pi$ degrees. The minus sign refers to the fact that his head turns in a clockwise direction.)

EXAMPLE 12. The voltage E at time t seconds in a 60-cycle ac of amplitude 110 volts is given by $E = 110 \sin(120\pi t + k)$ for some constant k. Find the maximum voltages and when they occur.

Solution. The function E is differentiable at all t, so the maximum and minimum values must occur when $dE/dt = 0$. Now

$$\frac{dE}{dt} = 120\pi \cdot 110 \cos(120\pi t + k),$$

which is zero when $\cos(120\pi t + k) = 0$. But $\cos x = 0$ only when $x = \pi/2 + n\pi$ for some integer n, and hence

$$120\pi t + k = \frac{\pi}{2} + n\pi,$$

or

$$t = \frac{2n+1}{240} - \frac{k}{120\pi}.$$

If n is even, then $\sin(120\pi t + k) = \sin(\pi/2 + n\pi) = 1$; and if n is odd, then $\sin(120\pi t + k) = \sin(\pi/2 + n\pi) = -1$. Thus the maximum $E = 110$ occurs when

$$t = \frac{2n+1}{240} - \frac{k}{120\pi} \quad (n \text{ even}),$$

and the minimum $E = -110$ occurs when

$$t = \frac{2n+1}{240} - \frac{k}{120\pi} \quad (n \text{ odd}).$$

PROBLEMS

1. The top of a mountain makes an angle of 15°18′ from one location and an angle of 15°14′ from a location 100 feet farther from the mountain. What is the elevation of the top of the mountain if the elevation at each of the two locations is 1000 feet above sea level?

244 CHAPTER 7: THE CIRCULAR FUNCTIONS

2. A paper cup is to be made by starting with a circular-shaped piece of paper of radius 10 centimeters, cutting along the dotted line, and then glueing into a cone-shaped cup (see Figure 7.21). What is the maximum volume of such a cup?

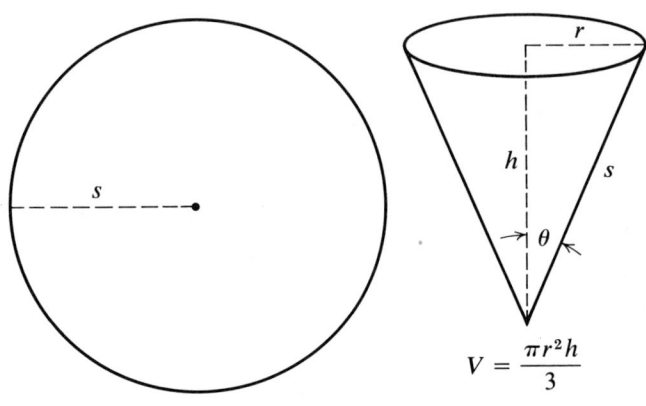

FIGURE 7.21

3. An outdoor movie screen 50 feet wide is perpendicular to and 500 feet away from a street. Cheap Charley wants to park his car beside the street and watch a free movie. Where should he park in order that his field of vision of the screen be as wide as possible?

4. A grandfather clock stands on the y-axis and the pendulum swings back and forth between $x = -a$ and $x = a$ in such a way that the horizontal component of velocity at the point x is $v = \sqrt{k/m}\sqrt{a^2 - x^2}$, where k and m are constants. Show that $x = a \sin(t\sqrt{k/m} + C)$ for some constant C. $\left[\textit{Hint:} \text{ Note that } \int \dfrac{dx}{\sqrt{a^2 - x^2}} = \sqrt{\dfrac{k}{m}} \int dt. \right]$

5. A weight on the end of a spring is moving up and down according to the function $y = 3 \sin(t/2 - 4)$, where t is time in seconds. How high and how low does the weight move? When does the weight reach the highest point?

6. The bottom of a 15-foot ladder is being pulled away from a house at a rate of 2 feet per second. How fast is the top of the ladder moving down the side of the house at an instant when the bottom of the ladder is 9 feet from the house?

7. A 2-meter fence stands one meter from a high wall. What is the length of the shortest ladder that can be set up outside the fence which will reach the wall over the top of the fence?

8. A rotating searchlight on a guard tower patrols the straight outside wall of a prison. The tower is 40 meters from the nearest point on the outside wall, and the light makes two complete revolutions per minute. How fast is the light beam moving along the wall at an

instant when the searchlight is aimed at a point on the wall 15 meters from the point nearest the tower?

9. A point (x, y) is moving at a constant rate of 3 revolutions per second around a circle of radius r with center at the origin. What is the rate of change of x at the point (x_0, y_0)? [*Hint*: $x = r \cos \theta$.]

10. In Example 10, which route should the man take if he can walk around on the beach at 3 miles per hour and across the island at 2 miles per hour?

11. Show that $\dfrac{d}{dx} \sin x° = \dfrac{\pi}{180} \cos x°$. Explain why radians are used instead of degrees in calculus.

THE HYPERBOLIC FUNCTIONS (OPTIONAL)

The functions $\tfrac{1}{2}(e^t - e^{-t})$ and $\tfrac{1}{2}(e^t + e^{-t})$ occur often enough that it is convenient to give names to them: the **hyperbolic sine** and the **hyperbolic cosine**. That is,

$$\sinh t = \frac{1}{2}(e^t - e^{-t}) \tag{36}$$

and

$$\cosh t = \frac{1}{2}(e^t + e^{-t}). \tag{37}$$

Notice that

$$\cosh^2 t - \sinh^2 t = \frac{1}{4}(e^t + e^{-t})^2 - \frac{1}{4}(e^t - e^{-t})^2$$

$$= \frac{1}{4}(e^{2t} + 2 + e^{-2t} - e^{2t} + 2 - e^{-2t})$$

$$= 1.$$

Thus the point $(\cosh t, \sinh t)$ lies on the hyperbola $x^2 - y^2 = 1$, whence the term "hyperbolic." [Recall that $\cos^2 t + \sin^2 t = 1$, and hence the point $(\cos t, \sin t)$ lies on the circle $x^2 + y^2 = 1$.]

There is a strong parallel between the theory of circular functions and the theory of hyperbolic functions. Since $\sinh(-t) = \tfrac{1}{2}[e^{-t} - e^{-(-t)}] = -\tfrac{1}{2}(e^t - e^{-t}) = -\sinh t$, it follows that hyperbolic sine is an odd function, just as sine is an odd function. Similarly, cosine and hyperbolic cosine are both even functions. Differentiation looks familiar, because

$$\frac{d}{dx} \sinh x = \frac{d}{dx} \frac{1}{2}(e^x - e^{-x})$$

$$= \frac{1}{2}[e^x - (-1)e^{-x}]$$

$$= \frac{1}{2}(e^x + e^{-x})$$

$$= \cosh x;$$

that is,

(38) $$\frac{d}{dx} \sinh x = \cosh x,$$

whereas (you will recall)

$$\frac{d}{dx} \sin x = \cos x.$$

Similarly,

$$\frac{d}{dx} \cosh x = \frac{d}{dx} \frac{1}{2}(e^x + e^{-x})$$

$$= \frac{1}{2}(e^x - e^{-x})$$

$$= \sinh x,$$

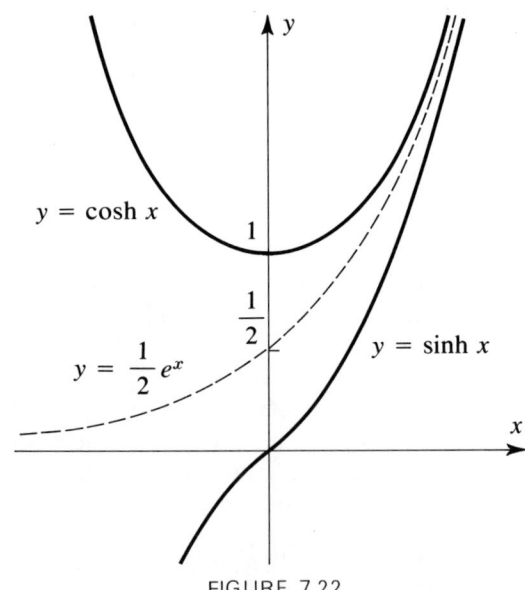

FIGURE 7.22

so that

(39) $$\frac{d}{dx} \cosh x = \sinh x,$$

while

$$\frac{d}{dx} \cos x = -\sin x.$$

There are, however, some fundamental differences between the circular functions and the hyperbolic functions, as can be seen by looking at Figure 7.22.

The circular functions are periodic with period 2π, so that $\sin(x + 2\pi) = \sin x$ for all x. On the other hand, if $\sinh(x + a) = \sinh x$ for any x, then $a = 0$. Also, $-1 \le \sin x \le 1$ for all x, whereas $\sinh x$ tends to infinity as x tends to infinity.

For completeness, we define the remaining hyperbolic functions, whose graphs are given in Figures 7.23 and 7.24.

(40) $$\tanh x = \frac{\sinh x}{\cosh x}$$

(41) $$\coth x = \frac{\cosh x}{\sinh x}$$

(42) $$\operatorname{sech} x = \frac{1}{\cosh x}$$

(43) $$\operatorname{csch} x = \frac{1}{\sinh x}$$

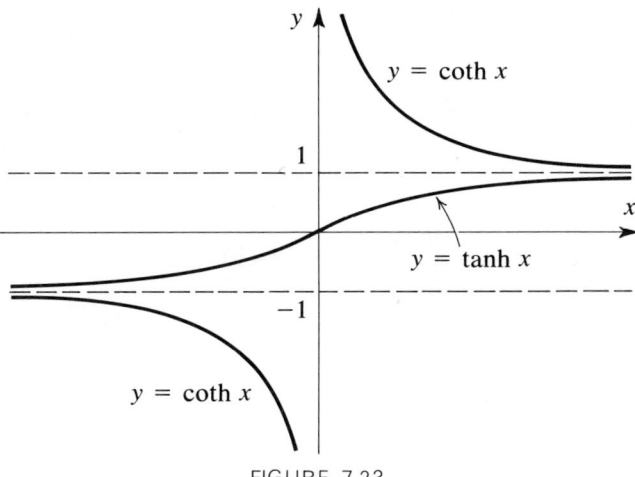

FIGURE 7.23

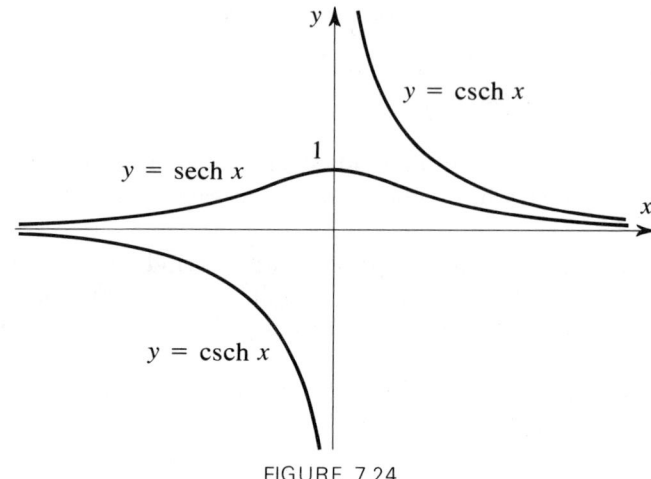

FIGURE 7.24

Using the quotient rule for derivatives, we obtain

$$\frac{d}{dx}\tanh x = \frac{\cosh x (D_x \sinh x) - \sinh x (D_x \cosh x)}{\cosh^2 x}$$

$$= \frac{\cosh^2 x - \sinh^2 x}{\cosh^2 x}$$

$$= \frac{1}{\cosh^2 x}$$

$$= \operatorname{sech}^2 x;$$

that is,

(44) $$\frac{d}{dx}\tanh x = \operatorname{sech}^2 x.$$

Similarly,

(45) $$\frac{d}{dx}\coth x = -\operatorname{csch}^2 x.$$

Also,

$$\frac{d}{dx}\operatorname{sech} x = \frac{d}{dx}\frac{1}{\cosh x}$$

$$= \frac{-\sinh x}{\cosh^2 x}$$

$$= -\frac{1}{\cosh x} \cdot \frac{\sinh x}{\cosh x}$$

$$= -\operatorname{sech} x \tanh x,$$

which shows that

(46) $$\frac{d}{dx} \operatorname{sech} x = -\operatorname{sech} x \tanh x.$$

In a similar manner,

(47) $$\frac{d}{dx} \operatorname{csch} x = -\operatorname{csch} x \coth x.$$

The parallel between the circular functions and the hyperbolic functions can be extended by considering the **inverse hyperbolic functions.** The derivatives of these functions are algebraic and look very similar to the corresponding derivatives of the inverse circular functions.

Since the hyperbolic functions are defined in terms of the exponential function, it should not be surprising that the inverse hyperbolic functions involve logarithms.

Since $\sinh x = \frac{1}{2}(e^x - e^{-x})$, we should be able to find an inverse function for $\sinh x$ by solving $y = \frac{1}{2}(e^x - e^{-x})$ for x in terms of y. Multiplying both sides by 2, we get

$$2y = e^x - e^{-x},$$

or

$$e^x - 2y - e^{-x} = 0.$$

Multiplying by e^x, this becomes

$$(e^x)^2 - 2ye^x - 1 = 0,$$

which is quadratic in e^x. The quadratic formula now yields

$$e^x = \frac{2y \pm \sqrt{4y^2 + 4}}{2}$$

$$= y \pm \sqrt{y^2 + 1}.$$

But e^x is always positive, and hence we must discard the minus sign. Finally, taking logarithms of both sides of

$$e^x = y + \sqrt{y^2 + 1}$$

yields

$$x = \ln(y + \sqrt{y^2 + 1}).$$

There are no restrictions on x or y, and so the inverse hyperbolic sine is given by arcsinh $y = \ln(y + \sqrt{y^2 + 1})$ for all numbers y. As usual, we now interchange the roles of x and y so that

(48) $$\text{arcsinh } x = \ln(x + \sqrt{x^2 + 1})$$

for all real numbers x.

The remainder of the inverse hyperbolic functions can be computed in a similar manner. It is left as an exercise to show that

(49) $$\text{arccosh } x = \ln(x + \sqrt{x^2 - 1}) \quad (x \geq 1),$$

(50) $$\text{arctanh } x = \frac{1}{2} \ln \frac{1+x}{1-x} \quad (-1 < x < 1),$$

(51) $$\text{arccoth } x = \frac{1}{2} \ln \frac{x+1}{x-1} \quad (|x| > 1),$$

(52) $$\text{arcsech } x = \ln \frac{1 + \sqrt{1-x^2}}{x} \quad (0 < x \leq 1),$$

and

(53) $$\text{arccsch } x = \ln \frac{1 + \sqrt{1+x^2}}{x} \quad (x \neq 0).$$

The alternate notation $\sinh^{-1} x$, $\cosh^{-1} x$, $\tanh^{-1} x$, $\coth^{-1} x$, $\text{sech}^{-1} x$, and $\text{csch}^{-1} x$ is also sometimes used for the preceding functions.

It is now easy to compute derivatives of the inverse hyperbolic functions. For example,

$$\frac{d}{dx} \text{arctanh } x = \frac{d}{dx} \frac{1}{2} \ln \frac{1+x}{1-x}$$

$$= \frac{d}{dx} \frac{1}{2} [\ln(1+x) - \ln(1-x)]$$

$$= \frac{1}{2} \left(\frac{1}{1+x} - \frac{-1}{1-x} \right)$$

$$= \frac{1}{1-x^2};$$

that is,

(54) $$\frac{d}{dx} \text{arctanh } x = \frac{1}{1-x^2},$$

where $-1 < x < 1$.

THE HYPERBOLIC FUNCTIONS   251

There is another method of computing such derivatives—namely, $y = \text{arctanh } x$ means that $x = \tanh y$. Thus differentiating with respect to x, we get

$$1 = \frac{d}{dx} \tanh y$$

$$= \text{sech}^2 y \frac{dy}{dx}.$$

But dividing both sides of $1 = \cosh^2 y - \sinh^2 y$ by $\cosh^2 y$ leads to $\text{sech}^2 y = 1 - \tanh^2 y$, and, consequently, solving for dy/dx yields

$$\frac{dy}{dx} = \frac{1}{\text{sech}^2 y}$$

$$= \frac{1}{1 - \tanh^2 y}$$

$$= \frac{1}{1 - x^2},$$

which agrees with Formula (54).

Either method can now be used to get the following list of differentiation formulas:

(55) $$\frac{d}{dx} \text{arcsinh } x = \frac{1}{\sqrt{1 + x^2}}$$

(56) $$\frac{d}{dx} \text{arccosh } x = \frac{1}{\sqrt{x^2 - 1}}$$

(57) $$\frac{d}{dx} \text{arccoth } x = \frac{1}{1 - x^2} \quad (|x| > 1)$$

(58) $$\frac{d}{dx} \text{arcsech } x = \frac{-1}{x\sqrt{1 - x^2}}$$

(59) $$\frac{d}{dx} \text{arccsch } x = \frac{-1}{|x|\sqrt{1 + x^2}}$$

Each of these formulas yields a corresponding integration formula. For example, Formula (56) gives

$$\int \frac{dx}{\sqrt{x^2 - 1}} = \text{arccosh } x + C.$$

Part of the parallel between the circular functions and the hyperbolic functions is summarized in Table 7.3.

Table 7.3

Circular Functions	Hyperbolic Functions
$\cos^2 x + \sin^2 x = 1$	$\cosh^2 x - \sinh^2 x = 1$
$1 + \tan^2 x = \sec^2 x$	$1 - \tanh^2 x = \text{sech}^2 x$
$\sin(a+b) = \sin a \cos b + \cos a \sin b$	$\sinh(a+b) = \sinh a \cosh b + \cosh a \sinh b$
$\cos(a+b) = \cos a \cos b - \sin a \sin b$	$\cosh(a+b) = \cosh a \cosh b + \sinh a \sinh b$
$y = \sin x$ satisfies $y + y'' = 0$, $y(0) = 0$, $y'(0) = 1$	$y = \sinh x$ satisfies $y - y'' = 0$, $y(0) = 0$, $y'(0) = 1$
$y = \cos x$ satisfies $y + y'' = 0$, $y(0) = 1$, $y'(0) = 0$	$y = \cosh x$ satisfies $y - y'' = 0$, $y(0) = 1$, $y'(0) = 0$
$D \sin x = \cos x$, $D \cos x = -\sin x$	$D \sinh x = \cosh x$, $D \cosh x = \sinh x$
$D \tan x = \sec^2 x$, $D \cot x = -\csc^2 x$	$D \tanh x = \text{sech}^2 x$, $D \coth x = -\text{csch}^2 x$
$D \sec x = \sec x \tan x$	$D \text{ sech } x = -\text{sech } x \tanh x$
$D \csc x = -\csc x \cot x$	$D \text{ csch } x = -\text{csch } x \coth x$
$D \arcsin x = \dfrac{1}{\sqrt{1-x^2}}$	$D \text{ arcsinh } x = \dfrac{1}{\sqrt{1+x^2}}$
$D \arccos x = \dfrac{-1}{\sqrt{1-x^2}}$	$D \text{ arccosh } x = \dfrac{1}{\sqrt{x^2-1}}$
$D \arctan x = \dfrac{1}{1+x^2}$	$D \text{ arctanh } x = \dfrac{1}{1-x^2} \quad (-1 < x < 1)$
$D \text{arccot } x = \dfrac{-1}{1+x^2}$	$D \text{ arccoth } x = \dfrac{1}{1-x^2} \quad (|x| > 1)$
$D \text{arcsec } x = \dfrac{1}{x\sqrt{x^2-1}}$	$D \text{ arcsech } x = \dfrac{-1}{x\sqrt{1-x^2}}$
$D \text{arccsc } x = \dfrac{-1}{x\sqrt{x^2-1}}$	$D \text{ arccsch } x = \dfrac{-1}{x\sqrt{1+x^2}}$

PROBLEMS

1. Verify that $\dfrac{d}{dx} \coth x = -\text{csch}^2 x$.

2. Verify that $D \text{ csch } x = -\text{csch } x \coth x$.

3. Show that $\text{sech}^2 x + \tanh^2 x = 1$.

4. Show that $\sinh(a+b) = \sinh a \cosh b + \cosh a \sinh b$.

5. Show that $\cosh(a+b) = \cosh a \cosh b + \sinh a \sinh b$.

6. If $\sinh x = -2$, find $\cosh x$, $\tanh x$, $\coth x$, $\text{sech } x$, and $\text{csch } x$.

7. If $\tanh x = \tfrac{3}{5}$, find $\text{sech } x$, $\cosh x$, $\sinh x$, $\coth x$, and $\text{csch } x$.

Compute each of the following derivatives.

8. $D_x \coth(x^2)$

9. $D(\coth^2 x - \text{csch}^2 x)$

10. $D \sinh(3x)$

11. $D(\sinh x \cosh x)$

12. $D(\cosh^2 x + \sinh^2 x)$

13. $\dfrac{d}{dx} \tanh \dfrac{1}{1-x^2}$

Compute each of the following integrals.

14. $\int x \sinh(x^2)\, dx$

15. $\int \operatorname{sech} x \tanh x \, dx$

16. $\int \sinh^3 x \, dx$

17. $\int \operatorname{sech}^2 x \tanh^2 x \, dx$

18. $\int \cosh^2 x \, dx$

19. $\int \operatorname{sech}^2 x \tanh x \, dx$

20. Verify that $\operatorname{arctanh} x = \dfrac{1}{2} \ln \dfrac{1+x}{1-x}$ $\quad (-1 < x < 1)$.

21. Verify that $\operatorname{arcsech} x = \ln \dfrac{1 + \sqrt{1-x^2}}{x}$ $\quad (0 < x \le 1)$.

22. Obtain the formula $D_x \operatorname{arcsinh} x = \dfrac{1}{\sqrt{1+x^2}}$.

23. Show that $\dfrac{d}{dx} \operatorname{arcsech} x = \dfrac{-1}{x\sqrt{1-x^2}}$.

Compute the following derivatives.

24. $D_x \operatorname{arctanh} \dfrac{x}{\sqrt{x^2 + 1}}$

25. $D_x \operatorname{arcsinh}(\cosh x)$

26. $\dfrac{d}{dx} \operatorname{arccoth}(\cosh x)$

27. $\dfrac{d}{dx} \cosh(\operatorname{arccoth} x)$

Compute the following integrals.

28. $\displaystyle\int_2^3 \dfrac{x\, dx}{1 - x^4}$

29. $\displaystyle\int \dfrac{\sinh x \, dx}{\sqrt{1 + \cosh^2 x}}$

30. $\displaystyle\int \dfrac{\operatorname{arcsinh} x \, dx}{\sqrt{1 + x^2}}$

31. $\displaystyle\int \dfrac{dx}{x\sqrt{1 - x^4}}$

32. $\displaystyle\int \dfrac{dx}{\sqrt{x^2 - 9}}$

33. $\displaystyle\int \dfrac{dx}{x\sqrt{9 - 4x^2}}$

FOR HAND-HELD CALCULATORS ONLY (OPTIONAL)

The following formulas are useful for approximating values of the trigonometric functions with a hand-held calculator. It is helpful, but not necessary, to have at least one memory register available.

(60) $\quad \sin x \approx ((x^2/5/4 - 1)x^2/3/2 + 1)x \quad \left(-\dfrac{\pi}{4} \le x \le \dfrac{\pi}{4}\right)$

(61) $\quad \cos x \approx ((x^2/6/(-5) + 1)x^2/4/3 - 1)x^2/2 + 1 \quad \left(-\dfrac{\pi}{4} \le x \le \dfrac{\pi}{4}\right)$

(62) $\quad \tan x \approx ((2x^2/5 + 1)x^2/3 + 1)x \quad \left(-\dfrac{\pi}{4} \le x \le \dfrac{\pi}{4}\right)$

(63) $\quad \arcsin x \approx ((9x^2/20 + 1)x^2/6 + 1)x \quad \left(-\dfrac{1}{2} \le x \le \dfrac{1}{2}\right)$

(64) $\quad \arctan x \approx ((3x^2/5 - 1)x^2/3 + 1)x \quad \left(-\dfrac{1}{2} \le x \le \dfrac{1}{2}\right)$

(65) $\quad \arccos x = \pi/2 - \arcsin x$

In the preceding formulas, expressions such as $x^2/5/4$ should be interpreted as $(x^2/5)/4$ and $3x^2/5 - 1$ as $(3x^2/5) - 1$.

These approximations are best for x close to zero. They are based on the use of Taylor series and will be studied in greater depth in Chapter 11.

The following formulas are somewhat simpler in case a reciprocal function $1/x$ is available.

(66) $\quad \sin x \approx ((x^2/20 + 1)^{-1} * 10 - 7)x/3 \quad \left(-\dfrac{\pi}{4} \le x \le \dfrac{\pi}{4}\right)$

(67) $\quad \cos x \approx ((x^2/30 + 1)^{-1} * 5 - 3)x^2/(-4) + 1 \quad \left(-\dfrac{\pi}{4} \le x \le \dfrac{\pi}{4}\right)$

(68) $\quad \tan x \approx (((-2)x^2/5 + 1)^{-1} * 5 + 1)x/6 \quad \left(-\dfrac{\pi}{4} \le x \le \dfrac{\pi}{4}\right)$

(69) $\quad \arcsin x \approx (((-9)x^2/20 + 1)^{-1} * 10 + 17)x/27 \quad \left(0 \le x \le \dfrac{1}{2}\right)$

(70) $\quad \arctan x \approx ((3x^2/5 + 1)^{-1} * 5 + 4)x/9 \quad \left(0 \le x \le \dfrac{1}{2}\right)$

For other values of x, Formulas (60) through (70) should be used together with the following identities.

$$\sin x = \cos\left(\dfrac{\pi}{2} - x\right) \qquad \cos x = \sin\left(\dfrac{\pi}{2} - x\right)$$

$$\arcsin x = \dfrac{\pi - 4\arcsin\sqrt{(1-x)/2}}{2}$$

$$\arccos x = \arcsin\sqrt{1 - x^2}$$

$$\arctan x = \arctan\left(\dfrac{2x - 1}{2 + x}\right) + \arctan\left(\dfrac{1}{2}\right)$$

The following constants may be useful.

$$\pi \approx 3.141592658979$$

$$\arctan\left(\dfrac{1}{2}\right) \approx 0.4636476$$

PROBLEMS

Use Formulas (60) through (70) to estimate the following expressions.

1. $\sin(0.1)$
2. $\cos(0.25)$
3. $\arctan(0.1)$
4. $\sec(0.15)$
5. $\sin(85°)$
6. $\cos[\cos(0.223)]$
7. $\tan(10°)$
8. $\arcsin(0.234)$
9. $\arcsin[\cos(0.45)]$
10. $\tan[\arcsin(0.56)]$
11. Compute $\cos(0.1)$ and $\sin(0.1)$; then compute $\cos^2(0.1) + \sin^2(0.1)$.
12. Repeat Problem 11 with $\cos(1)$ and $\sin(1)$.

REVIEW SECTION

Many definitions were given and many identities developed in this chapter. Familiarity with the following list will make life much easier during Chapter 8.

(a) $\sin(-x) = -\sin x$

(b) $\cos(-x) = \cos x$

(c) $\sin^2 x + \cos^2 x = 1$

(d) $\cos(a \pm b) = \cos a \cos b \mp \sin a \sin b$

(e) $\sin(a \pm b) = \sin a \cos b \pm \cos a \sin b$

(f) $\tan x = \dfrac{\sin x}{\cos x}$

(g) $\cot x = \dfrac{\cos x}{\sin x}$

(h) $\sec x = \dfrac{1}{\cos x}$

(i) $\csc x = \dfrac{1}{\sin x}$

(j) $\tan^2 x + 1 = \sec^2 x$

(k) $D \sin x = \cos x$

(l) $D \cos x = -\sin x$

(m) $D \tan x = \sec^2 x$

(n) $D \cot x = -\csc^2 x$

(o) $D \sec x = \sec x \tan x$

(p) $D \csc x = -\csc x \cot x$

(q) $D \arcsin x = \dfrac{1}{\sqrt{1-x^2}}$

(r) $D \arctan x = \dfrac{1}{1+x^2}$

(s) $D \operatorname{arcsec} x = \dfrac{1}{x\sqrt{x^2-1}}$

REVIEW PROBLEMS

Graph each of the following functions.

1. $f(x) = \sin x + \cos x$
2. $y = \sin x + \cos 2x$
3. $x \to \sin(\ln x)$
4. $y = \ln |\sin x|$
5. $f(x) = e^{\sin x}$
6. $f(x) = \sin(e^x)$
7. $f(x) = \tan x + \cot x$
8. $y = \arctan(e^x)$

Compute each of the following integrals.

9. $\displaystyle\int x \sin(x^2)\, dx$

10. $\displaystyle\int_0^1 e^x \cos(e^x)\, dx$

11. $\displaystyle\int x \sin(x^2) \cos(x^2)\, dx$

12. $\displaystyle\int e^{\cos x} \sin x\, dx$

13. $\displaystyle\int \dfrac{\tan \sqrt{t}}{\sqrt{t}}\, dt$

14. $\displaystyle\int \dfrac{dx}{\sqrt{x} \cos^2(\sqrt{x})}$

15. $\displaystyle\int x \sec(x^2+1) \tan(x^2+1)\, dx$

16. $\displaystyle\int \dfrac{\cos x \ln |\sin x|}{\sin x}\, dx$

17. $\displaystyle\int \dfrac{x \sin(x^2)}{\cos(x^2)}\, dx$

18. $\displaystyle\int \dfrac{dx}{\sqrt{e^{2x}-1}}$

19. $\displaystyle\int_0^1 \dfrac{e^x\, dx}{e^{2x}+1}$

20. $\displaystyle\int \dfrac{dx}{\cos x \sqrt{1-\sin^2 x}}$

21. $\displaystyle\int \dfrac{dx}{9x^2+16}$

22. $\displaystyle\int \dfrac{dx}{\sqrt{4-9x^2}}$

23. $\displaystyle\int \cos^2 x\, dx \quad \left[\text{Hint: } \cos^2 x = \dfrac{1}{2}(1+\cos 2x).\right]$

24. Show that one solution to the differential equation $y'' - 6y' + 10y = 0$ is given by $y = e^{3x} \sin x$.

25. A hanging cable, suspended between two points, satisfies the differential equation $\dfrac{d^2y}{dx^2} = a\sqrt{1+\left(\dfrac{dy}{dx}\right)^2}$. Show that $y = a \cosh(x/a)$ is a solution to this differential equation.

The equation of a hanging cable, which appears at first to look like a parabola, is actually an equation of the form $y = a \cosh(x/a)$, and it is called a *catenary* (after *catena*, Latin for chain).

26. Let $z = dy/dx$ and use separation of variables (twice) to solve the differential equation
$$\frac{d^2y}{dx^2} = a\sqrt{1 + \left(\frac{dy}{dx}\right)^2}$$
if $dy/dx = 0$ when $x = 0$ and $y = a$ at $x = 0$.

CHAPTER 8

METHODS OF INTEGRATION

In Chapter 1 methods were given to approximate the definite integral of any continuous function to any desired degree of accuracy. Since then techniques have been developed for computing some of these integrals exactly. In this chapter we will discover new techniques and revisit the old.

These methods for evaluating integrals will allow us to compute some integrals that we could not handle previously. There will always remain, however, a large collection of functions that can never be integrated exactly. It is for this reason that it is necessary to have techniques available, such as the trapezoidal rule and Simpson's rule (both of which are discussed in this chapter), which will allow us to approximate closely the definite integral of any continuous function. On the other hand, if we have a choice between Simpson's rule and a method that will yield an exact answer, the latter method will usually be preferred.

1. SUBSTITUTION

Many integrals can be found by *guessing*. Recall that this process involves making a thoughtful guess and then checking the guess by differentiating. The next most widely used method is **substitution**. We begin by looking at some examples.

EXAMPLE 1. Compute $\int_{-1}^{3} x\sqrt{x+1}\,dx$.

Solution. As a rule of thumb, the first substitution to try when square roots are involved is the *rationalizing substitution*. Let $u = \sqrt{x+1}$ so that $u^2 = x + 1$. Then $2u\,du = dx$, and hence

$$\int_{-1}^{3} x\sqrt{x+1}\,dx = \int_{x=-1}^{x=3} (u^2 - 1)u2u\,du$$

$$= \int_{x=-1}^{x=3} (2u^4 - 2u^2)\,du$$

$$= \frac{2}{5}u^5 - \frac{2}{3}u^3 \Big|_{x=-1}^{x=3}$$

$$= \frac{2}{5}(x+1)^{5/2} - \frac{2}{3}(x+1)^{3/2} \Big|_{-1}^{3}$$

$$= \frac{2}{5}(32 - 0) - \frac{2}{3}(8 - 0)$$

$$= \frac{112}{15}.$$

When using substitution to compute definite integrals, the following method may be used to find new limits of integration. If a substitution $u = g(x)$ is used to change $\int_a^b f(x)\,dx$ into some integral $\int_{x=a}^{x=b} h(u)\,du$, then $\int_{x=a}^{x=b} h(u)\,du = \int_{g(a)}^{g(b)} h(u)\,du$. In other words, to find the new limits of integration, it is sufficient to compute the value of $u = g(x)$ at $x = a$ and $x = b$. To see how this works, we give another solution to Example 1.

Alternate Solution. Let $u = g(x) = \sqrt{x+1}$, so that $u^2 = x + 1$, $2u\,du = dx$ and $x = u^2 - 1$ as before. At $x = -1$, $g(-1) = 0$; and at $x = 3$, $g(3) = 2$. Hence

$$\int_{-1}^{3} x\sqrt{x+1}\,dx = \int_{g(-1)}^{g(3)} (u^2 - 1)u2u\,du$$

$$= 2\int_0^2 (u^4 - u^2)\,du$$

$$= 2\left(\frac{u^5}{5} - \frac{u^3}{3}\right)\Big|_0^2$$

$$= 2\left(\frac{32}{5} - \frac{8}{3}\right)$$

$$= \frac{112}{15}.$$

Notice that the computations were simplified in the alternate solution because there was no need to convert the antiderivative back to a function of x. This fact is particularly useful if several substitutions need to be made in order to compute one integral.

EXAMPLE 2. Compute $\displaystyle\int_0^8 \frac{x\,dx}{1+\sqrt[3]{x}}$.

Solution. We try the rationalizing substitution $u = \sqrt[3]{x} = g(x)$, so that $u^3 = x$, $3u^2\,du = dx$, $g(0) = 0$, and $g(8) = 2$. It follows that

$$\int_0^8 \frac{x\,dx}{1+\sqrt[3]{x}} = \int_{g(0)}^{g(8)} \frac{u^3 3u^2\,du}{1+u}$$

$$= 3\int_0^2 \frac{u^5}{1+u}\,du.$$

Long division yields

$$\frac{u^5}{1+u} = u^4 - u^3 + u^2 - u + 1 - \frac{1}{1+u},$$

so that

$$\int_0^8 \frac{x\,dx}{1+\sqrt[3]{x}} = 3\int_0^2 \left(u^4 - u^3 + u^2 - u + 1 - \frac{1}{1+u}\right)du$$

$$= 3\left(\frac{u^5}{5} - \frac{u^4}{4} + \frac{u^3}{3} - \frac{u^2}{2} + u - \ln(1+u)\right)\bigg|_0^2$$

$$= 3\left(\frac{116}{5} - \ln 3\right)$$

$$\approx 66.30417.$$

EXAMPLE 3. Compute $\displaystyle\int_0^1 \sqrt{1-x^2}\,dx$.

Solution. We again try the rationalizing substitution $u = \sqrt{1-x^2}$, so that $u^2 = 1 - x^2$, $x = \sqrt{1-u^2}$, and $u\,du = -x\,dx$. Then

$$\int_0^1 \sqrt{1-x^2}\,dx = \int_0^1 \frac{\sqrt{1-x^2}\,x\,dx}{x}$$

$$= -\int_1^0 \frac{u^2\,du}{\sqrt{1-u^2}},$$

which is more complicated than the integral we started with.

There is a substitution, however, that will work. Let $x = \sin u$. Then $dx = \cos u\,du$, and hence

$$\int_0^1 \sqrt{1-x^2}\,dx = \int_0^{\pi/2} \sqrt{1-\sin^2 u}\,\cos u\,du$$

$$= \int_0^{\pi/2} \cos^2 u\,du.$$

Before we can compute this integral, we need to look at some trigonometric identities. Recall that

$$\cos 2A = \cos^2 A - \sin^2 A$$
$$= \cos^2 A - 1 + \cos^2 A$$
$$= 2\cos^2 A - 1,$$

and hence

$$\cos^2 A = \frac{1}{2}(1 + \cos 2A).$$

Using this identity, we obtain

$$\int_0^{\pi/2} \cos^2 u\,du = \int_0^{\pi/2} \frac{1}{2}(1 + \cos 2u)\,du$$

$$= \frac{1}{2}\int_0^{\pi/2} du + \frac{1}{2}\int_0^{\pi/2} \cos 2u\,du$$

$$= \frac{1}{2}\cdot u\,\Big|_0^{\pi/2} + \frac{1}{2}\cdot\frac{1}{2}\sin 2u\,\Big|_0^{\pi/2}$$

$$= \frac{\pi}{4} + \frac{1}{4}\sin\pi - \frac{1}{4}\sin 0$$

$$= \frac{\pi}{4}.$$

The integral $\int_0^1 \sqrt{1-x^2}\,dx$ can be interpreted as the first quadrant area of the circle $x^2 + y^2 = 1$. The area of this circle is π, so the first quadrant area is indeed $\pi/4$.

In Example 3 the expression $\sqrt{1-x^2}$ was simplified by the substitution $x = \sin u$. Table 8.1 lists other **trigonometric substitutions** that are sometimes useful. These substitutions are particularly useful where square roots such as $\sqrt{a^2 - x^2}$ are involved because the new expressions are perfect squares.

Table 8.1

Expression	Substitution	New Expression
$a^2 - b^2x^2$	$x = \dfrac{a}{b}\sin u$	$a^2 - b^2x^2 = a^2 \cos^2 u$
$b^2x^2 - a^2$	$x = \dfrac{a}{b}\sec u$	$b^2x^2 - a^2 = a^2 \tan^2 u$
$a^2 + b^2x^2$	$x = \dfrac{a}{b}\tan u$	$a^2 + b^2x^2 = a^2 \sec^2 u$

In making such substitutions, it is often necessary to simplify expressions such as $\sin(\arctan x)$ or even $\tan[\arccos(\sqrt{x^2-1}/x)]$. This step can be done most easily by thinking of the circular functions as trigonometric functions, so that

$$\sin A = \frac{\text{opposite side}}{\text{hypotenuse}},$$

$$\cos A = \frac{\text{adjacent side}}{\text{hypotenuse}},$$

and

$$\tan A = \frac{\text{opposite side}}{\text{adjacent side}},$$

as in Section 4 of Chapter 7.

To simplify $\sin(\arctan x)$, we think of $\arctan x$ as being the angle A whose tangent is $x = x/1$, as in Figure 8.1. Thus the opposite side has length x, the adjacent side has length 1, and by the Pythagorean Theorem the hypotenuse has length $\sqrt{x^2 + 1}$. Hence

$$\sin A = \frac{x}{\sqrt{x^2 + 1}}.$$

EXAMPLE 4. Compute $\displaystyle\int_0^1 \frac{dx}{\sqrt{9x^2 + 16}}$.

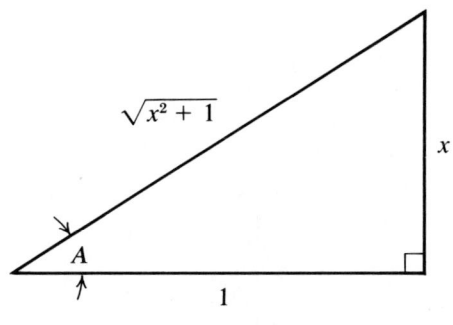

FIGURE 8.1

Solution. We let $x = \frac{4}{3} \tan u$, so that
$$\sqrt{9x^2 + 16} = \sqrt{16 \tan^2 u + 16} = 4\sqrt{1 + \tan^2 u} = 4 \sec u$$
and $dx = \frac{4}{3} \sec^2 u \, du$. Also, $u = 0$ when $x = 0$ and $u = \arctan \frac{3}{4}$ when $x = 1$. It follows that

$$\int_0^1 \frac{dx}{\sqrt{9x^2 + 16}} = \int_0^{\arctan(3/4)} \frac{(4/3) \sec^2 u \, du}{4 \sec u}$$

$$= \frac{1}{3} \int_0^{\arctan(3/4)} \sec u \, du.$$

Recall (see Example 4 of Chapter 7) that

$$\int \sec u \, du = \ln |\sec u + \tan u| + C.$$

Thus

$$\frac{1}{3} \int_0^{\arctan(3/4)} \sec u \, du = \frac{1}{3} \ln |\sec u + \tan u| \Big|_0^{\arctan 3/4}$$

$$= \frac{1}{3} \ln \left| \sec \left(\arctan \frac{3}{4} \right) + \tan \left(\arctan \frac{3}{4} \right) \right|$$

$$- \frac{1}{3} \ln |\sec (0) + \tan (0)|$$

$$= \frac{1}{3} \ln \left(\frac{5}{4} + \frac{3}{4} \right)$$

$$= \frac{1}{3} \ln 2$$

$$\approx 0.23105.$$

EXAMPLE 5. Compute $\int_0^1 \frac{dx}{(1+x^2)^2}$.

Solution. Let $x = \tan u$, so that $dx = \sec^2 u\, du$. Then

$$\int_0^1 \frac{dx}{(1+x^2)^2} = \int_0^{\pi/4} \frac{\sec^2 u\, du}{(1+\tan^2 u)^2}$$

$$= \int_0^{\pi/4} \frac{\sec^2 u\, du}{\sec^4 u}$$

$$= \int_0^{\pi/4} \cos^2 u\, du$$

$$= \frac{1}{2}\int_0^{\pi/4} (1 + \cos 2u)\, du$$

$$= \frac{u}{2} + \frac{1}{4}\sin 2u \Big|_0^{\pi/4}$$

$$= \frac{\pi}{8} + \frac{1}{4}$$

$$\approx 0.6427.$$

EXAMPLE 6. Compute $\int_0^{1/2} \frac{x\, dx}{\sqrt{1-x^2}}$.

Solution. Let $x = \sin u$, so that $dx = \cos u\, du$. Then

$$\int_0^{1/2} \frac{x\, dx}{\sqrt{1-x^2}} = \int_0^{\pi/6} \frac{\sin u}{\sqrt{1-\sin^2 u}} \cos u\, du$$

$$= \int_0^{\pi/6} \sin u\, du$$

$$= -\cos u \Big|_0^{\pi/6}$$

$$= -\frac{\sqrt{3}}{2} + 1$$

$$\approx 0.133975.$$

Notice that in Example 6 the substitution $u = 1 - x^2$ would also lead to a solution.

SECTION 1: SUBSTITUTION   265

EXAMPLE 7.   Find an antiderivative of $f(x) = \dfrac{1}{x(x^2 - 1)}$.

Solution.   Let $x = \sec u$. Then $dx = \sec u \tan u \, du$, and hence

$$\int \frac{dx}{x(x^2 - 1)} = \int \frac{\sec u \tan u}{\sec u (\sec^2 u - 1)} du$$

$$= \int \cot u \, du$$

$$= \int \frac{\cos u}{\sin u} du.$$

A second substitution is required. Let $v = \sin u$, which implies $dv = \cos u \, du$. Then

$$\int \frac{\cos u}{\sin u} du = \int \frac{dv}{v}$$

$$= \ln |v| + C$$

$$= \ln |\sin u| + C$$

$$= \ln |\sin (\text{arcsec } x)| + C$$

$$= \ln \left| \frac{\sqrt{x^2 - 1}}{x} \right| + C.$$

EXAMPLE 8.   Compute $\displaystyle\int_0^{\arctan 2} \frac{\sec^2 x \, dx}{4 + \tan^2 x}$.

Solution.   Let $\tan x = 2u$. Then $\sec^2 x \, dx = 2 \, du$, so that

$$\int_0^{\arctan 2} \frac{\sec^2 x \, dx}{4 + \tan^2 x} = \int_0^1 \frac{2 \, du}{4 + 4u^2}$$

$$= \frac{1}{2} \int_0^1 \frac{du}{1 + u^2}$$

$$= \frac{1}{2} \arctan u \bigg|_0^1$$

$$= \frac{\pi}{8}.$$

PROBLEMS

1. Rewrite each of the following integrals by making a rationalizing substitution.

   a. $\int (x+2)\sqrt{x+3}\, dx$
   b. $\int \dfrac{x\, dx}{\sqrt{x+1}}$
   c. $\int_0^1 x\sqrt{x^2+1}\, dx$
   d. $\int_{-1}^7 x\sqrt[3]{x+1}\, dx$
   e. $\int x^3 \sqrt{x^2-4}\, dx$
   f. $\int x^2 \sqrt[3]{x+5}\, dx$

2. Rewrite each of the following integrals by making a trigonometric substitution.

   a. $\int_0^1 x\sqrt{x^2+1}\, dx$
   b. $\int_2^{2\sqrt{2}} x^3\, dx$
   c. $\int_{\sqrt{3}}^3 \dfrac{\sqrt{9-x^2}}{x}\, dx$
   d. $\int \dfrac{dx}{(x^2+1)^{3/2}}$
   e. $\int \dfrac{dx}{x^2 \sqrt{9+4x^2}}$
   f. $\int \dfrac{dx}{\sqrt{e^{2x}-1}}$

Use the method of substitution to compute each of the following integrals.

3. $\int x^3 \sqrt{x^2-4}\, dx$

4. $\int_{\sqrt{3}}^3 \dfrac{\sqrt{9-x^2}}{x}\, dx$

5. $\int \dfrac{dx}{x^2 \sqrt{9+4x^2}}$

6. $\int \dfrac{dx}{\sqrt{e^{2x}-1}}$

7. $\int \dfrac{dx}{1+\cos x}$ (Let $x = 2u$.)

8. $\int \dfrac{dx}{1-\sin x}$

9. $\int \dfrac{x^2\, dx}{(4+x^2)^2}$

10. $\int \dfrac{dx}{x\sqrt{9x^2-1}}$

11. $\int \dfrac{dx}{\sqrt{9-4x^2}}$

12. $\int \dfrac{dx}{25+16x^2}$

13. $\int_0^{3/2} \dfrac{dx}{4x^2+9}$

14. $\int \dfrac{dx}{x-\sqrt{x}}$

15. $\int_{5/4}^{13/5} \dfrac{dx}{(x^2-1)^{3/2}}$

16. $\int_0^{\pi/2} \sin^3 x\, dx$

17. $\int_0^1 x\sqrt{16+9x^2}\, dx$

18. $\int_0^\pi \sin^4 x\, dx$

19. $\int_0^3 \dfrac{x}{\sqrt{x+1}}\, dx$

20. $\int_0^{\pi/2} (1-\sin^2 x)^{3/2}\, dx$

21. $\int_0^{\pi/2} \sqrt{1+\cos 2x}\, dx$

22. $\int_0^1 (1-x^2)^{3/2}\, dx$

23. $\int_0^{1/2} \dfrac{dx}{(1-x^2)^{3/2}}$

24. $\int_0^1 \dfrac{dx}{(x^2+1)^{5/2}}$

25. $\int_4^{25} \dfrac{dx}{\sqrt{x}\sqrt{\sqrt{x}-1}}$

26. $\int_0^1 \dfrac{\sqrt{x}}{x+1}\, dx$

27. $\int_0^{(1/2)\ln 3} \dfrac{e^x\, dx}{e^{2x}+1}$

28. $\int_0^{\pi/3} \tan^2 x\, dx$

29. $\int_1^5 \dfrac{\sqrt{x-1}}{x+1}\, dx$

30. $\int_0^1 \dfrac{\sqrt{x}\, dx}{(16x+9)^2}$

31. $\int_0^{81} \sqrt{16+\sqrt{x}}\, dx$

32. $\int \dfrac{dx}{e^x-1}$

33. Show that $\int \csc x \, dx = \ln |\tan (x/2)| + C$ by using the identity $\sin x = 2 \sin (x/2) \cos (x/2)$.
34. Show that $-\ln |\cot x + \csc x| = \ln |\tan(x/2)| + C$ for some constant C; then evaluate the constant.

2. INTEGRATION BY PARTS

A third useful technique of integration (after guessing and substitution) is called **integration by parts**. The method of substitution is a consequence of the composition rule for derivatives. The product rule for derivatives yields the integration-by-parts formula. In differential notation, the product rule reads:

$$d(uv) = u \, dv + v \, du.$$

Taking antiderivatives on both sides, we obtain

$$uv = \int d(uv) = \int u \, dv + \int v \, du$$

or solving for $\int u \, dv$ yields

(1) $$\int u \, dv = uv - \int v \, du.$$

By picking u and v cleverly, we will find this formula very useful.

EXAMPLE 9. Compute $\int \ln x \, dx$.

Solution. Let $u = \ln x$, $dv = dx$. Then $du = (1/x) \, dx$ and $v = x$. (There is no need to worry about constants of integration until the very end. See Problem 27.) Using the integration-by-parts formula, we get

$$\int \ln x \, dx = \int u \, dv = uv - \int v \, du$$

$$= x \ln x - \int x \cdot \frac{1}{x} \, dx$$

$$= x \ln x - \int dx$$

$$= x \ln x - x + C$$

$$= x (\ln x - 1) + C.$$

EXAMPLE 10. Compute $\int x \ln x \, dx$.

Solution. Let $u = \ln x$ and $dv = x \, dx$. Then $du = (1/x) \, dx$ and $v = x^2/2$, so that

$$\int x \ln x \, dx = \frac{x^2}{2} \ln x - \int \frac{x^2}{2} \cdot \frac{1}{x} \, dx$$

$$= \frac{x^2}{2} \ln x - \frac{1}{2} \int x \, dx$$

$$= \frac{x^2}{2} \ln x - \frac{1}{2} \cdot \frac{x^2}{2} + C$$

$$= \frac{x^2}{2} \left(\ln x - \frac{1}{2} \right) + C.$$

There are no hard and fast rules for picking u and v. The basic goal is to obtain a new integral that is easy to evaluate. It follows that dv should be chosen so that v is simple, and u should be a function that has a simple derivative. In Example 10 the choice $u = \ln x$ was made because $\ln x$ has a simple derivative. Another rough guideline is to let dv be the largest part that can be easily integrated.

EXAMPLE 11. Compute $\int x^2 e^x \, dx$.

Solution. Let $u = x^2$, $dv = e^x \, dx$. Then $du = 2x \, dx$ and $v = e^x$. Hence

$$\int x^2 e^x \, dx = x^2 e^x - \int 2x e^x \, dx$$

$$= x^2 e^x - 2 \int x e^x \, dx.$$

We must integrate by parts again. This time we let $u = x$ and $dv = e^x \, dx$. Then $du = dx$ and $v = e^x$, so that

$$\int x^2 e^x \, dx = x^2 e^x - 2 \int x e^x \, dx$$

$$= x^2 e^x - 2 \left(x e^x - \int e^x \, dx \right)$$

$$= x^2 e^x - 2x e^x + 2 e^x + C$$

$$= e^x (x^2 - 2x + 2) + C.$$

SECTION 2: INTEGRATION BY PARTS

EXAMPLE 12. Compute $\int e^x \sin x\, dx$.

Solution. The functions e^x and $\sin x$ can be integrated and differentiated with equal ease. Let $u = \sin x$ and $dv = e^x\, dx$. (However, the choices $u = e^x$ and $dv = \sin x\, dx$ would work just as well. See Problem 25.) Then $du = \cos x\, dx$ and $v = e^x$, so

$$\int e^x \sin x\, dx = e^x \sin x - \int e^x \cos x\, dx.$$

We use integration by parts again, this time with $u = \cos x$ and $dv = e^x\, dx$. (Notice that $u = e^x$ and $dv = \cos x\, dx$ will *not* work here. See Problem 26.) Then $du = -\sin x\, dx$ and $v = e^x$, which leads to

$$\int e^x \sin x\, dx = e^x \sin x - \left[e^x \cos x - \int e^x(-\sin x)\, dx \right]$$

$$= e^x \sin x - e^x \cos x - \int e^x \sin x\, dx.$$

At first glance it may seem as if we are no better off than before, since we still have the original integral as a term on the right. However, there is a minus sign in front of the integral, so by taking this integral to the left-hand side of the equation we obtain

$$2\int e^x \sin x\, dx = e^x \sin x - e^x \cos x + C_1,$$

where C_1 is an arbitrary constant. Dividing both sides by 2, we observe that

$$\int e^x \sin x\, dx = \frac{1}{2} e^x (\sin x - \cos x) + C,$$

where $C = C_1/2$.

EXAMPLE 13. Compute $\int \sec x \tan^2 x\, dx$.

Solution. Guessing and substitution do not seem to work, so we try integration by parts. (This method also may not work.) Let $u = \tan x$ and $dv = \sec x \tan x\, dx$. Then $du = \sec^2 x\, dx$ and $v = \sec x$. Hence

$$\int \sec x \tan^2 x\, dx = \sec x \tan x - \int \sec^3 x\, dx$$

$$= \sec x \tan x - \int \sec x (1 + \tan^2 x)\, dx$$

$$= \sec x \tan x - \int \sec x \, dx - \int \sec x \tan^2 x \, dx$$

$$= \sec x \tan x - \ln |\sec x + \tan x| - \int \sec x \tan^2 x \, dx.$$

As in Example 12, we can now solve this equation for the original integral to get

$$\int \sec x \tan^2 x \, dx = \frac{1}{2} (\sec x \tan x - \ln |\sec x + \tan x|) + C.$$

The integration-by-parts formula also applies to definite integrals. Given $u(x)$ and $v(x)$, we have $du = u'(x) \, dx$ and $dv = v'(x) \, dx$. Hence

(2) $$\int_a^b u(x) v'(x) \, dx = u(x) v(x) \Big|_a^b - \int_a^b v(x) u'(x) \, dx.$$

EXAMPLE 14. Compute $\int_0^1 x e^{-x} \, dx$.

Solution. Let $u(x) = x$ and $v'(x) = e^{-x}$, so that $u'(x) = 1$ and $v(x) = -e^{-x}$. Then

$$\int_0^1 x e^{-x} \, dx = \int_0^1 u(x) v'(x) \, dx$$

$$= u(x) v(x) \Big|_0^1 - \int_0^1 v(x) u'(x) \, dx$$

$$= -x e^{-x} \Big|_0^1 + \int_0^1 e^{-x} \, dx$$

$$= -e^{-1} + (-e^{-x}) \Big|_0^1$$

$$= -e^{-1} - e^{-1} + 1$$

$$= \frac{e - 2}{e}$$

$$\approx 0.2642411.$$

PROBLEMS

Compute each of the following integrals.

1. $\int x^2 \ln x \, dx$.

2. $\int (\ln x)^2 \, dx$

3. $\int_0^1 xe^{2x}\,dx$

4. $\int_0^1 \arcsin x\,dx$

5. $\int \sqrt{x}\ln x\,dx$

6. $\int x^3\sqrt{x^2+1}\,dx$
(Let $dv = x\sqrt{x^2+1}\,dx$.)

7. $\int \sin(\ln x)\,dx$

8. $\int_0^1 \arctan x\,dx$

9. $\int x^3 e^{-2x}\,dx$

10. $\int x^3 e^{x^2}\,dx$
(Let $dv = xe^{x^2}\,dx$.)

11. $\int \dfrac{x^3\,dx}{9+x^4}$

12. $\int_0^1 \dfrac{x^3\,dx}{1+x^2}$

13. $\int \dfrac{x^3\,dx}{\sqrt{x^2+1}}$

14. $\int x\sec x\tan x\,dx$

15. $\int x^9\sqrt{x^5+1}\,dx$

16. $\int x\,\text{arcsec}\,x\,dx$

17. $\int x^2 \ln x^2\,dx$

18. $\int \dfrac{\ln x}{x^2}\,dx$

19. $\int (\arcsin x)^2\,dx$

20. $\int x\tan^2 x\,dx$

21. $\int x\arctan x\,dx$

22. $\int_0^{\pi/4} \sec^3 x\,dx$

23. $\int \ln(x^2+1)\,dx$

24. $\int x^n \ln x\,dx$, n a positive integer

25. Compute the integral $\int e^x \sin x\,dx$ of Example 12 by first using the substitution $u = e^x$ and $dv = \sin x\,dx$ to get $\int e^x \sin x\,dx = e^x(-\cos x) - \int e^x(-\cos x)\,dx$.

26. What happens in Example 12 if the substitutions $u = e^x$ and $dv = \cos x\,dx$ are used in attempting to evaluate $\int e^x \sin x\,dx = e^x \sin x - \int e^x \cos x\,dx$?

27. Rework Example 9 with $v = x + a$ for some constant a. Show that the final answer differs from $x(\ln x - 1) + C$ by a constant.

28. Compute the volume generated by rotating the graph of $y = xe^x$ about the x-axis, $0 \le x \le 1$.

3. COMPLETING THE SQUARE

Tables of integration formulas can be a wonderful tool of assistance, but usually some work is still required before the integral is in a form that will appear in a standard table of integrals. This work often involves completing a square or making a substitution.

272 CHAPTER 8: METHODS OF INTEGRATION

EXAMPLE 15. Compute $\int \dfrac{dx}{\sqrt{x^2 + 2x}}$.

Solution. Since $x^2 + 2x = x^2 + 2x + 1 - 1 = (x + 1)^2 - 1$, we see that

$$\int \frac{dx}{\sqrt{x^2 + 2x}} = \int \frac{dx}{\sqrt{(x + 1)^2 - 1}}.$$

Let $u = x + 1$ so that $du = dx$. Then

$$\int \frac{dx}{\sqrt{x^2 + 2x}} = \int \frac{du}{\sqrt{u^2 - 1}}.$$

We can now use the trigonometric substitution $u = \sec t$, $du = \sec t \tan t\, dt$:

$$\int \frac{dx}{\sqrt{x^2 + 2x}} = \int \frac{\sec t \tan t\, dt}{\tan t}$$

$$= \int \sec t\, dt$$

$$= \ln|\sec t + \tan t| + C$$

$$= \ln|u + \sqrt{u^2 - 1}| + C$$

$$= \ln|x + 1 + \sqrt{x^2 + 2x}| + C.$$

Notice that in order to compute the integral

$$\int \frac{dx}{\sqrt{x^2 + 2x}},$$

it was necessary to rewrite $x^2 + 2x$ as $(x + 1)^2 - 1$. This step is called *completing the square*. In general,

(3) $\qquad ax^2 + bx + c = a\left(x^2 + \dfrac{b}{a}x + \dfrac{b^2}{4a^2}\right) - \dfrac{b^2}{4a} + c$

$$= a\left[\left(x + \frac{b}{2a}\right)^2 + \frac{1}{a}\left(c - \frac{b^2}{4a}\right)\right].$$

To complete the square, factor out the coefficient of x^2; then add and subtract the square of half the coefficient of x. Thus

$$2x^2 + 3x - 1 = 2\left(x^2 + \frac{3}{2}x + \frac{9}{16}\right) - \frac{9}{8} - 1$$

SECTION 3: COMPLETING THE SQUARE

$$= 2\left(x + \frac{3}{4}\right)^2 - \frac{17}{8}$$

$$= 2\left[\left(x + \frac{3}{4}\right)^2 - \frac{17}{16}\right],$$

and

$$x^2 + 4x = x^2 + 4x + 4 - 4$$
$$= (x + 2)^2 - 4.$$

EXAMPLE 16. Compute $\displaystyle\int_{-2}^{-1} \frac{dx}{x^2 + 4x + 5}$.

Solution.

$$\int_{-2}^{-1} \frac{dx}{x^2 + 4x + 5} = \int_{-2}^{-1} \frac{dx}{x^2 + 4x + 4 + 1}$$

$$= \int_{-2}^{-1} \frac{dx}{(x + 2)^2 + 1}$$

$$= \int_{0}^{1} \frac{du}{u^2 + 1}$$

$$= \arctan u \Big|_0^1$$

$$= \frac{\pi}{4}$$

$$\approx 0.7854.$$

EXAMPLE 17. Compute $\displaystyle\int_0^1 \frac{2\,dx}{4x^2 + 8x + 3}$.

Solution. Completing the square, we obtain

$$\int_0^1 \frac{2\,dx}{4x^2 + 8x + 3} = \int_0^1 \frac{2\,dx}{4(x^2 + 2x + 1) - 1}$$

$$= \int_0^1 \frac{2\,dx}{4(x + 1)^2 - 1}.$$

Let $u = 2(x + 1)$. Then $du = 2\,dx$, and hence

$$\int_0^1 \frac{2\,dx}{4x^2 + 8x + 3} = \int_2^4 \frac{du}{u^2 - 1}.$$

Now let $u = \sec w$, so that $du = \sec w \tan w \, dw$. Then

$$\int_0^1 \frac{2 \, dx}{4x^2 + 8x + 3} = \int_{\text{arcsec } 2}^{\text{arcsec } 4} \frac{\tan w \sec w \, dw}{\tan^2 w}$$

$$= \int_{\text{arcsec } 2}^{\text{arcsec } 4} \csc w \, dw$$

$$= -\ln |\csc w + \cot w| \Big|_{\text{arcsec } 2}^{\text{arcsec } 4}$$

$$= -\ln \left| \frac{4}{\sqrt{15}} + \frac{1}{\sqrt{15}} \right| + \ln \left| \frac{2}{\sqrt{3}} + \frac{1}{\sqrt{3}} \right|$$

$$= -\ln 5 + \frac{1}{2} \ln 15 + \ln 3 - \frac{1}{2} \ln 3$$

$$= -\ln 5 + \frac{1}{2} \ln 3 + \frac{1}{2} \ln 5 + \ln 3 - \frac{1}{2} \ln 3$$

$$= \ln 3 - \frac{1}{2} \ln 5$$

$$\approx 0.29389.$$

PROBLEMS

Compute the following integrals.

1. $\displaystyle\int_1^{1+(\sqrt{2}/2)} \frac{dx}{\sqrt{2x - x^2}}$

2. $\displaystyle\int \frac{2 \, dx}{(2x + 3)\sqrt{4x^2 + 12x + 8}}$

3. $\displaystyle\int \frac{dx}{\sqrt{x}\sqrt{1 - x}}$

4. $\displaystyle\int \frac{dx}{\sqrt{x(1 + x)}}$

5. $\displaystyle\int_{-1}^1 (x + 1)\sqrt{x^2 + 2x + 2} \, dx$

6. $\displaystyle\int \frac{dx}{(3x + 6)\sqrt{9x^2 + 36x + 32}}$

7. $\displaystyle\int \sqrt{x^2 + 4x} \, dx$

8. $\displaystyle\int_1^3 \frac{3 \, dx}{x^2 + 6x + 10}$

9. $\displaystyle\int_0^4 \sqrt{4x - x^2} \, dx$

10. $\displaystyle\int \frac{dx}{2x^2 + 8x}$

11. $\displaystyle\int_0^2 (x^2 + 6x)^{3/2} \, dx$

12. $\displaystyle\int \frac{dx}{(x^2 + 4x + 5)^{3/2}}$

4. PARTIAL FRACTIONS

A simpler solution to Example 17 is possible.

Solution. Notice that

$$\frac{2}{4x^2 + 8x + 3} = \frac{1}{2x + 1} - \frac{1}{2x + 3}.$$

Hence

$$\int_0^1 \frac{2\,dx}{4x^2 + 8x + 3} = \int_0^1 \frac{dx}{2x + 1} - \int_0^1 \frac{dx}{2x + 3}$$

$$= \frac{1}{2}\ln(2x+1)\Big|_0^1 - \frac{1}{2}\ln(2x+3)\Big|_0^1$$

$$= \frac{1}{2}\ln\left(\frac{2x+1}{2x+3}\right)\Big|_0^1$$

$$= \frac{1}{2}\left(\ln\frac{3}{5} - \ln\frac{1}{3}\right)$$

$$= \frac{1}{2}(\ln 3 - \ln 5 + \ln 3)$$

$$= \ln 3 - \frac{1}{2}\ln 5$$

$$\approx 0.29389.$$

This method works fine if it is known that

$$\frac{2}{4x^2 + 8x + 3} = \frac{1}{2x + 1} - \frac{1}{2x + 3}.$$

Discovering such identities can be done by a technique known as *partial fractions*. The purpose of this technique is to replace a quotient of polynomials with a sum of simpler functions. Partial fractions, roughly speaking, is merely the process of undoing the idea of least common denominator.

EXAMPLE 18. Rewrite $\dfrac{x^3 + 2}{x^2 - x}$ as a sum of simpler functions.

Solution 1. Since degree $(x^3 + 2) \geq$ degree $(x^2 - x)$, we begin by using long division to rewrite the quotient. The computation

$$
\begin{array}{r}
x + 1 \\
x^2 - x \overline{\smash{\big)} x^3 + 2} \\
\underline{x^3 - x^2 } \\
x^2 \\
\underline{x^2 - x} \\
x + 2
\end{array}
$$

leads to

$$\frac{x^3 + 2}{x^2 - x} = x + 1 + \frac{x + 2}{x^2 - x}.$$

The denominator $x^2 - x$ factors into the product $x(x - 1)$, and neither x nor $x - 1$ can be further factored. Is it possible to further simplify the expression on the right? In particular, is it possible to find real numbers A and B such that

$$\frac{x + 2}{x^2 - x} = \frac{A}{x} + \frac{B}{x - 1}?$$

Multiplying both sides by x, we get

$$\frac{x + 2}{x - 1} = A + \frac{Bx}{x - 1},$$

and evaluation at $x = 0$ yields

$$A = -2.$$

Now multiplying both sides by $x - 1$, we get

$$\frac{x + 2}{x} = \frac{A(x - 1)}{x} + B,$$

and by evaluating both sides at $x = 1$, we get

$$B = 3.$$

This leads to

$$\frac{x^3 + 2}{x^2 - x} = x + 1 - \frac{2}{x} + \frac{3}{x - 1}.$$

Checking, we have

$$x + 1 - \frac{2}{x} + \frac{3}{x-1} = \frac{(x+1)x(x-1) - 2(x-1) + 3x}{x(x-1)}$$

$$= \frac{x^3 - x - 2x + 2 + 3x}{x(x-1)}$$

$$= \frac{x^3 + 2}{x^2 - x}.$$

This method works whenever the denominator factors as a product of distinct linear factors. (A **linear** factor is of the form $ax + b$.) A second method of solution is also possible.

Solution 2. We again wish to find A and B such that

$$\frac{x+2}{x^2-x} = \frac{A}{x} + \frac{B}{x-1}.$$

Adding the fractions on the right-hand side gives

$$\frac{x+2}{x^2-x} = \frac{A(x-1) + Bx}{x(x-1)}$$

$$= \frac{(A+B)x - A}{x^2 - x}.$$

Since the denominators are identical, it follows from the theory of equations that the numerators must also be identical. Thus

$$A + B = 1$$

and
$$-A = 2.$$

Replacing A by -2 in the first equation, we see that

$$B = 1 - (-2)$$
$$= 3.$$

Hence

$$\frac{x+2}{x^2-x} = -\frac{2}{x} + \frac{3}{x-1},$$

which agrees with Solution 1.

CHAPTER 8: METHODS OF INTEGRATION

EXAMPLE 19. Rewrite $\dfrac{x^2 + x - 2}{x^3 + x}$ as a sum.

Solution. The degree of $x^2 + x - 2$ is already less than the degree of $x^3 + x$, so there is no need for long division. The denominator factors as $x(x^2 + 1)$, and the quadratic $x^2 + 1$ cannot be further factored. (A *quadratic* is a polynomial of the form $ax^2 + bx + c$, $a \neq 0$.) Is it possible to find A and B so that

$$\frac{x^2 + x - 2}{x^3 + x} = \frac{A}{x} + \frac{B}{x^2 + 1}?$$

Adding the fractions on the right, we get

$$\frac{x^2 + x - 2}{x^3 + x} = \frac{Ax^2 + A + Bx}{x(x^2 + 1)},$$

so that

$$A = 1, \quad B = 1, \quad \text{and} \quad A = -2,$$

which is sheer nonsense. (The number A cannot simultaneously equal both 1 and -2.) We conclude that such a solution is not possible. However, if a linear factor $Bx + C$ is allowed, then

$$\frac{x^2 + x - 2}{x^3 + x} = \frac{A}{x} + \frac{Bx + C}{x^2 + 1}$$

$$= \frac{Ax^2 + A + Bx^2 + Cx}{x(x^2 + 1)}$$

$$= \frac{(A + B)x^2 + Cx + A}{x^3 + x}$$

leads to

$$A + B = 1$$
$$C = 1$$
$$A = -2,$$

and hence $A = -2$, $C = 1$, and $B = 1 - (-2) = 3$. Checking, we have

$$-\frac{2}{x} + \frac{3x + 1}{x^2 + 1} = \frac{-2x^2 - 2 + 3x^2 + x}{x(x^2 + 1)}$$

$$= \frac{x^2 + x - 2}{x^3 + x}.$$

What about fractions such as

$$\frac{3x+4}{(x-2)^2}$$

that have repeated factors in the denominator? Certainly we could write

$$\frac{3x+4}{(x-2)^2} = \frac{A}{x-2} + \frac{Bx+C}{(x-2)^2}$$

by taking $A = 0$, $B = 3$, and $C = 4$, but then we would be back where we started. How about something simpler; how about

$$\frac{3x+4}{(x-2)^2} = \frac{A}{x-2} + \frac{B}{(x-2)^2}?$$

As before, we write

$$\frac{3x+4}{(x-2)^2} = \frac{A(x-2) + B}{(x-2)^2}$$

$$= \frac{Ax + (-2A + B)}{(x-2)^2},$$

which implies $A = 3$, $-2A + B = 4$, so that $B = 4 + 6 = 10$. Checking, we get

$$\frac{3}{x-2} + \frac{10}{(x-2)^2} = \frac{3(x-2) + 10}{(x-2)^2}$$

$$= \frac{3x+4}{(x-2)^2}.$$

This result works in general. If the degree of $p(x)$ is less than n, then

(4) $$\frac{p(x)}{(ax+b)^n} = \frac{A_1}{ax+b} + \frac{A_2}{(ax+b)^2} + \frac{A_3}{(ax+b)^3} + \cdots + \frac{A_n}{(ax+b)^n}$$

for some numbers $A_1, A_2, A_3, \ldots, A_n$.

EXAMPLE 20. Use the method of partial fractions to write $\dfrac{x^4 + 2x^2}{(x+1)^3}$ as a sum.

Solution. Since the degree of the numerator exceeds the degree of the denominator, we use long division to obtain

$$\frac{x^4 + 2x^2}{(x+1)^3} = x - 3 + \frac{8x^2 + 8x + 3}{(x+1)^3}.$$

280 CHAPTER 8: METHODS OF INTEGRATION

Now by Formula (4),

$$\frac{8x^2 + 8x + 3}{(x+1)^3} = \frac{A}{x+1} + \frac{B}{(x+1)^2} + \frac{C}{(x+1)^3}$$

$$= \frac{A(x+1)^2 + B(x+1) + C}{(x+1)^3}$$

$$= \frac{Ax^2 + (2A+B)x + (A+B+C)}{(x+1)^3}$$

Comparing coefficients, we obtain

$$A + B + C = 3$$
$$2A + B \quad\quad = 8$$
$$A \quad\quad\quad\quad = 8.$$

Hence $A = 8$, $B = -8$, and $C = 3$. Thus

$$\frac{x^4 + 2x^2}{(x+1)^3} = x - 3 + \frac{8}{x+1} - \frac{8}{(x+1)^2} + \frac{3}{(x+1)^3}.$$

EXAMPLE 21. Write $\dfrac{x^3 + x^2 + 3x + 2}{(x^2+1)^2}$ as a sum.

Solution. The degree of the numerator is less than the degree of the denominator, so there is no need for long division. An equation of the form

$$\frac{x^3 + x^2 + 3x + 2}{(x^2+1)^2} = \frac{Ax + B}{x^2+1} + \frac{Cx^3 + Dx^2 + Ex + F}{(x^2+1)^2}$$

would lead to nothing new, for we could simply take $A = B = 0$, $C = D = 1$, $E = 3$, and $F = 2$. As before, we try to get by with less, say

$$\frac{x^3 + x^2 + 3x + 2}{(x^2+1)^2} = \frac{Ax + B}{x^2+1} + \frac{Cx + D}{(x^2+1)^2}.$$

Then

$$\frac{x^3 + x^2 + 3x + 2}{(x^2+1)^2} = \frac{(Ax+B)(x^2+1) + Cx + D}{(x^2+1)^2}$$

$$= \frac{Ax^3 + Bx^2 + (A+C)x + (B+D)}{(x^2+1)^2}.$$

Comparing coefficients, we see that $A = 1$, $B = 1$, $A + C = 3$, and $B + D = 2$, which leads to $A = 1$, $B = 1$, $C = 2$, and $D = 1$. Hence

$$\frac{x^3 + x^2 + 3x + 2}{(x^2+1)^2} = \frac{x+1}{x^2+1} + \frac{2x+1}{(x^2+1)^2}.$$

SECTION 4: PARTIAL FRACTIONS

In general, if $ax^2 + bx + c$ cannot be factored as a product of linear polynomials and the degree of $p(x)$ is less than $2n$, then

(5) $$\frac{p(x)}{(ax^2 + bx + c)^n} = \frac{A_1 x + B_1}{ax^2 + bx + c} + \frac{A_2 x + B_2}{(ax^2 + bx + c)^2} + \cdots$$
$$+ \frac{A_n x + B_n}{(ax^2 + bx + c)^n}$$

for some numbers $A_1, A_2, \ldots, A_n$ and $B_1, B_2, \ldots, B_n$.

EXAMPLE 22. Write $\dfrac{4x^3 + 2x^2 + 3x + 2}{x^2(x^2 + x + 1)}$ as a sum.

Solution. We must combine Formulas (4) and (5) to write

$$\frac{4x^3 + 2x^2 + 3x + 2}{x^2(x^2 + x + 1)} = \frac{A}{x} + \frac{B}{x^2} + \frac{Cx + D}{x^2 + x + 1}$$

$$= \frac{Ax(x^2 + x + 1) + B(x^2 + x + 1) + (Cx + D)x^2}{x^2(x^2 + x + 1)}$$

$$= \frac{(A + C)x^3 + (A + B + D)x^2 + (A + B)x + B}{x^2(x^2 + x + 1)}$$

Comparing coefficients, we see that

$$A \quad + C \quad \quad = 4$$
$$A + B \quad + D = 2$$
$$A + B \quad \quad = 3$$
$$B \quad \quad = 2.$$

Solving, we get $B = 2$, $A = 1$, $C = 3$, $D = -1$, and hence

$$\frac{4x^3 + 2x^2 + 3x + 2}{x^2(x^2 + x + 1)} = \frac{1}{x} + \frac{2}{x^2} + \frac{3x - 1}{x^2 + x + 1}.$$

EXAMPLE 23. Compute $\displaystyle\int \frac{4x^3 + 2x^2 + 3x + 2}{x^2(x^2 + x + 1)} dx$.

Solution. By Example 22, we have

$$\int \frac{4x^3 + 2x^2 + 3x + 2}{x^2(x^2 + x + 1)} dx = \int \frac{1}{x} dx + \int \frac{2}{x^2} dx + \int \frac{3x - 1}{x^2 + x + 1} dx$$

$$= \ln|x| - \frac{2}{x} + \int \frac{3x - 1}{x^2 + x + 1} dx.$$

If $u = x^2 + x + 1$, then $du = (2x + 1) dx$. Thus

$$\int \frac{3x - 1}{x^2 + x + 1} dx = \int \frac{(3/2)(2x + 1) - (3/2) - 1}{x^2 + x + 1} dx$$

$$= \frac{3}{2} \int \frac{du}{u} - \frac{5}{2} \int \frac{dx}{x^2 + x + 1}$$

$$= \frac{3}{2} \ln |u| - \frac{5}{2} \int \frac{dx}{x^2 + x + (1/4) + (3/4)}$$

$$= \frac{3}{2} \ln |x^2 + x + 1| - \frac{5}{2} \int \frac{dx}{[x + (1/2)]^2 + (3/4)}.$$

Let $x + 1/2 = (\sqrt{3}/2)w$. Then $dx = (\sqrt{3}/2) dw$, and hence

$$-\frac{5}{2} \int \frac{(\sqrt{3}/2) dw}{(3/4)w^2 + (3/4)} = -\frac{5}{2} \cdot \frac{\sqrt{3}}{2} \cdot \frac{4}{3} \int \frac{dw}{w^2 + 1}$$

$$= -\frac{5\sqrt{3}}{3} \arctan w + C$$

$$= -\frac{5\sqrt{3}}{3} \arctan \left[\frac{2\sqrt{3}}{3}\left(x + \frac{1}{2}\right)\right] + C.$$

Thus

$$\int \frac{4x^3 + 2x^2 + 3x + 2}{x^2(x^2 + x + 1)} dx = \ln |x| - \frac{2}{x} + \frac{3}{2} \ln (x^2 + x + 1)$$

$$- \frac{5\sqrt{3}}{3} \arctan \left[\frac{2\sqrt{3}}{3}\left(x + \frac{1}{2}\right)\right] + C.$$

PROBLEMS

Rewrite the following fractions as a sum, using the method of partial fractions.

1. $\dfrac{x^2 + x + 1}{x + 1}$

2. $\dfrac{x^3 - 3x^2 + 4x - 1}{(x - 1)^2}$

3. $\dfrac{3x^2 + 2x + 2}{(x - 2)(x + 1)^2}$

4. $\dfrac{1 - x}{x(x^2 + 1)}$

5. $\dfrac{x^2 + 2x + 2}{(x + 1)^3}$

6. $\dfrac{2x^2 + 2x + 2}{x^3 + 3x^2 + 2x}$

7. $\dfrac{1 + x - x^2}{x^3(x + 1)}$

8. $\dfrac{7x^2 + 7x + 1}{(x + 2)(2x^2 + 2x + 1)}$

9. $\dfrac{2x^3 + x^2 + 3x + 2}{(x^2 + 1)^2}$

10. $\dfrac{x^5 + 3x^4 + 2x^3 - x^2 - 2x - 2}{x^4(x^2 + 2x + 2)}$

11. $\dfrac{x^7 + 2x^5 - x^4 + 2x^3 - 2x^2 - 1}{x^2(x^2 + 1)^2}$

12. $\dfrac{4x}{(x^2 - 1)^2}$

Compute the following integrals.

13. $\displaystyle\int_1^2 \dfrac{x^2 + x + 1}{x^3 + x}\,dx$

14. $\displaystyle\int_2^3 \dfrac{dx}{x^2 - 1}$

15. $\displaystyle\int \dfrac{4x^3 + 3x^2 + x + 2}{x^2(x^2 + 1)}\,dx$

16. $\displaystyle\int \dfrac{x^4 + 2x^3 - 2x^2 + 5x - 2}{x^2 + 2x - 3}\,dx$

17. $\displaystyle\int \dfrac{6 + 3x - x^2}{(1 - x)(x^2 + 2x + 5)}\,dx$

18. $\displaystyle\int \dfrac{2x^2 + 3x - 2}{x^3 - x^2 - 2x}\,dx$

19. $\displaystyle\int \dfrac{5x^2 - 8x + 12}{(x - 1)(4x^2 - 8x + 13)}\,dx$

20. $\displaystyle\int_{1/2}^{3/2} \dfrac{x^3 - 2x^2 + x + 2}{x^2 - 2x}\,dx$

21. $\displaystyle\int_{-1}^1 \dfrac{2x^3 + x^2 + 3x + 2}{(x^2 + 1)^2}\,dx$

22. $\displaystyle\int \dfrac{x^3 - 2x^2 + 2x}{(x - 1)^3}\,dx$

5. IMPROPER INTEGRALS

Consider the integral

$$\int_0^1 \dfrac{dx}{\sqrt{x}}.$$

The function $y = 1/\sqrt{x}$ has a vertical asymptote at $x = 0$, and it is not continuous on $[0, 1]$. However, for each $t > 0$, the integral $\int_t^1 dx/\sqrt{x}$ exists, since $y = 1/\sqrt{x}$ is continuous on $[t, 1]$. In fact,

$$\int_t^1 \dfrac{dx}{\sqrt{x}} = 2\sqrt{x}\,\Big|_t^1$$

$$= 2(1 - \sqrt{t}),$$

so that

$$\lim_{t \to 0^+} \int_t^1 \dfrac{dx}{\sqrt{x}} = \lim_{t \to 0^+} 2(1 - \sqrt{t})$$

$$= 2,$$

where $\lim_{t \to 0^+}$ means that t approaches 0 from the right-hand side. It is then natural to write

$$\int_0^1 \dfrac{dx}{\sqrt{x}} = 2.$$

The integral $\int_0^1 dx/\sqrt{x}$ can be interpreted as the area of the (unbounded) region in Figure 8.2. This region is approximated by the (bounded) region in Figure 8.3.

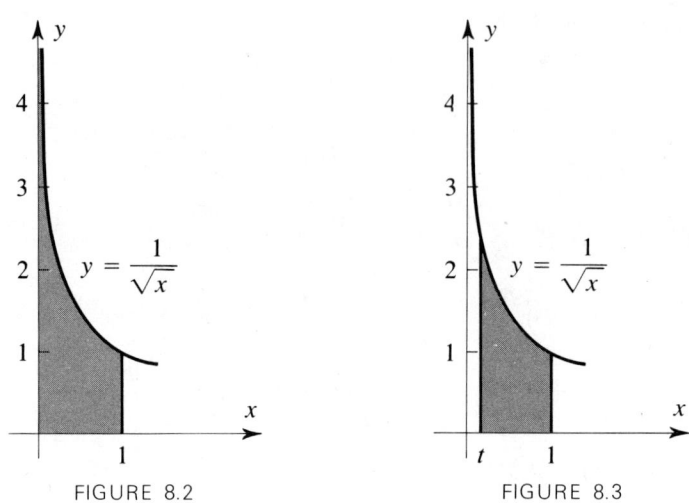

FIGURE 8.2   FIGURE 8.3

In general, if f is continuous on $[t, b]$ for each t such that $a < t < b$ but is not continuous at $x = a$, then we write

(6) $$\int_a^b f(x)\, dx = \lim_{t \to a^+} \int_t^b f(x)\, dx$$

if this one-sided limit exists.

Similarly, if f is continuous on $[a, s]$ for each s such that $a < s < b$ but is not continuous at $x = b$, then

(7) $$\int_a^b f(x)\, dx = \lim_{s \to b^-} \int_a^s f(x)\, dx$$

if this one-sided limit exists, where $\lim_{s \to b^-}$ means that s approaches b from the left-hand side.

If f has a discontinuity at some point c between a and b, then

$$\int_a^b f(x)\, dx = \lim_{s \to c^-} \int_a^s f(x)\, dx + \lim_{t \to c^+} \int_t^b f(x)\, dx$$

if both of these one-sided limits exist.

SECTION 5: IMPROPER INTEGRALS

EXAMPLE 24. Compute $\int_{-1}^{1} \frac{1}{x} dx$.

Solution. The following is *not* valid:

$$\int_{-1}^{1} \frac{1}{x} dx = \ln |x| \Big|_{-1}^{1} = \ln |1| - \ln |-1| = 0.$$

There is a vertical asymptote at $x = 0$, and we must evaluate the one-sided limits

$$\lim_{s \to 0^-} \int_{-1}^{s} \frac{1}{x} dx \quad \text{and} \quad \lim_{t \to 0^+} \int_{t}^{1} \frac{1}{x} dx.$$

Both limits fail to exist. For example,

$$\lim_{t \to 0^+} \int_{t}^{1} \frac{1}{x} dx = \lim_{t \to 0^+} \ln x \Big|_{t}^{1}$$

$$= \lim_{t \to 0^+} (\ln 1 - \ln t)$$

$$= +\infty.$$

Thus the integral $\int_{-1}^{1} (1/x) dx$ does not exist.

EXAMPLE 25. Evaluate $\int_{-1}^{8} \frac{dx}{\sqrt[3]{x}}$.

Solution. There is a vertical asymptote at $x = 0$. However,

$$\lim_{s \to 0^-} \int_{-1}^{s} \frac{dx}{\sqrt[3]{x}} = \lim_{s \to 0^-} \frac{3}{2} x^{2/3} \Big|_{-1}^{s}$$

$$= \frac{3}{2} \lim_{s \to 0^-} [s^{2/3} - (-1)^{2/3}]$$

$$= -\frac{3}{2},$$

and

$$\lim_{t \to 0^+} \int_{t}^{8} \frac{dx}{\sqrt[3]{x}} = \lim_{t \to 0^+} \frac{3}{2} x^{2/3} \Big|_{t}^{8}$$

$$= \frac{3}{2} \lim_{t \to 0^+} (8^{2/3} - t^{2/3})$$

$$= 6.$$

Therefore the integral $\int_{-1}^{8} dx/\sqrt[3]{x}$ exists and equals $-3/2 + 6 = 9/2$.

The function $f(x) = 1/(x^2 + 1)$ is continuous on the interval $[0, t]$ for every t. In fact,

$$\lim_{t \to \infty} \int_0^t \frac{dx}{x^2 + 1} = \lim_{t \to \infty} \arctan x \Big|_0^t$$

$$= \lim_{t \to \infty} (\arctan t - \arctan 0)$$

$$= \frac{\pi}{2}.$$

It is natural to write

$$\int_0^\infty \frac{dx}{x^2 + 1} = \frac{\pi}{2}.$$

In general, we define

(8) $$\int_a^\infty f(x) \, dx = \lim_{t \to \infty} \int_a^t f(x) \, dx$$

if this limit exists. Integrals of the type $\int_{-\infty}^b f(x) \, dx$ are defined similarly. An integral $\int_{-\infty}^\infty f(x) \, dx$ exists if and only if both of the integrals $\int_{-\infty}^a f(x) \, dx$ and $\int_a^\infty f(x) \, dx$ exist for some real number a, in which case

(9) $$\int_{-\infty}^\infty f(x) \, dx = \int_{-\infty}^a f(x) \, dx + \int_a^\infty f(x) \, dx.$$

EXAMPLE 26. Evaluate $\int_0^\infty \frac{dx}{\sqrt{x}(x + 1)}$.

Solution. This integral is improper for two reasons. There is a vertical asymptote at $x = 0$, and ∞ appears as a limit of integration. The integral exists if and only if each of the integrals

$$\int_0^1 \frac{dx}{\sqrt{x}(x + 1)} \quad \text{and} \quad \int_1^\infty \frac{dx}{\sqrt{x}(x + 1)}$$

exists. Let $t > 0$, $x = u^2$, and $dx = 2u \, du$. Then

$$\int_t^1 \frac{dx}{\sqrt{x}(x + 1)} = \int_{\sqrt{t}}^1 \frac{2u \, du}{u(u^2 + 1)}$$

$$= 2 \int_{\sqrt{t}}^1 \frac{du}{u^2 + 1}$$

$$= 2 \arctan u \Big|_{\sqrt{t}}^1.$$

Thus
$$\int_0^1 \frac{dx}{\sqrt{x}(x+1)} = \lim_{t \to 0^+} 2\left[\frac{\pi}{4} - \arctan(\sqrt{t})\right] = \frac{\pi}{2}.$$

Similarly,
$$\int_1^\infty \frac{dx}{\sqrt{x}(x+1)} = \lim_{s \to \infty} 2[\arctan(\sqrt{s}) - \arctan(1)]$$
$$= 2\left(\frac{\pi}{2} - \frac{\pi}{4}\right) = \frac{\pi}{2}.$$

Hence
$$\int_0^\infty \frac{dx}{\sqrt{x}(x+1)} = \frac{\pi}{2} + \frac{\pi}{2} = \pi.$$

EXAMPLE 27. Evaluate $\int_2^\infty \frac{dx}{x\sqrt{x^2-1}}$.

Solution.
$$\int_2^t \frac{dx}{x\sqrt{x^2-1}} = \operatorname{arcsec} x \Big|_2^t$$
$$= \operatorname{arcsec} t - \operatorname{arcsec} 2,$$

so that
$$\int_2^\infty \frac{dx}{x\sqrt{x^2-1}} = \lim_{t \to \infty} (\operatorname{arcsec} t - \operatorname{arcsec} 2)$$
$$= \frac{\pi}{2} - \frac{\pi}{3}$$
$$= \frac{\pi}{6}.$$

PROBLEMS

Decide which of the following improper integrals exist and evaluate the ones that do exist.

1. $\int_1^\infty \frac{1}{x} dx$

2. $\int_0^1 \frac{dx}{x-1}$

3. $\int_0^1 \frac{dx}{\sqrt{1-x}}$

4. $\int_1^\infty \frac{dx}{\sqrt{x}}$

5. $\int_0^1 \frac{dx}{x^2}$

6. $\int_1^\infty \frac{dx}{x^2}$

7. $\int_{-\infty}^0 x^2 e^x \, dx$

8. $\int_0^\infty e^{-x} dx$

9. $\int_0^{1/2} \frac{dx}{\sqrt{1-4x^2}}$

10. $\displaystyle\int_0^\infty xe^{-x^2}\,dx$

11. $\displaystyle\int_e^\infty \frac{dx}{x\ln x}$

12. $\displaystyle\int_e^\infty \frac{dx}{x(\ln x)^2}$

13. $\displaystyle\int_{-\infty}^\infty \frac{x\,dx}{(x^2+1)^2}$

14. $\displaystyle\int_5^6 \frac{dx}{(x-5)(x-6)}$

15. $\displaystyle\int_0^2 \frac{dx}{(2-x)^3}$

16. $\displaystyle\int_1^2 \frac{dx}{\sqrt[3]{x-1}}$

17. $\displaystyle\int_0^1 x\ln x\,dx$

18. $\displaystyle\int_0^\infty e^{-x}\cos x\,dx$

19. $\displaystyle\int_0^1 \frac{dx}{\sqrt{x}\sqrt{1-x}}$

20. $\displaystyle\int_0^\infty \frac{dx}{(x^2+1)^{3/2}}$

21. What is the volume of the solid generated by rotating the graph of $y = 1/x$ about the x-axis, $1 < x < \infty$?

22. What is the area of the region bounded above by the graph of $y = 5/\sqrt{x}$ and below by the graph of $y = -1/\sqrt{1-x}$, $0 < x < 1$?

6. THE TRAPEZOIDAL RULE

We now have an arsenal of techniques for attacking integrals $\int_a^b f(x)\,dx$. Occasionally, however, all of these techniques fail (or we fail to use them cleverly enough). When this happens, we must resort to some type of numerical technique.

Suppose that f is continuous on $[a, b]$. Since the definite integral is defined by

$$\int_a^b f(x)\,dx = \lim_{n\to\infty} \sum_{k=1}^n f(\bar{x}_k)\frac{b-a}{n},$$

it is possible to approximate $\int_a^b f(x)\,dx$ as closely as desired with sums $\sum_{k=1}^n f(\bar{x}_k)(b-a)/n$ by taking n sufficiently large, but doing so may require much time and effort. We will now look at some more efficient techniques.

One such technique for approximating integrals is a technique known as the trapezoidal rule. Recall that with the Riemann sum method we used a rectangle of area (or negative of area) $f(\bar{x}_k)(x_k - x_{k-1})$ as a building block to estimate the integral $\int_a^b f$. With the trapezoidal rule we use, as the name implies, the area of a trapezoid as our fundamental approximating area.

As usual, subdivide the interval $[a, b]$ into n subintervals, each of length $(b-a)/n$, and label the endpoints as in Figure 8.4. The area of the kth trapezoid in Figure 8.4 is

$$\frac{1}{2}[f(x_{k-1}) + f(x_k)]\frac{b-a}{n}.$$

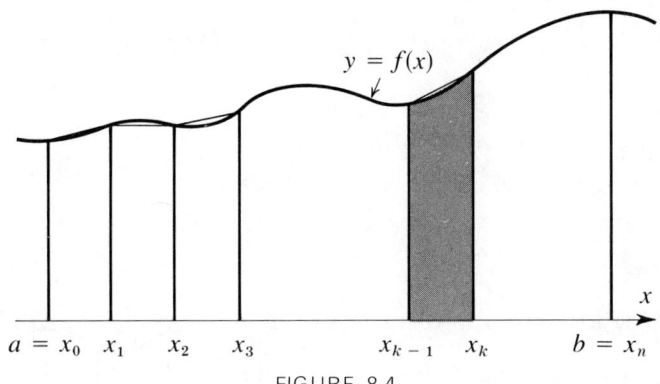

FIGURE 8.4

(Recall that the area of a trapezoid is one-half the sum of the lengths of the parallel sides times the distance between them.) The sum of the areas of these trapezoids is

$$\sum_{k=1}^{n} \frac{1}{2}[f(x_{k-1}) + f(x_k)]\frac{b-a}{n},$$

which can be rewritten

$$\frac{1}{2}\sum_{k=1}^{n} f(x_{k-1})\frac{b-a}{n} + \frac{1}{2}\sum_{k=1}^{n} f(x_k)\frac{b-a}{n}.$$

But $\sum_{k=1}^{n} f(x_{k-1})(b-a)/n$ is an approximation to $\int_a^b f$ (using left-hand endpoints) and $\sum_{k=1}^{n} f(x_k)(b-a)/n$ is also an approximation to $\int_a^b f$ (using right-hand endpoints). By taking n large enough, these approximations can be made as close as desired to the integral $\int_a^b f$. Thus

$$\int_a^b f(x)\,dx = \frac{1}{2}\int_a^b f(x)\,dx + \frac{1}{2}\int_a^b f(x)\,dx$$

$$\approx \frac{1}{2}\sum_{k=1}^{n} f(x_{k-1})\frac{b-a}{n} + \frac{1}{2}\sum_{k=1}^{n} f(x_k)\frac{b-a}{n}$$

$$= \frac{1}{2}\sum_{k=1}^{n} [f(x_{k-1}) + f(x_k)]\frac{b-a}{n}$$

$$= \frac{b-a}{2n}\sum_{k=1}^{n} [f(x_{k-1}) + f(x_k)]$$

$$= \frac{b-a}{2n}[f(x_0) + 2f(x_1) + 2f(x_2) + \cdots + 2f(x_{n-1}) + f(x_n)],$$

since each of the terms $f(x_1), f(x_2), \ldots, f(x_{n-1})$ appears twice.

CHAPTER 8: METHODS OF INTEGRATION

The approximation

(10) $$\int_a^b f(x)\,dx \approx \frac{b-a}{2n}[f(x_0) + 2f(x_1) + 2f(x_2) + 2f(x_3) + \cdots + 2f(x_{n-1}) + f(x_n)]$$

is known as the *trapezoidal rule*. It is usually a better approximation than either the left-hand or right-hand endpoint approximation.

EXAMPLE 28. Approximate $\int_1^2 \frac{1}{x}\,dx$ by using the trapezoidal rule with $n = 4$.

Solution.

$$\int_1^2 \frac{1}{x}\,dx \approx \frac{2-1}{2\cdot 4}\left(\frac{1}{1} + 2\cdot\frac{4}{5} + 2\cdot\frac{2}{3} + 2\cdot\frac{4}{7} + \frac{1}{2}\right)$$

$$\approx \frac{1}{8}(1 + 1.6 + 1.3333 + 1.1428 + 0.5)$$

$$\approx 0.6970.$$

Recall that $\int_1^2 (1/x)\,dx = \ln 2 \approx 0.69315$, so the error in the preceding approximation is less than 0.004.

The trapezoidal rule as it now stands has a serious flaw: it does not give any indication about how good the approximation is. The following theorem overcomes this flaw. The proof involves a clever use of integration by parts and will not be given here.

THEOREM 1. Suppose that $|f''(x)| \leq M$ for $a \leq x \leq b$. Subdivide the interval $[a, b]$ into n subintervals, each of length $(b - a)/n$, and label the endpoints as in Figure 8.4. Then

$$\int_a^b f = \frac{b-a}{2n}[f(x_0) + 2f(x_1) + 2f(x_2) + \cdots + 2f(x_{n-1}) + f(x_n)] + E,$$

where
$$|E| \leq \frac{M(b-a)^3}{12n^2}$$

$$\leq \frac{M(b-a)}{12}\left(\frac{b-a}{n}\right)^2.$$

SECTION 6: THE TRAPEZOIDAL RULE   291

This theorem gives a useful estimate for the error $|E|$. Using this estimate, it is possible to figure out beforehand how large n must be in order to be confident that the error will be smaller than some predetermined permissible error.

The following examples demonstrate how to use Theorem 1.

EXAMPLE 29. Estimate the error in the approximation given in Example 28.

Solution. Let $f(x) = 1/x$. Then

$$f'(x) = -\frac{1}{x^2} \quad \text{and} \quad f''(x) = \frac{2}{x^3}.$$

Since $1 \leq x \leq 2$, it follows that

$$|f''(x)| = \left|\frac{2}{x^3}\right| \leq 2,$$

and hence

$$|E| \leq \frac{M(b-a)^3}{12n^2} = \frac{2 \cdot 1^3}{12 \cdot 4^2} = \frac{1}{96} \approx 0.0104167.$$

It follows that

$$0.6865833 \leq \int_1^2 \frac{1}{x} dx \leq 0.7074167.$$

EXAMPLE 30. Approximate $\int_1^9 \sqrt{x}\, dx$ by using the trapezoidal rule with $n = 4$. Use Theorem 1 to estimate the error, then use the Fundamental Theorem of Calculus to compute the integral exactly, and from this result compute the actual error when using the trapezoidal rule. Compare the actual error with the estimated error.

Solution. Looking in a book of mathematical tables, we see that $\sqrt{3} \approx 1.732$, $\sqrt{5} \approx 2.236$, and $\sqrt{7} \approx 2.646$. Hence

$$\int_1^9 \sqrt{x}\, dx \approx \frac{9-1}{2 \cdot 4}(\sqrt{1} + 2\sqrt{3} + 2\sqrt{5} + 2\sqrt{7} + \sqrt{9})$$

$$\approx 17.228.$$

If $f(x) = \sqrt{x}$, then

$$f'(x) = \frac{1}{2\sqrt{x}} \quad \text{and} \quad f''(x) = -\frac{1}{4x\sqrt{x}}.$$

Therefore

$$|f''(x)| = \left|\frac{1}{4x\sqrt{x}}\right| \le \frac{1}{4}$$

for x between 1 and 9. It follows from Theorem 1 that

$$|E| \le \frac{(1/4) \cdot 8^3}{12 \cdot 4^2}$$

$$= \frac{2}{3}.$$

By the Fundamental Theorem of Calculus,

$$\int_1^9 \sqrt{x}\, dx = \frac{2}{3} x^{3/2} \Big|_1^9$$

$$= \frac{2}{3}(27 - 1)$$

$$= \frac{52}{3}$$

$$\approx 17.333.$$

Thus $|E| \approx |17.333 - 17.228| = 0.105$, which is significantly smaller than the estimate $\frac{2}{3}$.

PROBLEMS

Use the trapezoidal rule to approximate the following integrals. Use Theorem 1 to estimate the error.

1. $\int_{-1}^{2} (x^2 + 2x)\, dx,\ n = 3$

2. $\int_0^2 (4x^3 + 2x)\, dx,\ n = 2$

3. $\int_1^3 \frac{1}{x}\, dx$ with $n = 4$

4. $\int_0^2 \frac{dx}{x+1}$ with $n = 4$

5. $\int_1^4 \sqrt{x}\, dx$ with $n = 3$

6. $\int_1^4 \sqrt{x^2 + 1}\, dx$ with $n = 3$

7. $\int_{0.5}^1 \frac{1}{x^2}\, dx$ with $n = 2$

8. $\int_5^9 \frac{1}{x^2}\, dx$ with $n = 4$

9. $\int_1^5 \frac{dx}{x^2 + 1}$ with $n = 4$

10. $\int_1^4 \frac{dx}{x^2 + x + 1}$ with $n = 3$

11. $\int_0^{1/2} \frac{dx}{\sqrt{1 - x^2}},\ n = 4$

12. $\int_0^1 \sqrt{1 - x^2}\, dx,\ n = 4$

7. SIMPSON'S RULE

Another method that is often convenient for computing the integral $\int_a^b f(x)\,dx$ depends on approximating the area under the graph of $y = f(x)$ by areas under parabolas. (See Section 3 of Chapter 10 for a discussion of arbitrary parabolas. In this section we consider only parabolas of the form $y = ax^2 + bx + c$.) This method usually gives a closer approximation than the trapezoidal rule, but with about the same amount of effort.

We again partition $[a, b]$ into subintervals of length $(b - a)/n$, but this time we choose n to be even. It can be shown that for the points $P_0 = (x_0, f(x_0))$, $P_1 = (x_1, f(x_1))$, and $P_2 = (x_2, f(x_2))$, we may find constants A_0, B_0, and C_0 such that the parabola $y = A_0 x^2 + B_0 x + C_0$ passes through the points P_0, P_1, and P_2. Similarly, we may find constants A_2, B_2, and C_2 such that the parabola $y = A_2 x^2 + B_2 x + C_2$ passes through the points P_2, P_3, and P_4, where $P_i = (x_i, f(x_i))$. In this manner, we can construct a piecewise parabola over the interval $[a, b]$ that approximates the graph of $y = f(x)$, as in Figure 8.5.

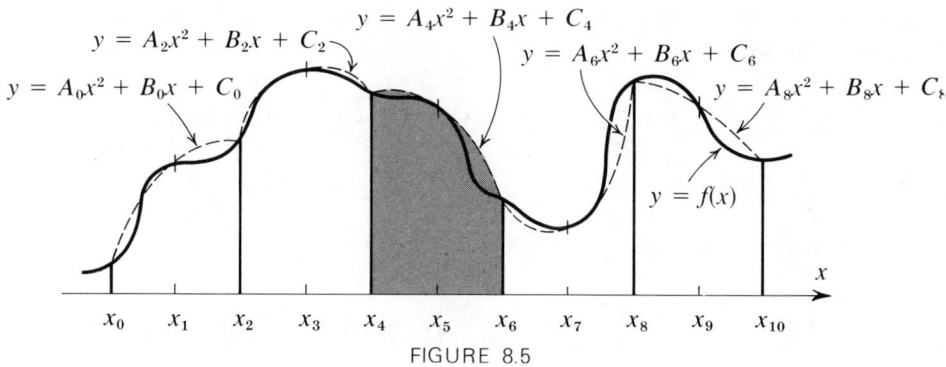

FIGURE 8.5

CHAPTER 8: METHODS OF INTEGRATION

We will approximate the area under the graph of $y = f(x)$ by computing exactly the area under the piecewise parabola. To do so, we first compute the area under the parabola $y = A_i x^2 + B_i x + C_i$ between x_i and x_{i+2}. This area is independent of the location of the y-axis, so, for notational purposes, we will assume that $x_i = -h$, $x_{i+1} = 0$, and $x_{i+2} = h$. The equation of the parabola would now become $y = Ax^2 + Bx + C$ for some A, B, and C. The area is given by

$$\int_{-h}^{h} (Ax^2 + Bx + C)\,dx = \frac{A}{3}[h^3 - (-h)^3] + \frac{B}{2}[h^2 - (-h)^2] + C[h - (-h)]$$

$$= \frac{2A}{3}h^3 + 2Ch.$$

Since the parabola $y = Ax^2 + Bx + C$ passes through the points $(-h, f(x_i))$, $(0, f(x_{i+1}))$, and $(h, f(x_{i+2}))$, it follows that

$$A(-h)^2 + B(-h) + C = f(x_i),$$
$$A(0)^2 + B(0) + C = f(x_{i+1}),$$

and
$$A(h)^2 + B(h) + C = f(x_{i+2}).$$

From the second equation, we get $C = f(x_{i+1})$. By adding the first and third equations, we see that

$$A = \frac{1}{2h^2}[f(x_i) + f(x_{i+2}) - 2f(x_{i+1})].$$

Thus

$$\int_{-h}^{h} (Ax^2 + Bx + C)\,dx = \frac{2Ah^3}{3} + 2Ch$$

$$= \frac{2h^3}{3}\frac{1}{2h^2}[f(x_i) + f(x_{i+2}) - 2f(x_{i+1})] + 2hf(x_{i+1})$$

$$= \frac{h}{3}[f(x_i) + 4f(x_{i+1}) + f(x_{i+2})].$$

But $h = (b - a)/n$, and hence the total area under the piecewise parabola equals

$$\frac{b-a}{3n}[f(x_0) + 4f(x_1) + f(x_2)] + \frac{b-a}{3n}[f(x_2) + 4f(x_3) + f(x_4)]$$

$$+ \cdots + \frac{b-a}{3n}[f(x_{n-2}) + 4f(x_{n-1}) + f(x_n)]$$

$$= \frac{b-a}{3n}[f(x_0) + 4f(x_1) + 2f(x_2) + 4f(x_3) + 2f(x_4)$$

$$+ \cdots + 2f(x_{n-2}) + 4f(x_{n-1}) + f(x_n)].$$

We can now state **Simpson's rule**:
If f is continuous on $[a, b]$, then

(11) $\quad \int_a^b f(x)\,dx \approx \dfrac{b-a}{3n}[f(x_0) + 4f(x_1) + 2f(x_2) + \cdots + 4f(x_{n-1}) + f(x_n)].$

EXAMPLE 31. Approximate $\int_1^2 \dfrac{1}{x}\,dx$ by using Simpson's rule with $n = 4$.

Solution.

$$\int_1^2 \frac{1}{x}\,dx \approx \frac{2-1}{3 \cdot 4}\left(\frac{1}{1} + 4 \cdot \frac{4}{5} + 2 \cdot \frac{2}{3} + 4 \cdot \frac{4}{7} + \frac{1}{2}\right)$$

$$\approx \frac{1}{12}(1 + 3.2 + 1.3333 + 2.2857 + 0.5)$$

$$\approx 0.6932,$$

which is very close to the actual value of $\ln 2 = 0.69315\ldots$. (Compare this with Example 28.)

EXAMPLE 32. Approximate $\int_0^2 x^3\,dx$, using Simpson's rule with $n = 2$.

Solution.

$$\int_0^2 x^3\,dx \approx \frac{2}{3 \cdot 2}(0^3 + 4 \cdot 1^3 + 2^3)$$

$$= \frac{1}{3}(4 + 8)$$

$$= 4,$$

which is actually *exact*. In fact, Simpson's rule is exact for all polynomial functions of degree 3 or less.

As with the trapezoidal rule, it is also possible to estimate the error involved in using Simpson's rule. The following theorem is stated without proof.

THEOREM 2. Suppose that $|f^{(4)}(x)| \leq M$ for $a \leq x \leq b$. Subdivide the interval $[a, b]$ into n subintervals (n even) of equal length and label as in Figure 8.5. Then

$$\int_a^b f = \frac{b-a}{3n}[f(x_0) + 4f(x_1) + 2f(x_2) + \cdots + 4f(x_{n-1}) + f(x_n)] + E,$$

where

$$|E| \le \frac{M(b-a)}{180}\left(\frac{b-a}{n}\right)^4$$

and where $f^{(4)}(x)$ denotes the fourth derivative of f at x.

As the following examples illustrate, the preceding error is often very small.

EXAMPLE 33. Use Theorem 2 to estimate the error in the approximation given in Example 31.

Solution. Let $f(x) = 1/x$. Then $f'(x) = -x^{-2}$, $f''(x) = 2x^{-3}$, $f^{(3)}(x) = -6x^{-4}$, and $f^{(4)}(x) = 24x^{-5}$. Since $1 \le x \le 2$, it follows that

$$|f^{(4)}(x)| = \left|\frac{24}{x^5}\right| \le 24.$$

Thus

$$|E| \le \frac{M(b-a)}{180}\left(\frac{b-a}{n}\right)^4$$

$$= \frac{24 \cdot 1}{180}\left(\frac{1}{4}\right)^4$$

$$\approx 0.0005208.$$

It follows that

$$0.6927 \le \int_1^2 \frac{1}{x} dx \le 0.6937.$$

EXAMPLE 34. Approximate $\int_1^9 \sqrt{x}\, dx$, using Simpson's rule with $n = 4$. Estimate the error.

Solution.

$$\int_1^9 \sqrt{x}\, dx \approx \frac{9-1}{3 \cdot 4}(\sqrt{1} + 4\sqrt{3} + 2\sqrt{5} + 4\sqrt{7} + \sqrt{9})$$

$$\approx \frac{2}{3}[1 + 4(1.732) + 2(2.236) + 4(2.646) + 3]$$

$$\approx 17.322.$$

If $f(x) = x^{1/2}$, then $f'(x) = \frac{1}{2}x^{-1/2}$, $f''(x) = -\frac{1}{4}x^{-3/2}$, $f^{(3)}(x) = \frac{3}{8}x^{-5/2}$, and $f^{(4)}(x) = -\frac{15}{16}x^{-7/2}$, so that

$$|f^{(4)}(x)| = \frac{15}{16}x^{-7/2} \le \frac{15}{16}$$

whenever $1 \le x \le 9$. From Theorem 2,

$$|E| \le \frac{15}{16} \frac{8}{180} \left(\frac{8}{4}\right)^4$$

$$= \frac{2}{3}.$$

The actual error is much smaller, since

$$\int_1^9 \sqrt{x}\, dx = 17.\overline{3},$$

using the Fundamental Theorem of Calculus. (The bar over the 3 indicates an infinite repeating decimal: $17.\overline{3} = 17.333\ldots$.)

The function $x \to e^{x^2}$ does not have an antiderivative that is an elementary function. In computing a definite integral of $x \to e^{x^2}$, it is necessary to use some numerical techniques.

EXAMPLE 35. Evaluate $\int_0^1 e^{x^2}\, dx$.

Solution. Using Simpson's rule with $n = 4$, we obtain

$$\int_0^1 e^{x^2}\, dx \approx \frac{1}{3 \cdot 4} [e^0 + 4e^{(1/4)^2} + 2e^{(1/2)^2} + 4e^{(3/4)^2} + e^1]$$

$$\approx \frac{1}{12} [1 + 4(1.646) + 2(1.284) + 4(1.755) + 2.7183]$$

$$\approx 1.4638.$$

PROBLEMS

Use Simpson's rule to approximate the following integrals.

1. $\int_0^1 \sin x^2\, dx,\ n = 4$
2. $\int_{-\pi/2}^0 \sin(\cos x)\, dx,\ n = 2$
3. $\int_0^{\pi/2} e^{\sin x}\, dx,\ n = 2$
4. $\int_0^1 \ln(x^2 + 1)\, dx,\ n = 6$

Use Simpson's rule to approximate the following integrals and use Theorem 2 to estimate the error. Use the Fundamental Theorem of Calculus to compute the integrals exactly.

5. $\int_{1/2}^1 \frac{dx}{x^2},\ n = 2$
6. $\int_{-2}^2 (x^2 + 2x)\, dx,\ n = 4$

7. $\int_1^5 \sqrt{x}\, dx, n = 4$

8. $\int_1^5 \dfrac{dx}{x^2+1}, n = 4$

9. $\int_0^{1/2} \dfrac{dx}{\sqrt{1-x^2}}, n = 4$

10. $\int_0^6 \sqrt{x^2+1}\, dx, n = 6$

11. Approximate $\int_0^4 x^4\, dx$ using
    a. the left-hand endpoint method with $n = 4$;
    b. the trapezoidal rule with $n = 4$; and
    c. Simpson's rule with $n = 4$.
    d. Use the Fundamental Theorem of Calculus to compute this integral exactly.

12. Repeat Problem 11 for the integral $\int_1^9 (1/\sqrt{x})\, dx$.

13. At the beginning of fall term, a campus bookstore starts selling books at a rate of $1 per minute at 9 A.M., increases to $5 per minute at 10 A.M., $7 at 11 A.M., $8 at noon, $5 at 1 P.M., $10 at 2 P.M., $11 at 3 P.M., $10 at 4 P.M., and closes at $4 per minute at 5 P.M. Use Simpson's rule to estimate how many dollars worth of books were sold that day. [*Answer:* $3660.]

14. Using remote television cameras, a traffic control officer can monitor traffic at various intersections. During a typical rush hour, he counts the cars passing through a certain intersection in 60 seconds, and he repeats this every 10 minutes from 7:00 until 8:01 A.M. He gets the following table:

Time Interval	Number of Cars
7:00–7:01	10
7:10–7:11	15
7:20–7:21	20
7:30–7:31	30
7:40–7:41	35
7:50–7:51	20
8:00–8:01	10

Use Simpson's rule to estimate the number of cars passing through the intersection between 7:00 and 8:00 A.M. [*Answer:* 1300 cars.]

REVIEW PROBLEMS

Compute each of the following integrals.

1. $\int x^2 \sin(x^3) \cos(x^3)\, dx$

2. $\int_0^3 \dfrac{x\, dx}{2\sqrt{x+1}}$

3. $\int \dfrac{dx}{\sqrt{9-x^2}}$

4. $\int_{-3}^3 \sqrt{9-x^2}\, dx$

5. $\int_1^{\sqrt{5}} \dfrac{x\, dx}{\sqrt{x^2-1}}$

6. $\int_2^\infty \dfrac{dx}{x\sqrt{4x^2+9}}$

SECTION 7: SIMPSON'S RULE

7. $\displaystyle\int \frac{x\,dx}{\sqrt{4x^2+9}}$

8. $\displaystyle\int \frac{\sqrt{x^2-4}}{x^2}\,dx$

9. $\displaystyle\int x \sin x \cos x\,dx$

10. $\displaystyle\int_0^2 x^2 e^{-x}\,dx$

11. $\displaystyle\int \ln(x^2)\,dx$

12. $\displaystyle\int x^3 \ln x\,dx$

13. $\displaystyle\int e^x \ln(e^x+1)\,dx$

14. $\displaystyle\int_{-1}^0 x\sqrt{x+1}\,dx$

15. $\displaystyle\int_0^1 \frac{\sqrt{x+1}}{x}\,dx$

16. $\displaystyle\int \ln\left|\frac{x+1}{x-1}\right|\,dx$

17. $\displaystyle\int_4^\infty \frac{x\,dx}{x^2-9}$

18. $\displaystyle\int_0^1 \sqrt{\frac{x}{x+1}}\,dx$

19. $\displaystyle\int \sqrt{x^2-4x}\,dx$

20. $\displaystyle\int_0^2 \frac{dx}{\sqrt{4x-x^2}}$

21. $\displaystyle\int_{-2}^0 \frac{dx}{x^2+2x}$

22. $\displaystyle\int x\sqrt{x^2-4x}\,dx$

23. $\displaystyle\int \frac{dx}{x^3+2x^2+x}$

24. $\displaystyle\int \sqrt{x^2+2x}\,dx$

25. $\displaystyle\int \frac{(x^4+2x)\,dx}{x^3+2x^2+x+2}$

26. $\displaystyle\int_0^1 \frac{x^2+2x}{\sqrt{x}}\,dx$

27. $\displaystyle\int \frac{dx}{(x^2+3x-4)^2}$

28. $\displaystyle\int \frac{dx}{\sqrt{x^2+2x}}$

29. $\displaystyle\int_{-2/3}^0 \frac{x\,dx}{\sqrt{3x+2}}$

30. $\displaystyle\int \sqrt{\sin^2 x + 4\sin x \cos x}\,dx$

31. Find the area of the region bounded by $y = \ln x$ and $y = 0$ between $x = 2$ and $x = 5$.

32. Find the area of the region between the graphs of $y = x \sin x$ and $y = xe^x$, $0 \le x \le 1$.

33. Find the volume of the solid formed by rotating the graph of $y = \ln x$ about the x-axis, $1 \le x \le 2$.

34. Find the volume of the solid generated by rotating the graph of $y = \sqrt{\sin xe^x}$ about the x-axis, $0 \le x \le 1$.

35. Find the average value of the function $y = 2\sin 3x\, e^{2x}$ on the interval $[0, \pi/3]$.

Solve each of the following differential equations.

36. $e^{2x} + 1 + \sin y\,\dfrac{dy}{dx} = 0$

37. $e^{x-y}\dfrac{dy}{dx} + 1 = 0$

38. $y + x\dfrac{dy}{dx} = 0$

39. $\sec x + \tan y\,\dfrac{dy}{dx} = 0$.

CHAPTER 9

FURTHER APPLICATIONS

It has already been seen how the definite integral can be interpreted as an area and as a volume. In this chapter additional interpretations of the integral will be introduced in order to solve various types of problems. (As an added bonus, there will be ample opportunity to apply the techniques of integration learned in Chapter 8.) We begin by taking a close look at arc length.

1. ARC LENGTH

We have assumed that a circle of radius 1 has a perimeter of length 2π and that it makes sense to consider the length of any arc of a circle. But what do we really mean by the length of an arc? Is it possible to start with the formula for the distance between two points in the plane, and from this formula derive a meaning for arc length? Starting with rectangles as basic building blocks of area, we were able to obtain areas for other regions. We will treat arc length similarly.

Consider the graph of $y = f(x)$ between $x = a$ and $x = b$. Subdivide the interval $[a, b]$ into n equal subintervals, as in Figure 9.1. If n is large, so that $x_{i+1} - x_i = (b - a)/n$ is small, then it appears that the length of the chord should

SECTION 1: ARC LENGTH

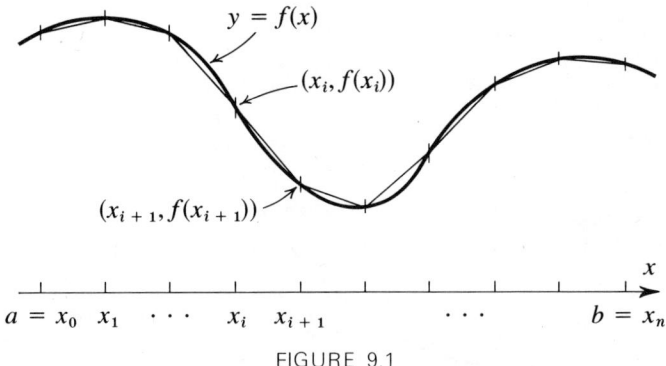

FIGURE 9.1

approximate the length of the arc between $(x_i, f(x_i))$ and $(x_{i+1}, f(x_{i+1}))$. The chord has length

$$\sqrt{(x_{i+1} - x_i)^2 + (f(x_{i+1}) - f(x_i))^2} = \sqrt{1 + \left[\frac{f(x_{i+1}) - f(x_i)}{x_{i+1} - x_i}\right]^2} (x_{i+1} - x_i).$$

If f is differentiable on an interval containing $[x_i, x_{i+1}]$, then by the Mean Value Theorem for Derivatives (see Chapter 3),

$$\frac{f(x_{i+1}) - f(x_i)}{x_{i+1} - x_i} = f'(\bar{x}_i)$$

for some $\bar{x}_i$ between x_i and x_{i+1}. The arc length from $x = a$ to $x = b$ should be approximated by

$$\sum_{i=0}^{n-1} \sqrt{1 + (f'(\bar{x}_i))^2} \, \frac{b-a}{n}.$$

If the limit

$$\lim_{n \to \infty} \sum_{i=0}^{n-1} \sqrt{1 + (f'(\bar{x}_i))^2} \, \frac{b-a}{n}$$

exists, then this limit is the definite integral

$$\int_a^b \sqrt{1 + (f'(x))^2} \, dx.$$

We define the *arc length* of the graph of $y = f(x)$ between a and b to be the number

(1) $$\int_a^b \sqrt{1 + (f'(x))^2} \, dx$$

if this integral exists.

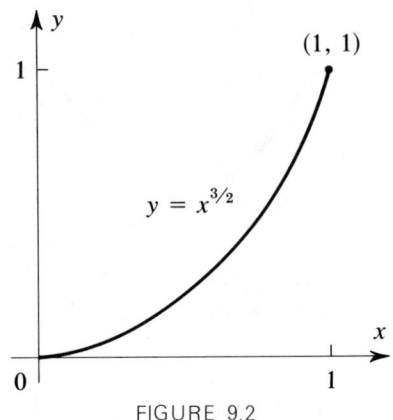

FIGURE 9.2

EXAMPLE 1. Find the arc length of $f(x) = x^{3/2}$ between $x = 0$ and $x = 1$ (see Figure 9.2).

Solution. We must compute the integral

$$\int_0^1 \sqrt{1 + (f'(x))^2}\, dx = \int_0^1 \sqrt{1 + \left(\frac{3}{2}x^{1/2}\right)^2}\, dx$$

$$= \int_0^1 \sqrt{1 + \frac{9}{4}x}\, dx.$$

Let $u = 1 + (9/4)x$. Then $dx = (4/9)\, du$, so that

$$\int_0^1 \sqrt{1 + \frac{9}{4}x}\, dx = \frac{4}{9}\int_1^{13/4} u^{1/2}\, du$$

$$= \frac{4}{9}\frac{2}{3}u^{3/2}\Big|_1^{13/4}$$

$$= \frac{8}{27}\left[\left(\frac{13}{4}\right)^{3/2} - 1\right]$$

$$= \frac{(13)^{3/2} - 8}{27}$$

$$\approx 1.43971.$$

SECTION 1: ARC LENGTH

EXAMPLE 2. Find the arc length of the function
$$f(x) = \frac{1}{2}x\sqrt{x^2 - 1} - \frac{1}{2}\ln(x + \sqrt{x^2 - 1})$$
between $x = 1$ and $x = 3$.

Solution. It can be seen that $f'(x) = \sqrt{x^2 - 1}$. Hence $(f'(x))^2 = x^2 - 1$, so that

$$\int_1^3 \sqrt{1 + (f'(x))^2}\, dx = \int_1^3 \sqrt{1 + x^2 - 1}\, dx$$

$$= \int_1^3 x\, dx$$

$$= \frac{1}{2}x^2 \Big|_1^3$$

$$= \frac{1}{2}(9 - 1)$$

$$= 4.$$

It is traditional to let s denote arc length. Thus

$$s = \int \sqrt{1 + (f'(x))^2}\, dx$$

$$= \int \sqrt{1 + \left(\frac{dy}{dx}\right)^2}\, dx$$

$$= \int \sqrt{(dx)^2 + (dy)^2}.$$

We usually write dx^2 instead of $(dx)^2$, and we define the differential of arc length to be

(2) $$ds = \sqrt{dx^2 + dy^2}.$$

Equation (2) often appears in the equivalent form
$$ds^2 = dx^2 + dy^2.$$
Formula (2) is also valid in case x is a function of y.

EXAMPLE 3. Find the length of the arc given by $x = (1 - y^{2/3})^{3/2}$, $0 \le y \le 1$.

Solution. Note that
$$dx = \frac{3}{2}(1 - y^{2/3})^{1/2}\left(-\frac{2}{3}y^{-1/3}\right)dy,$$

so that
$$s = \int_{y=0}^{y=1} \sqrt{dx^2 + dy^2}$$
$$= \int_0^1 \sqrt{\frac{1 - y^{2/3}}{y^{2/3}}dy^2 + dy^2}$$
$$= \int_0^1 \frac{dy}{y^{1/3}}$$
$$= \frac{3}{2}y^{2/3}\Big|_0^1$$
$$= 1.5.$$

PROBLEMS

Compute the arc length.

1. $y = \ln(\cos x),\ 0 \le x \le \frac{\pi}{4}$

2. $y = x,\ 0 \le x \le 1$

3. $y = \sqrt{1 - x^2},\ 0 \le x \le 1$

4. $y = \ln(\sin x),\ \frac{\pi}{4} \le x \le \frac{\pi}{3}$

5. $y = \frac{2}{3}(x - 1)^{3/2},\ 1 \le x \le 4$

6. $y = \frac{1}{3}(x^2 + 2)^{3/2},\ 0 \le x \le 1$

7. $y = \frac{1}{2}x^2,\ 0 \le x \le 6$

8. $y = \ln x,\ a \le x \le b$

9. $y = 2\sqrt{x},\ 1 \le x \le 4$

10. $y = \frac{2}{3}x^{3/2},\ 0 \le x \le 6$

11. $y = \frac{x^3}{3} + \frac{1}{4x},\ 1 \le x \le 3$

2. CENTER OF MASS

Consider a weightless bar resting on a fulcrum with weights of mass m_1 and m_2, as in Figure 9.3. The bar will balance if $a_2 m_2 = a_1 m_1$; that is, if $a_2 m_2 - a_1 m_1 = 0$. More generally, if a mass m_i is placed at the point x_i on the x-axis as in Figure 9.4, then the *moment* about the point $\bar{x}$ is defined to be the sum

(3) $$\sum_{i=1}^{n} (x_i - \bar{x}) m_i.$$

Note that the coefficient of m_i is positive if $x_i > \bar{x}$ and negative if $x_i < \bar{x}$. If the moment about $\bar{x}$ is zero, then the system would be in balance if a fulcrum were placed at $\bar{x}$.

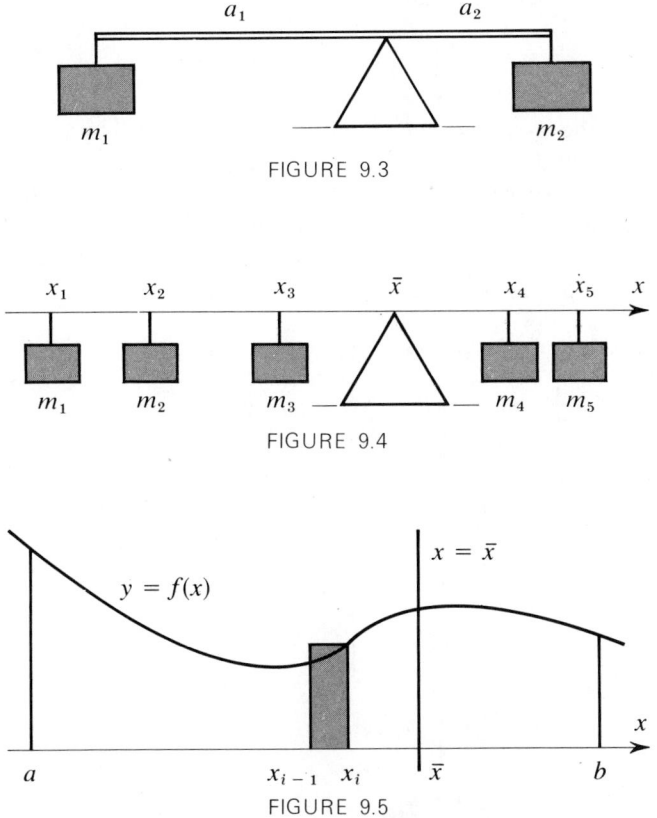

FIGURE 9.3

FIGURE 9.4

FIGURE 9.5

Consider a region in the plane bounded by $y = f(x)$, $y = 0$, $x = a$, and $x = b$, as in Figure 9.5. If mass is proportional to area, what is the moment of this region about the line $x = \bar{x}$? Since mass is proportional to area, the *density* (mass per unit of area) must be some constant k. We can estimate the moment by subdividing the interval $[a, b]$ into n equal subintervals. The mass of the slice between x_{i-1} and x_i is approximately $k f(x_i)(b - a)/n$, since mass equals area times density. Thus the contribution of this slice to the moment about $x = \bar{x}$ is approximately equal to $(x_i - \bar{x})kf(x_i)(b - a)/n$. The total moment is roughly

$$\sum_{i=1}^{n} (x_i - \bar{x})kf(x_i) \frac{b-a}{n},$$

which we recognize as being a Riemann sum for the integral $k\int_a^b (x - \bar{x})f(x)\,dx$. The **moment of the region about the line** $x = \bar{x}$ is defined to be the integral

(4) $$k\int_a^b (x - \bar{x})f(x)\,dx.$$

If $\bar{x}$ is a number such that

(5) $$k\int_a^b (x - \bar{x})f(x)\,dx = 0,$$

then $\bar{x}$ is called the **x-component of the center of mass**. The y-component $\bar{y}$ is defined similarly, and the point $(\bar{x}, \bar{y})$ is then the **center of mass**. The flight characteristics of an airplane, the performance of an athlete, and the stability of a tall building during a wind storm all depend on the location of a center of mass. If a mass has uniform density, then the center of mass is called a **centroid**.

Equation (5) can be solved for $\bar{x}$:

$$0 = k\int_a^b (x - \bar{x})f(x)\,dx$$

$$= k\int_a^b xf(x)\,dx - k\bar{x}\int_a^b f(x)\,dx,$$

and hence

(6) $$\bar{x} = \frac{\int_a^b xf(x)\,dx}{\int_a^b f(x)\,dx}.$$

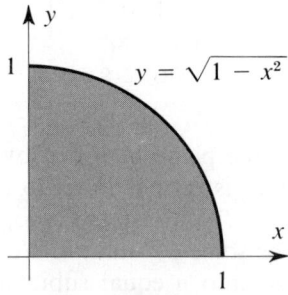

FIGURE 9.6

Notice that the integral that appears in the denominator is just the area of the region.

EXAMPLE 4. Find the centroid of the first quadrant section of the circle with radius 1 and center at the origin.

Solution. The region is bounded by $y = \sqrt{1-x^2}$, $x = 0$, $x = 1$, and $y = 0$, as in Figure 9.6. Thus

$$\bar{x} = \frac{\int_0^1 x\sqrt{1-x^2}\,dx}{\int_0^1 \sqrt{1-x^2}\,dx}.$$

The integral in the denominator is just the area of the first quadrant of the unit circle—that is, $\pi/4$. In the numerator, we use the substitution $u = 1 - x^2$, so that $du = -2x\,dx$. It follows that

$$\bar{x} = \frac{-(1/2)\int_1^0 \sqrt{u}\,du}{\pi/4}$$

$$= -\frac{1}{2} \cdot \frac{4}{\pi} \cdot \frac{2}{3} u^{3/2} \Big|_1^0$$

$$= \frac{4}{3\pi}.$$

The region is symmetric in x and y, and hence the centroid is the point $(4/3\pi, 4/3\pi)$.

EXAMPLE 5. Find the center of mass of the triangular region in Figure 9.7 bounded by $x = 0$, $y = 0$, and $y = -2x + 2$.

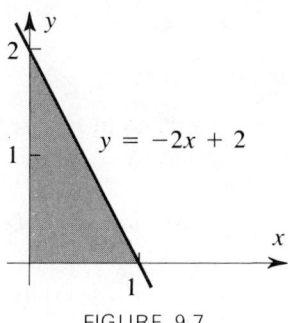

FIGURE 9.7

Solution.

$$\bar{x} = \frac{\int_0^1 x(-2x+2)\,dx}{\int_0^1 (-2x+2)\,dx}$$

$$= \frac{-2\int_0^1 x^2\,dx + 2\int_0^1 x\,dx}{-2\int_0^1 x\,dx + 2\int_0^1 dx}$$

$$= \frac{-(2/3)x^3\,|_0^1 + x^2\,|_0^1}{-x^2\,|_0^1 + 2x\,|_0^1}$$

$$= \frac{1-(2/3)}{-1+2}$$

$$= \frac{1}{3}.$$

The area of the triangular region is 1, which corresponds to the integral in the denominator. This area need not be computed again in the computation of $\bar{y}$. Also, since $y = -2x + 2$, it follows that $-2x = y - 2$, so that $x = (2-y)/2$. Hence

$$\bar{y} = \int_0^2 y\left(\frac{2-y}{2}\right) dy$$

$$= \int_0^2 y\,dy - \frac{1}{2}\int_0^2 y^2\,dy$$

$$= \frac{1}{2}y^2\,\Big|_0^2 - \frac{1}{6}y^3\,\Big|_0^2$$

$$= 2 - \frac{4}{3}$$

$$= \frac{2}{3}.$$

Thus the center of mass is the point $(\frac{1}{3}, \frac{2}{3})$.

A more general region is sketched in Figure 9.8. The x-component of the center of mass of such a region is given by

$$\bar{x} = \frac{\int_a^b x[f(x) - g(x)]\,dx}{\int_a^b [f(x) - g(x)]\,dx}.$$

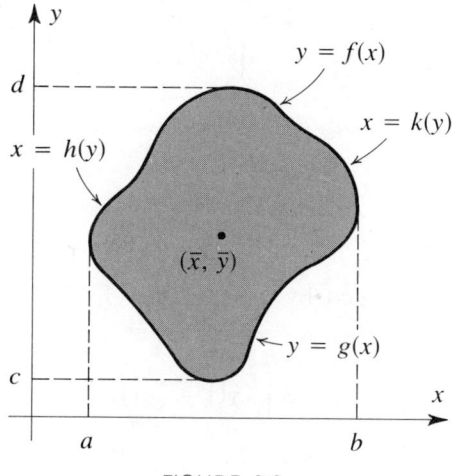

FIGURE 9.8

On the other hand, in order to compute $\bar{y}$, we must solve for x as a function of y so that the region is bounded by $x = h(y)$ and $x = k(y)$. Then

$$\bar{y} = \frac{\int_c^d y[k(y) - h(y)]\,dy}{\int_c^d [k(y) - h(y)]\,dy}.$$

EXAMPLE 6. Find the center of mass of the region in Figure 9.9 bounded by $y = x^2$, $x = 1$, and $y = 0$.

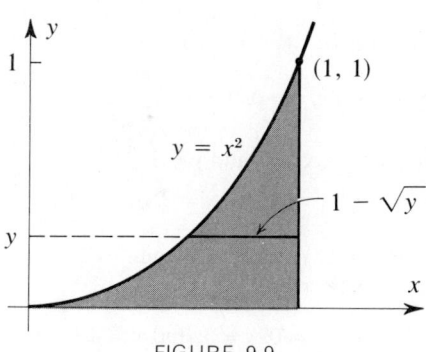

FIGURE 9.9

Solution.

$$\bar{x} = \frac{\int_0^1 x(x^2)\, dx}{\int_0^1 x^2\, dx}$$

$$= \frac{(1/4)x^4 \big|_0^1}{(1/3)x^3 \big|_0^1}$$

$$= \frac{3}{4}.$$

The region is bounded by $x = \sqrt{y}$, $x = 1$, and $y = 0$. The area of the region is $\frac{1}{3}$, and hence

$$\bar{y} = 3 \int_0^1 y(1 - \sqrt{y})\, dy$$

$$= \frac{3}{2} y^2 \Big|_0^1 - \frac{6}{5} y^{5/2} \Big|_0^1$$

$$= \frac{3}{2} - \frac{6}{5}$$

$$= \frac{3}{10}.$$

Thus the center of mass is the point $(\frac{3}{4}, \frac{3}{10})$.

PROBLEMS

Find the center of mass of the following regions.

1. The region bounded by $y = \ln x$, $y = 0$, and $x = e$.
2. The region bounded by $y = \frac{1}{\sqrt{1 - x^2}}$, $x = 0$, $x = \frac{1}{2}$, and $y = 0$.
3. The region bounded by $y = \sin x$, $y = 0$, $0 \leq x \leq \frac{\pi}{2}$.
4. The region bounded by $y = \sqrt{x}$, $x = 0$, and $y = 1$.
5. The region bounded by $y = x^3$, $y = 0$, and $x = 2$.
6. The region bounded by $y = \frac{1}{\sqrt{x}}$, $y = 0$, $x = 1$, and $x = 2$.
7. The region bounded by $y = \frac{1}{x^3}$, $y = 0$, $x = 1$, and $x = 4$.

8. The region bounded by $y = \dfrac{x}{x^2+1}$, $y = 0$, and $x = 1$.

9. The region bounded by $y = (1 - \sqrt{x})^2$, $x = 0$, and $y = 0$.

10. The region bounded by $y = (1 + \sqrt[3]{x})^3$, $y = 0$, $x = 0$, and $x = 1$.

11. The region bounded by $y = \ln(x^2)$, $y = 0$, and $x = e$.

12. The region bounded by $y = \ln(\sqrt{x})$, $y = 0$, and $x = e$.

3. WORK

The **work** W performed by a constant force f acting through a distance d is defined to be

(7) $$W = fd.$$

Thus it takes 600 foot-pounds of work to lift a 200-pound weight from the floor onto a 3-foot table.

If the force f is not constant, the situation is somewhat more complicated. Assume that $f(x)$ is the force acting on an object at the point x. What is the work required to move the object from a to b?

To solve this problem, subdivide the interval $[a, b]$ into n equal subintervals $[x_{i-1}, x_i]$, $i = 1, 2, \ldots, n$, where $x_0 = a$ and $x_n = b$. If n is large, then f is nearly constant on each of the subintervals $[x_{i-1}, x_i]$, and so the work done in moving the object from x_{i-1} to x_i is approximately equal to $f(x_i)(x_i - x_{i-1})$. Hence the total work done in moving the object from a to b is approximately equal to

$$\sum_{i=1}^{n} f(x_i)(x_i - x_{i-1}).$$

Taking a limit as n tends to infinity, we get the definite integral

$$\lim_{n \to \infty} \sum_{i=1}^{n} f(x_i)(x_i - x_{i-1}) = \int_a^b f(x)\,dx.$$

Thus the total work done in moving the object from a to b is given by

(8) $$\text{Work} = \int_a^b f(x)\,dx.$$

EXAMPLE 7. How much work is done in stretching a coil spring 7 centimeters, starting with the spring in its natural position, if a one-gram weight will stretch the spring one centimeter?

312 CHAPTER 9: FURTHER APPLICATIONS

Solution. According to **Hooke's law**, in stretching a spring, the restoring force is proportional to elongation. (For this reason, a spring can be used to measure weight.) Since $f(1) = 1$, it follows that $f(x) = x$. Hence

$$\text{Work} = \int_0^7 f(x)\, dx$$
$$= \int_0^7 x\, dx$$
$$= \frac{x^2}{2}\bigg|_0^7$$
$$= 24.5 \text{ gram-centimeters.}$$

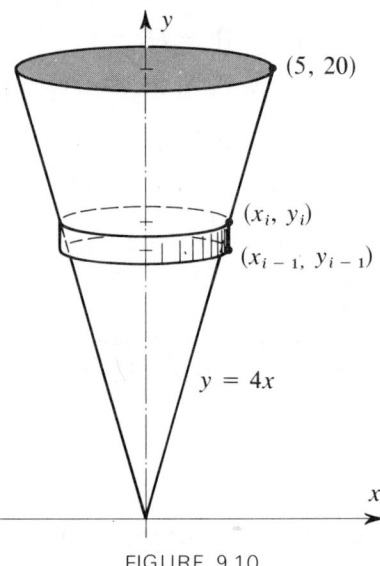

FIGURE 9.10

EXAMPLE 8. A water storage tank in the form of a right circular cone of height 20 feet and base radius 5 feet, as in Figure 9.10, is full of water. The valve at the bottom of the tank is stuck, and so the water must be pumped out the top of the tank. How much work is required to empty the tank?

Solution. The density δ of water is approximately 62.5 pounds per cubic foot. To compute the work done, we first subdivide the interval $[0, 20]$ on the y-axis into n equal subintervals. The volume of the ith slab is equal to $\pi x_i^2 (y_i - y_{i-1})$, where (by looking at similar triangles) $x_i = y_i/4$. Thus the weight of the ith slab is $\delta\pi(y_i/4)^2(y_i - y_{i-1})$. This slab must be pumped a distance of $20 - y_i$ feet, and hence the work done to remove this slab is

$$(20 - y_i)\,\delta\pi\left(\frac{y_i}{4}\right)^2(y_i - y_{i-1}).$$

The total work done in removing all the water is approximately

$$\sum_{i=1}^{n}(20 - y_i)\,\delta\pi\left(\frac{y_i}{4}\right)^2(y_i - y_{i-1}).$$

This is a Riemann sum for the definite integral

$$\int_0^{20}(20 - y)\,\delta\pi\left(\frac{y}{4}\right)^2 dy,$$

which represents the total amount of work done. But

$$\int_0^{20}(20 - y)\,\delta\pi\left(\frac{y}{4}\right)^2 dy = \delta\pi\int_0^{20}\left(\frac{5}{4}y^2 - \frac{y^3}{16}\right)dy$$

$$= \delta\pi\left(\frac{5}{12}y^3 - \frac{1}{64}y^4\right)\bigg|_0^{20}$$

$$= \delta\pi(833.3)$$

$$\approx 163{,}624.61 \text{ foot-pounds.}$$

PROBLEMS

A particle is acted on by a force $f(x)$. Find the work done in moving the particle from $x = a$ to $x = b$.

1. $f(x) = x^2$, $a = 0$, $b = 3$
2. $f(x) = \sqrt{x}$, $a = 1$, $b = 4$
3. $f(x) = \dfrac{1}{x}$, $a = 1$, $b = 2$
4. $f(x) = \dfrac{1}{x^2}$, $a = 1$, $b = 2$
5. $f(x) = \sin x$, $a = 0$, $b = \dfrac{\pi}{2}$
6. $f(x) = \ln x$, $a = 1$, $b = 2$
7. $f(x) = x \sin x^2$, $a = 0$, $b = \sqrt{\dfrac{\pi}{2}}$
8. $f(x) = xe^{x^2}$, $a = 0$, $b = 1$
9. $f(x) = \begin{cases} x^2 & \text{if } 0 \le x \le 1 \\ 1 & \text{if } 1 \le x \le 2, \\ \dfrac{1}{x-1} & \text{if } x \ge 2 \end{cases}$ $a = 0$, $b = 5$

10. $f(x) = |x|$, $a = -1$, $b = 2$

11. $f(x) = \begin{cases} x & \text{if } 0 \le x \le 1 \\ \dfrac{1}{x} & \text{if } 1 \le x \end{cases}$, $a = 0, b = 3$

12. $f(x) = \begin{cases} -\sin x & \text{if } x \le 0 \\ 1 - \cos x & \text{if } x \ge 0 \end{cases}$, $a = -\dfrac{\pi}{2}, b = \dfrac{\pi}{2}$

13. A water storage tank in the form of a sphere of radius 10 meters is half full of water. How much work is required to pump the water out the top?

14. Force is equal to mass times acceleration. How much work is required to halt a 4000-pound automobile going at 88 feet per second if the acceleration due to applying the brakes is -10 feet/sec^2?

15. How much work is required to stretch a spring 1 meter if a 1-gram force will stretch the spring 2 centimeters?

16. During the power stroke of a piston, the pressure inside the cylinder is given by $P = K/V^c$, where K and c are constants and V is the volume inside the cylinder. As the gas in the cylinder expands from volume V_1 to volume V_2, the total work done by the expanding gas is $W = \int_{V_1}^{V_2} P\, dV$. Evaluate this integral if $0 < c < 1$.

4. MISCELLANEOUS APPLICATIONS

We have repeated the following four-step process many times during this chapter.

1. Subdivide the interval $[a, b]$ into a large number (say n) of subintervals $[x_{k-1}, x_k]$.
2. Estimate the quantity (work, force, arc length, area, volume) between x_{k-1} and x_k by a product of the form $f(x_k)(x_k - x_{k-1})$.
3. Estimate the total quantity (work, force, arc length, area, volume) by taking a sum of the form

$$\sum_{k=1}^{n} f(x_k)(x_k - x_{k-1}).$$

4. Compute the limit

$$\lim_{n \to \infty} \sum_{k=1}^{n} f(x_k)(x_k - x_{k-1}) = \int_a^b f(x)\, dx.$$

This four-step process can be used to solve many additional types of problems.

EXAMPLE 9. A parachutist falls vertically with a velocity $v(t) = 32t$, $0 \le t \le 2$ seconds. How many feet vertically does he fall during the first 2 seconds?

SECTION 4: MISCELLANEOUS APPLICATIONS

Solution.
(i) Subdivide the interval $[0, 2]$ into n subintervals of the form $[t_{k-1}, t_k]$, where $t_k = 2k/n$.

(ii) The equation $d = rt$ (distance = rate × time) is valid if r is constant, and it is a good approximation if r is allowed to change only slightly. If n is chosen large enough, then between t_{k-1} and t_k the velocity is roughly equal to $32t_k$, and hence

(iii) the distance fallen during this time interval is approximately

$$\sum_{k=1}^{n} 32t_k(t_k - t_{k-1}).$$

(iv) The exact total distance fallen is

$$\lim_{n \to \infty} \sum_{k=1}^{n} 32t_k(t_k - t_{k-1}) = \int_0^2 32t \, dt = 16t^2 \Big|_0^2 = 64 \text{ feet}.$$

Thus the parachutist falls vertically a distance of 64 feet during the first 2 seconds.

The preceding four-step process can also be used to compute the area of the surface generated by rotating the graph of $y = f(x)$ about the x-axis, $a \le x \le b$.

1. We first subdivide the interval $[a, b]$ into n subintervals $[x_{k-1}, x_k]$, where $x_k = k(b - a)/n$, $0 \le k \le n$.

2. The chord connecting the points $(x_{k-1}, f(x_{k-1}))$ and $(x_k, f(x_k))$ has length

$$\sqrt{(x_{k-1} - x_k)^2 + (f(x_{k-1}) - f(x_k))^2} = \sqrt{1 + (f'(\bar{x}_k))^2} \frac{b-a}{n},$$

as in Section 1. This arc is rotated through a distance of approximately $2\pi f(\bar{x}_k)$, and hence the surface area generated by the arc between x_{k-1} and x_k is roughly equal to

$$2\pi f(\bar{x}_k)\sqrt{1 + (f'(\bar{x}_k))^2} \frac{b-a}{n}.$$

3. Adding these surface areas as k goes from 1 to n gives

$$\sum_{k=1}^{n} 2\pi f(\bar{x}_k)\sqrt{1 + (f'(\bar{x}_k))^2} \frac{b-a}{n}.$$

316   CHAPTER 9: FURTHER APPLICATIONS

4. The exact total surface area between $x = a$ and $x = b$ is the limit

$$\lim_{n \to \infty} \sum_{k=1}^{n} 2\pi f(\bar{x}_k)\sqrt{1 + (f'(\bar{x}_k))^2}\, \frac{b-a}{n},$$

which we recognize as being the definite integral

(9) $$S = \int_a^b 2\pi f(x)\sqrt{1 + f'(x)^2}\, dx.$$

In general, the **area of a surface** generated by rotating the graph of a function about some line is given by

(10) $$S = \int_\alpha^\beta 2\pi \rho\, ds,$$

where $ds = \sqrt{dx^2 + dy^2}$ is the differential of arc length and ρ is the radius through which the graph is rotated. The **differential of surface area** is $dS = 2\pi \rho\, ds$ (see Figure 9.11).

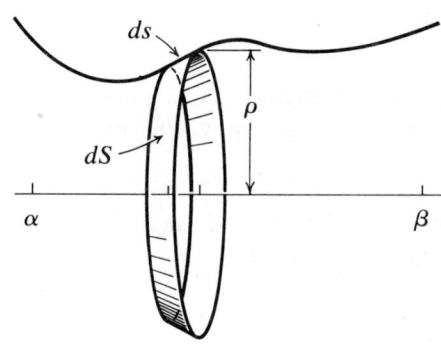

FIGURE 9.11

EXAMPLE 10. Find the area of the surface generated by rotating the graph of $y = \sqrt{x}$ about the x-axis, $0 \le x \le 1$.

Solution. In this example, $\rho = y = \sqrt{x}$, and hence the surface area is given by

$$2\pi \int_0^1 \sqrt{x}\sqrt{dx^2 + dy^2} = 2\pi \int_0^1 \sqrt{x}\sqrt{1 + \left(\frac{dy}{dx}\right)^2}\, dx$$

$$= 2\pi \int_0^1 \sqrt{x}\sqrt{1 + \frac{1}{4x}}\, dx$$

$$= \pi \int_0^1 \sqrt{4x+1}\,dx$$

$$= \pi \frac{2}{3} \cdot \frac{1}{4}(4x+1)^{3/2}\Big|_0^1$$

$$= \pi \frac{5\sqrt{5}-1}{6}.$$

EXAMPLE 11. Find the lateral surface area of a right circular cone of height h and base radius r.

Solution. The surface of a cone of height h and base radius r is generated by rotating the graph of $y = rx/h$ about the x-axis, $0 \le x \le h$, as in Figure 9.12. The surface area is given by

$$2\pi \int_{x=0}^{x=h} y\sqrt{dx^2+dy^2} = 2\pi \int_0^h \frac{rx}{h}\sqrt{1+\left(\frac{dy}{dx}\right)^2}\,dx$$

$$= \frac{2\pi r}{h}\int_0^h x\sqrt{1+\frac{r^2}{h^2}}\,dx$$

$$= \frac{2\pi r\sqrt{r^2+h^2}}{h^2}\int_0^h x\,dx$$

$$= \pi r\sqrt{r^2+h^2}.$$

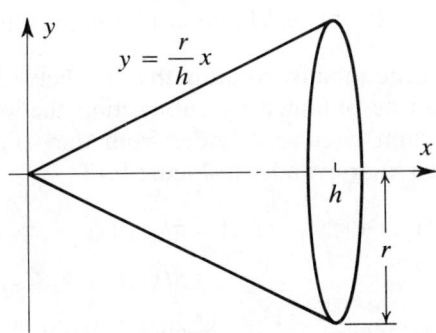

FIGURE 9.12

EXAMPLE 12. Find the surface area of a sphere of radius a.

Solution. A sphere of radius a is generated by rotating the graph of $y = \sqrt{a^2-x^2}$ about the x-axis, $-a \le x \le a$. The surface area is then given by

$$2\pi \int_{-a}^{a} \sqrt{a^2 - x^2} \sqrt{1 + \left(\frac{x}{\sqrt{a^2 - x^2}}\right)^2} \, dx = 2\pi \int_{-a}^{a} \sqrt{a^2 - x^2} \frac{\sqrt{a^2 - x^2 + x^2}}{\sqrt{a^2 - x^2}} \, dx$$

$$= 2\pi \int_{-a}^{a} a \, dx$$

$$= 4\pi a^2.$$

As another application of the four-step process, consider the volume generated by rotating the region bounded by the graph of $y = f(x)$, and the x-axis, $a \le x \le b$, about the y-axis, as in Figure 9.13.

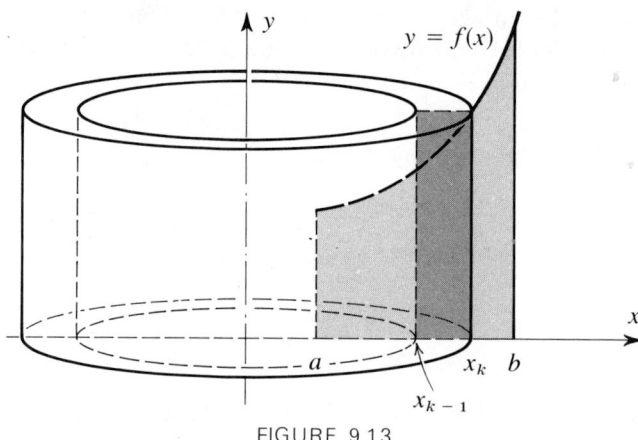

FIGURE 9.13

1. Subdivide the interval $[a, b]$ into n subintervals $[x_{k-1}, x_k]$.

2. The volume generated by rotating the slice between x_{k-1} and x_k about the y-axis can be obtained by subtracting the volume $\pi x_{k-1}^2 f(x_k)$ of the smaller right circular cylinder from the volume $\pi x_k^2 f(x_k)$ of the larger right circular cylinder in Figure 9.13:

$$\pi x_k^2 f(x_k) - \pi x_{k-1}^2 f(x_k) = \pi f(x_k)(x_k^2 - x_{k-1}^2)$$
$$= \pi f(x_k)(x_k + x_{k-1})(x_k - x_{k-1})$$
$$\approx 2\pi x_k f(x_k)(x_k - x_{k-1}),$$

using $2x_k \approx x_k + x_{k-1}$.

3. The total volume is roughly

$$\sum_{k=1}^{n} 2\pi x_k f(x_k)(x_k - x_{k-1}).$$

SECTION 4: MISCELLANEOUS APPLICATIONS

4. The exact volume is the limit

$$\lim_{n \to \infty} \sum_{k=1}^{n} 2\pi x_k f(x_k)(x_k - x_{k-1}) = 2\pi \int_a^b x f(x)\, dx.$$

Thus the volume is given by the integral

(11) $$V = 2\pi \int_a^b x f(x)\, dx.$$

EXAMPLE 13. Find the volume generated by rotating the region bounded by $y = \sqrt{x}$ and $y = 0$, $0 \le x \le 1$, about the y-axis.

Solution. We must compute the integral $2\pi \int_0^1 x\sqrt{x}\, dx$, which is given by

$$2\pi \int_0^1 x\sqrt{x}\, dx = 2\pi \int_0^1 x^{3/2}\, dx$$

$$= 2\pi \frac{2}{5} x^{5/2} \Big|_0^1$$

$$= \frac{4\pi}{5}.$$

The foregoing technique is called the **cylindrical shell** method for computing volumes of revolution.

EXAMPLE 14. Use the cylindrical shell method to compute the volume of a sphere of radius a.

Solution. We can rotate the graph of $y = \sqrt{a^2 - x^2}$ about the y-axis to get the top half of a sphere, so the total volume is

$$4\pi \int_0^a x\sqrt{a^2 - x^2}\, dx = -\pi \frac{4}{3}(a^2 - x^2)^{3/2} \Big|_0^a$$

$$= \frac{4\pi a^3}{3}.$$

EXAMPLE 15. Find the **hydrostatic force** on a vertical dam.

Solution. Let h be the depth of water backed up by the dam. We will again use the familiar four-step process.
 (i) We subdivide the interval $[0, h]$ into n subintervals $[x_{k-1}, x_k]$, each of width h/n, as in Figure 9.14.

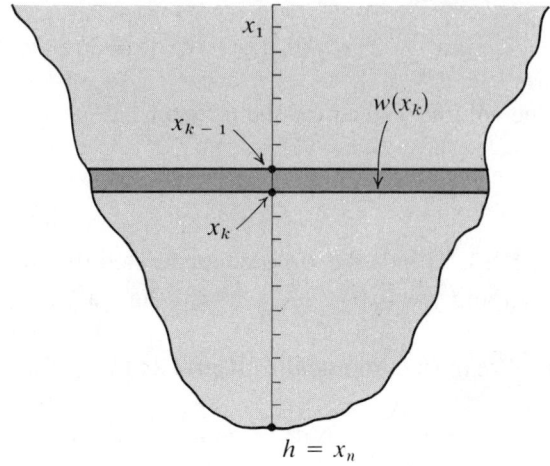

FIGURE 9.14

(ii) From physics, if pressure is constant, then force = pressure × area. Pressure is certainly much greater at the bottom of the dam than at the top; in fact, pressure = density × depth. (Pressure does *not* depend on the shape of a container, and it is independent of direction. The pressure on a vertical surface is the same as the pressure on a horizontal surface.) The density of water is $\delta \approx 62.5$ pounds per cubic foot. If n is large, then the pressure is nearly constant on the kth narrow strip, so the force on the kth strip is approximately $\delta x_k w(x_k)(x_k - x_{k-1})$.

(iii) We can estimate the total force on the dam by taking the sum of the forces on the narrow strips:

$$\sum_{k=1}^{n} \delta x_k w(x_k)(x_k - x_{k-1}).$$

(iv) The exact total force on the dam is the limit

$$\lim_{n \to \infty} \sum_{k=1}^{n} \delta x_k w(x_k)(x_k - x_{k-1}) = \delta \int_0^h x w(x) \, dx.$$

PROBLEMS

Find the surface area generated by rotating the given arcs about the x-axis.
1. $y = e^x, 0 \le x \le 2$
2. $y = \sin x, 0 \le x \le \pi$

3. $y = \ln x$, $1 \le x \le 2$ [Use Simpson's rule with $n = 4$.]

4. $y = 2\sqrt{x}$, $0 \le x \le 3$

5. $y = \frac{1}{2}(e^x + e^{-x})$, $0 \le x \le 1$

6. $y = x^3$, $0 \le x \le 2$

7. $x^{2/3} + y^{2/3} = 1$, $0 \le x \le 1$, $0 \le y \le 1$

In Problems 8 to 12, the given region is rotated about the given line. Find the volume of the solid generated.

8. Region bounded by $y = x^2$ and $y = 0$, $1 \le x \le 2$, about $x = -1$.

9. Region bounded by $y = \sqrt{x}$ and $y = 0$, $0 \le x \le 4$, about the x-axis.

10. Region bounded by $y = \sin x$ and $y = 0$, $0 \le x \le \pi$, about the y-axis.

11. Region bounded by $y = x - 2$, $y = 0$, and $x = 10$, about $x = 1$.

12. Region bounded by the circle $x^2 + (y - 3)^2 = 1$ about the x-axis.

Find the length of each of the following arcs.

13. $y = e^{-x}$, $0 \le x \le 1$

14. $y = \frac{1}{2\sqrt{2}} x^{3/2}$, $0 \le x \le 2$

15. $x^{2/3} + y^{2/3} = 1$, $0 \le x \le 1$, $0 \le y \le 1$

16. $y = \frac{2}{3} x^{3/2} - \frac{1}{2} x^{1/2}$, $0 \le x \le 1$

REVIEW SECTION

By interpreting the definite integral

$$\int_a^b f(x)\,dx = \lim_{n \to \infty} \sum_{k=1}^n f(x_i) \frac{b-a}{n}$$

in various ways, we can obtain numerous applications of the definite integral. We summarize a few of the applications.

The **arc length** of the graph of $y = f(x)$ between $x = a$ and $x = b$ is the number

(a) $$\int_a^b \sqrt{1 + (f'(x))^2}\,dx$$

if this integral exists. The **differential of arc length** is given by

(b) $$ds = \sqrt{dx^2 + dy^2}.$$

If a mass m_i is placed at the point x_i on the x-axis, then the *moment* about the point $x = \bar{x}$ is defined to be the sum

(c) $$\sum_{i=1}^{n}(x_i - \bar{x})m_i.$$

The *centroid* $(\bar{x}, \bar{y})$ of the plane region in Figure 9.8 is given by

(d) $$\bar{x} = \frac{\int_a^b x[f(x) - g(x)]\,dx}{A}$$

and

(e) $$\bar{y} = \frac{\int_c^d y[k(y) - h(y)]\,dy}{A},$$

where $A = \int_a^b [f(x) - g(x)]\,dx$ is the area of the region.

The amount of *work* done by a force $f(x)$ from $x = a$ to $x = b$ is the integral

(f) $$\text{Work} = \int_a^b f(x)\,dx.$$

The *differential of surface area* for a surface of revolution is given by

(g) $$dS = 2\pi\rho\,ds,$$

where ds is the differential of arc length and ρ is the radius of rotation. The *surface area* is then

(h) $$S = \int_\alpha^\beta 2\pi\rho\,ds.$$

Using the *cylindrical shell method*, the *volume* generated by rotating the region bounded by $y = f(x)$ and $y = 0$, $a \le x \le b$, about the y-axis is the integral

(i) $$V = 2\pi \int_a^b xf(x)\,dx.$$

Using the *disk method*, the *volume* generated by rotating the region bounded by $y = f(x)$ and $y = 0$, $a \le x \le b$, about the x-axis is the integral

(j) $$V = \pi \int_a^b (f(x))^2\,dx.$$

The *average value* of a function f on an interval $[a, b]$ is

(k) $$f_{av} = \frac{1}{b-a}\int_a^b f(x)\,dx.$$

REVIEW PROBLEMS

1. Find the length of each of the following arcs.

    a. $y = mx + b$, $x_1 \leq x \leq x_2$
    b. $f(x) = \frac{1}{2}(e^x + e^{-x})$, $0 \leq x \leq 1$
    c. $f(x) = \ln(\sec x)$, $0 \leq x \leq \frac{\pi}{4}$
    d. $f(x) = \frac{1}{3}(x^2 + 2)^{3/2}$, $0 \leq x \leq 1$
    e. $x \to e^x$, $0 \leq x \leq 1$
    f. $y = \ln x$, $0 < a \leq x \leq b$

2. Find the area of the surface generated by rotating the given arc about the x-axis.

    a. $f(x) = r$, $0 \leq x \leq h$, where r and h are positive constants.
    b. $y = mx + b$, $0 \leq x \leq h$, where m, b, and h are positive constants.
    c. $f(x) = \sqrt{x}$, $0 < a \leq x \leq b$
    d. $x^2 + y^2 = r^2$, $-r \leq x \leq r$
    e. $y = \sin x$, $\frac{\pi}{4} \leq x \leq \frac{\pi}{2}$
    f. $y = \frac{e^x + e^{-x}}{2}$, $0 \leq x \leq 1$

3. Find the centroid of the following regions.

    a. The region between $y = \sqrt{x}$ and $y = x^2$, $0 \leq x \leq 1$.
    b. The triangle with vertices $(0, 0)$, $(a, 0)$, and $(0, b)$.
    c. The region bounded by $y = \sin x$, and $y = 0$, $0 \leq x \leq \frac{\pi}{2}$.
    d. The region bounded by $y = \cos x$, and $y = 0$, $0 \leq x \leq \frac{\pi}{2}$.
    e. The region common to both $x^2 + y^2 = 3^2$ and $x^2 + (y - 3)^2 = 3^2$.
    f. The region bounded by $y = x^2$ and $y = 9$, $-3 \leq x \leq 3$.
    g. The region bounded by $y = x^2$ and $y = x^3$, $0 \leq x \leq 1$.

4. Find the average value of the function $f(x) = \sqrt{x^2 + 1}$ on the interval $[-1, 1]$.

5. Find the area of the surface generated by rotating the arc $x = 2\ln t$, $y = t^2$, $1 \leq t \leq 2$, about the x-axis.

6. Use the shell method to find the volume of the solid generated by rotating the given region about the y-axis.

    a. $y = \sqrt{x}$, $y = 0$, $0 \leq x \leq 4$
    b. $y = x\sqrt{x}$, $y = 0$, $0 \leq x \leq 4$
    c. $y = \frac{1}{x}$, $y = 0$, $1 \leq x \leq 4$
    d. $y = \frac{1}{\sqrt{x}}$, $y = 0$, $1 \leq x \leq 4$

7. Find the natural length of a spring if the work required to stretch the spring from 4 to 5 centimeters is equal to twice the work required to stretch the spring from 3 to 4 centimeters.

8. Find the volume of the solid generated by rotating the triangle with vertices $(1, 1)$, $(2, 4)$, and $(3, 3)$ about

    a. the x-axis, and
    b. the y-axis.

324   CHAPTER 9: FURTHER APPLICATIONS

9. Two hallways meet at right angles. Find the length of the longest thin ladder that can be carried horizontally from one hallway, which is 2 meters wide, to a second hallway, which is 3 meters wide.

10. A particle is moving along the x-axis by being acted upon by a force $F(x) = 3x + 4/x^2$. How much work is required to move the particle from $x = 1$ to $x = 10$?

11. A right circular cone is generated by rotating the graph of $y = (b/a)x$ about the x-axis, $0 \le x \le a$. Show that the volume of a right circular cone is $\frac{1}{3}Ah$, where A is the area of the base and $h = a$ is the height of the cone.

12. A force of 40 grams is required to compress a spring of natural length 10 centimeters to a length of 8 centimeters. Find the work required to compress the spring from 10 to 9 centimeters; from 9 to 8 centimeters; from 8 to 7 centimeters.

13. A 3000-foot steel cable weighing 4 pounds per foot is suspended in the shaft of an oil well. How much work is required to lift the cable out of the well? How much work is required if there is a 500-pound drill on the end of the cable?

14. A vertical cylindrical tank 20 feet in diameter and 30 feet deep sits on top of a 25-foot tower. How much work is required to fill the tank from the bottom with water from a holding pond at the base of the tower?

15. The base of a solid is the region bounded by $x^2 + y^2 = 1$, and each plane perpendicular to the x-axis intersects the solid in an equilateral triangle. Find the volume of the solid.

16. The base of a solid is the region bounded by $x^2 + y^2 = 1$, and each plane perpendicular to the x-axis intersects the solid in a square. Find the volume of the solid.

17. Impurities in a 5-gallon fish tank measure 1000 parts per million (ppm). A new filter is put in the tank that recycles 1 gallon per hour and removes 90% of the impurities. How long will it take before the impurity count is down to 100 ppm? [*Hint:* Let $x(t)$ be the amount of impurities in the tank at time t and find an expression for dx/dt in terms of the amount of impurities taken out and the amount returned per hour at time t.]

18. If the force of gravity is inversely proportional to the square of the distance of an object from the center of the Earth, how much work is required to move a satellite from a circular orbit of radius a to a circular orbit of radius b?

19. A farmer is digging a vertical hole of diameter 2 feet for a fence post when he strikes a large horizontal root of an old oak tree, also of diameter 2 feet. How much wood must he chop away if the axes of the two cylinders lie in the same plane?

20. A hole of radius a is drilled through the center of a sphere of radius r, $a \le r$. Find the volume of the solid remaining.

21. Assume that the Earth is a perfect sphere. Show that the surface area (including water and land) between 0 and 30 degrees latitude north is equal to the area above the 30th parallel.

22. Imagine the Earth cut by a pair of parallel planes, a distance h apart. Show that the surface area between these two planes is $2\pi rh$, where r is the radius of the Earth. Show that Problem 21 is a special case.

23. In free fall, a parachutist is acted on by gravity and air friction, and the resultant force is $mg - cv$, where m is mass, g is acceleration due to gravity, v is velocity, and c is a constant

determined by flight characteristics of the parachutist. From physics, force equals mass times acceleration, so that $\dfrac{dv}{dt} = a = g - c\dfrac{v}{m}$.

a. Show that $v = \dfrac{mg}{c}(1 - e^{-ct/m})$.

b. Find c if a parachutist in free fall attains a maximum velocity $\bar{v}$.

24. A net full of sponges, which weighs 500 pounds when soaking wet, is lifted 60 feet above the water level onto a ship. How much work is required to do this if the net is raised at a rate of 0.5 foot per second and the sponges lose water at a rate of $7e^{-t/60}$ pounds per second?

25. Find the volume generated by rotating the circle $x^2 + (y - c)^2 = r^2$, $r < c$, about the x-axis. Show that this volume is equal to the product of the area of the region bounded by $x^2 + (y - c)^2 = r^2$ with the distance $2\pi c$ traveled by the center of the circle.

26. Let R be a region entirely above the x-axis and consider the volume generated by rotating the region about the x-axis. Use the shell method to show that the volume generated is equal to the area of Region R multiplied by the distance $2\pi \bar{y}$ traveled by the centroid $(\bar{x}, \bar{y})$ of R. (This is actually a theorem due to Pappus, a mathematician who lived about A.D. 300.)

CHAPTER 10

TOPICS IN ANALYTIC GEOMETRY

We now pause in our development of calculus to consider some topics in geometry. Many problems in mathematics can be greatly simplified by looking at them from the proper geometric perspective. We begin by looking at parametric equations, a natural setting for studying motion in two or three dimensions. This topic is followed by a study of polar coordinates, a coordinate system designed to take advantage of the natural symmetry that is involved in many physical and geometric problems. The chapter is concluded with a study of the conic sections (circles, ellipses, parabolas, and hyperbolas) and related topics, such as the translation and rotation of axes, required to simplify the study of the conic sections.

1. PARAMETRIC EQUATIONS

One car is traveling east on State Route 14 and a second car is traveling north on State Route 47. The two roads meet at an uncontrolled intersection, so that the path of the first car crosses the path of the second car. Does this mean that the two cars will collide at the uncontrolled intersection? Certainly not. No such conclusion can be made until we know *when* the two cars go through the intersection. It is perfectly all right for the paths of the two cars to cross, as long as they do so at different times.

In order to keep track of the additional information needed for such problems, we introduce parametric equations. Such equations describe the location of a point (or perhaps a car or a plane or an atomic particle) in terms of time (or some other convenient parameter).

The equations

(1) $$x(t) = 5t \qquad y(t) = 64t - 16t^2$$

might describe the location $(x(t), y(t))$ of a ball thrown into the air from the point $(0, 0)$. Here the x-coordinate refers to the distance the ball has traveled to the east, and the y-coordinate is the altitude of the ball, as in Figure 10.1.

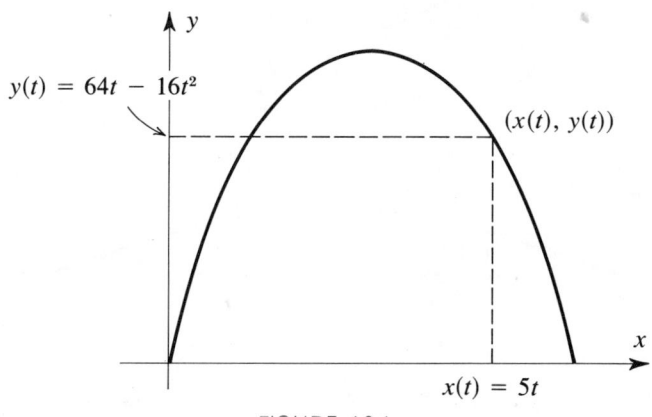

FIGURE 10.1

A point $(x(t), y(t))$ on a circle of radius 1 can be described by the equations

(2) $$x(t) = \cos t \qquad y(t) = \sin t.$$

Since $x^2 + y^2 = \cos^2 t + \sin^2 t = 1$, the point $(x(t), y(t))$ does indeed lie on the unit circle, as in Figure 10.2.

Pairs of equations like (1) and (2) are called **parametric equations** with **parameter** t.

EXAMPLE 1. How far will Lee Terrific's golf ball travel (neglecting roll and air friction) if the ball leaves the tee at a velocity of 200 feet per second at a 20° angle? (See Figure 10.3.)

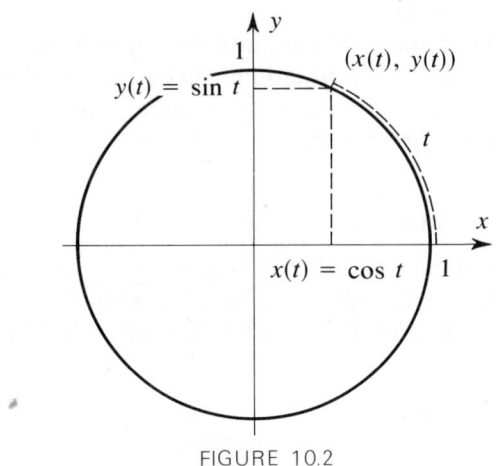

FIGURE 10.2

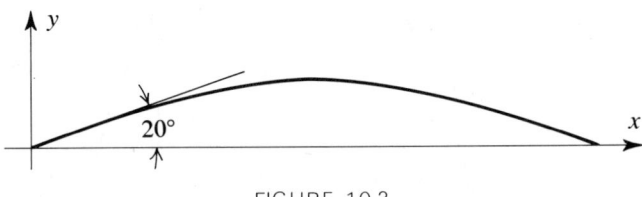

FIGURE 10.3

Solution. The initial vertical component of velocity is $y'(0) = 200 \sin 20°$ ≈ 69.8 feet per second. Thus, at any time t, $y'(t) = 69.8 - 32t$, so that the altitude is given by $y(t) = 69.8t - 16t^2$. The horizontal component of velocity is given by $x'(t) = 200 \cos 20° \approx 188$, so that $x(t) = 188t$. The ball hits the ground when $y(t) = 0$. Solving for t, we get

$$t \approx \frac{69.8}{16} \approx 4.36 \text{ seconds.}$$

The distance traveled is thus

$$x(4.36) \approx 820 \text{ feet.}$$

Another method of solution is possible. The parametric equations $y(t) = 69.8t - 16t^2$ and $x(t) = 188t$ can be solved for y in terms of x as follows:

$$y = 69.8t - 16t^2$$

$$= 69.8 \frac{x}{188} - 16\left(\frac{x}{188}\right)^2$$

$$= 0.371x - 0.00045x^2.$$

When $y = 0$, either $x = 0$ or else

$$x \approx \frac{0.371}{0.00045}$$

$$\approx 824 \text{ feet.}$$

(The discrepancy between the answers 820 and 824 comes from the roundoff error in the denominator 0.00045.) Notice that the parametric equations $x(t) = 188t$ and $y(t) = 69.8t - 16t^2$ contain much more information than the single equation $y = 0.371x - 0.00045x^2$. From the parametric equations, not only is it possible to find out how far the ball travels, but it is also possible to find out when it gets there and exactly where it is at each instant while it is in the air.

EXAMPLE 2. Find the equation of the line tangent to the graph of $x(t) = t^2 + 1$, $y(t) = \sqrt{t+1}$ at the point where $t = 3$.

Solution 1. By the chain rule,

$$\frac{dy}{dt} = \frac{dy}{dx}\frac{dx}{dt}.$$

Solving for dy/dx, we get

$$\frac{dy}{dx} = \left(\frac{dy}{dt}\right) \bigg/ \left(\frac{dx}{dt}\right).$$

But

$$\frac{dy}{dt} = \frac{1}{2\sqrt{t+1}} \quad \text{and} \quad \frac{dx}{dt} = 2t.$$

Evaluating at $t = 3$, we see that the slope of the tangent line is 1/24. The tangent line is then given by

$$y - 2 = \frac{1}{24}(x - 10),$$

since $y(3) = 2$ and $x(3) = 10$.

EXAMPLE 3. Eliminate the parameter t and sketch the graph of $x = a \cos t$, $y = b \sin t$.

Solution. Notice that $\cos t = x/a$ and $\sin t = y/b$. It follows that

$$\frac{x^2}{a^2} + \frac{y^2}{b^2} = \cos^2 t + \sin^2 t$$

$$= 1.$$

This is the equation of an ellipse, and the graph is given in Figure 10.4. (See Section 3 for a discussion of an ellipse.)

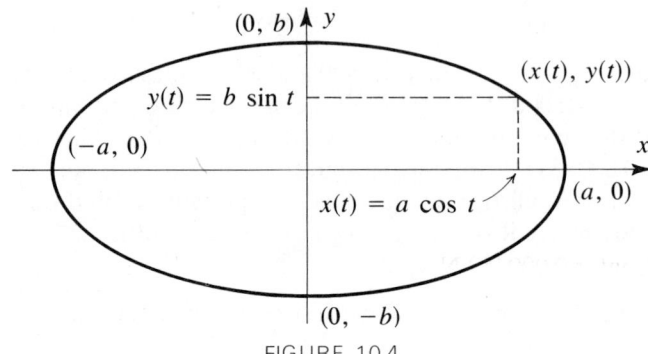

FIGURE 10.4

Recall that the differential of arc length is given by

(3) $$ds = \sqrt{dx^2 + dy^2}.$$

(See Chapter 9.) Formula (3) is also valid if x and y are given in parametric equations.

EXAMPLE 4. Find the length of the arc given by
$$x = 2t^3 \qquad y = 3t^2 \qquad (0 \leq t \leq \sqrt{3}).$$

Solution. The length is given by

$$\int_{t=0}^{t=\sqrt{3}} \sqrt{dx^2 + dy^2} = \int_0^{\sqrt{3}} \sqrt{(6t^2)^2\, dt^2 + (6t)^2\, dt^2}$$
$$= \int_0^{\sqrt{3}} 6t\sqrt{t^2 + 1}\, dt.$$

Let $u = t^2 + 1$. Then $du = 2t\, dt$, and hence

$$6\int_0^{\sqrt{3}} t\sqrt{t^2 + 1}\, dt = 3\int_1^4 \sqrt{u}\, du$$
$$= 3 \cdot \frac{2}{3} u^{3/2} \Big|_1^4$$
$$= 2(8 - 1)$$
$$= 14.$$

PROBLEMS

Compute dy/dx at the indicated points. Sketch the graph.

1. $y = b \sin t$, $x = a \cos t$, $t = \dfrac{\pi}{4}$

2. $y = t^3 + \sqrt{t}$, $x = \dfrac{1}{t}$, $t = 4$

3. $y = e^t$, $x = \ln t$, $t = 1$

4. $y = b \tan t$, $x = a \sec t$, $t = \dfrac{\pi}{4}$

5. $y = t + 1$, $x = t - 1$, $t = 5$

6. $y = t^3 + 1$, $x = t^3 - 1$, $t = 5$

7. $y = \sqrt{t^3 + 1}$, $x = \sqrt[3]{t^5 - 5}$, $t = 2$

8. $y = \cos 2t$, $x = \cos t$, $t = \dfrac{\pi}{3}$

Sketch the graph of each of the following. Eliminate the parameter, if necessary.

9. $y = b \tan t$, $x = a \sec t$

10. $y = t + 1$, $x = t - 1$

11. $y = \cos t$, $x = \sin^2 t$

12. $y = e^t$, $x = \ln t$

13. $y = t^3 + 1$, $x = t^2$

14. $y = \sqrt{t + 1}$, $x = \sqrt{t - 1}$

15. $y = a(1 - \cos t)$, $x = a(t - \sin t)$

16. $y = 3 \sin t \sin 3t$, $x = 3 \cos t + \cos 3t$

17. At what angle should a ball be thrown into the air in order to maximize the horizontal distance traveled by the ball in the air? Assume that the ball is thrown from ground level and ignore air friction.

18. Compute the length of the arc $y = \sin t$, $x = \cos t$, $0 \le t \le \pi$.

19. Compute the arc length: $x = e^t \cos t$, $y = e^t \sin t$, $0 \le t \le \ln 2$.

2. POLAR COORDINATES

The correspondence between ordered pairs of real numbers and points in the plane has been of central significance throughout our study of calculus. This correspondence was made by relating a pair of numbers to a pair of directed distances. We will now give a second useful method for associating pairs of numbers with points. This method relates a pair of numbers with one directed distance and one angle. This new coordinate system is particularly appropriate where a lot of symmetry exists about one point, such as with circles and spirals.

Consider the point P in Figure 10.5. The line L forms an angle θ with the positive x-axis. Given the angle θ and the distance $r = \sqrt{a^2 + b^2}$ from the point P to the origin, we can again locate the point P. The numbers r and θ are called **polar coordinates** of the point P. The description of a point in polar coordinates is not unique, for the point $P(r, \theta)$ is the same as the point $P(r, \theta + 2\pi)$. For $r < 0$, we define $P(r, \theta)$ to be the point $P(-r, \theta + \pi)$. Several sample points are plotted in Figure 10.6. Notice that $P(r, \theta)$ and $P(-r, \theta)$ lie in opposite directions from the origin.

332 CHAPTER 10: TOPICS IN ANALYTIC GEOMETRY

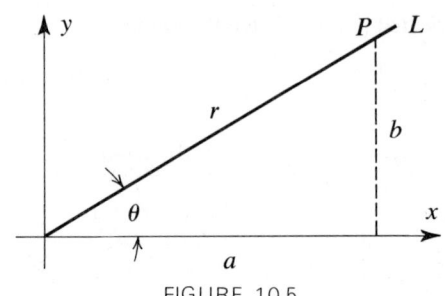

FIGURE 10.5

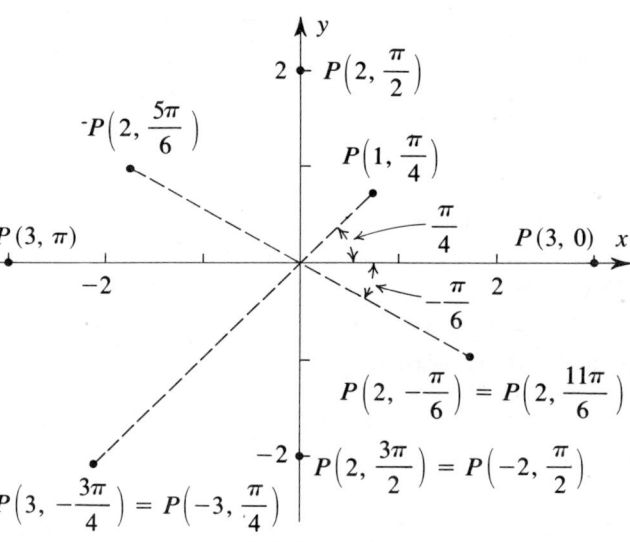

FIGURE 10.6

If $P(r, \theta)$ is a point in polar coordinates, then the rectangular coordinates of the same point are given by

(4) $$x = r \cos \theta$$

and

(5) $$y = r \sin \theta.$$

Given $x = r \cos \theta$ and $y = r \sin \theta$, notice that

$$x^2 + y^2 = r^2 \cos^2 \theta + r^2 \sin^2 \theta = r^2(\cos^2 \theta + \sin^2 \theta) = r^2$$

and

$$\frac{y}{x} = \frac{r \sin \theta}{r \cos \theta} = \frac{\sin \theta}{\cos \theta} = \tan \theta;$$

that is,

(6) $$r^2 = x^2 + y^2$$

and

(7) $$\tan \theta = \frac{y}{x}.$$

In particular, the point $P(2, \pi/6)$ in polar coordinates is the same as the point $(2 \cos(\pi/6), 2 \sin(\pi/6)) = (\sqrt{3}, 1)$ in rectangular coordinates. The point $(2,2)$ in rectangular coordinates is the same as the point $(\sqrt{4+4}, \arctan 1) = P(2\sqrt{2}, \pi/4)$ in polar coordinates.

Notice that $r = 2$ is an equation in polar coordinates of a circle with center at the origin and with radius 2, since the equation $r = 2$ is the same as $\sqrt{x^2 + y^2} = 2$ or $x^2 + y^2 = 4$ in rectangular coordinates. The equation $r = -2$ has the same graph.

EXAMPLE 5. Sketch the polar graph of $r = 1 - \cos \theta$. The graph is called a *cardioid* because of its heart shape.

Solution. We first plot a few points. When $\theta = 0$, we get $r = 0$, and hence the graph passes through the origin. Similarly, when $\theta = \pi/6$,

$$r = 1 - \cos \frac{\pi}{6} = 1 - \frac{\sqrt{3}}{2} \approx 0.134.$$

In a similar manner, we can compute the remaining entries in Table 10.1.

Table 10.1

θ	r
0	0
$\frac{\pi}{6}$	$1 - \frac{\sqrt{3}}{2} \approx 0.134$
$\frac{\pi}{4}$	$1 - \frac{\sqrt{2}}{2} \approx 0.293$
$\frac{\pi}{3}$	0.5
$\frac{\pi}{2}$	1
$\frac{2\pi}{3}$	1.5
$\frac{3\pi}{4}$	$1 + \frac{\sqrt{2}}{2} \approx 1.707$
$\frac{5\pi}{6}$	$1 + \frac{\sqrt{3}}{2} \approx 1.866$
π	2

These points are then plotted and connected with a smooth curve. Since r is an even function of θ, we get a graph as in Figure 10.7.

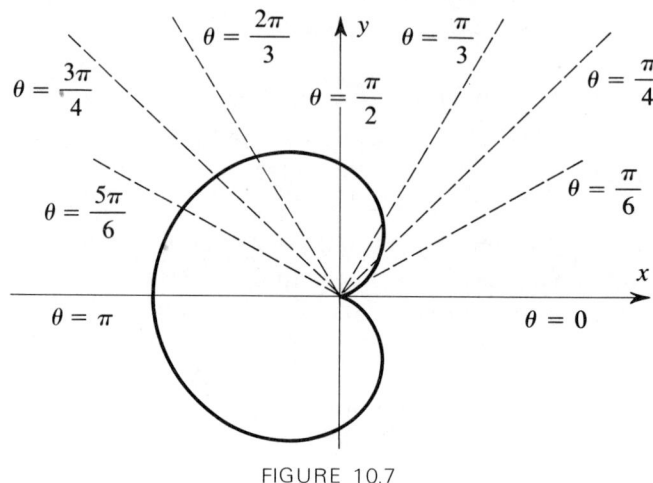

FIGURE 10.7

Suppose that we wish to compute the area of the region bounded by the graph of $r = f(\theta)$, $a \leq \theta \leq b$. As is usually done in problems of this type, subdivide the interval $[a, b]$ into n parts, each of length $(b - a)/n$. Label the points $\theta_0 = a, \theta_1, \theta_2, \ldots, \theta_n = b$. The area of the ith triangular-shaped region in Figure 10.8 is approximately equal to

$$\frac{1}{2} \frac{b-a}{n} (f(\theta_i))^2$$

(see Appendix). Thus the total area is approximately equal to

$$\sum_{i=1}^{n} \frac{1}{2} \frac{b-a}{n} (f(\theta_i))^2.$$

Taking the limit as n tends to infinity yields the integral

(8) $$\frac{1}{2} \int_a^b (f(\theta))^2 \, d\theta.$$

EXAMPLE 6. Find the area of the cardioid $r = 1 - \cos \theta$.

Solution. From Figure 10.7, it is clear that the limits of integration are from zero to 2π.

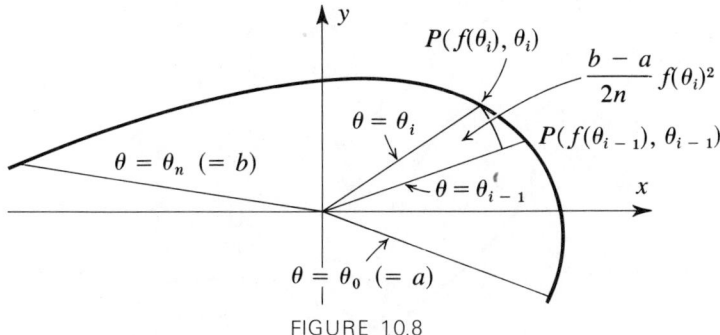

FIGURE 10.8

Thus

$$\frac{1}{2}\int_0^{2\pi} (1-\cos\theta)^2\, d\theta = \frac{1}{2}\int_0^{2\pi} (1 - 2\cos\theta + \cos^2\theta)\, d\theta$$

$$= \frac{1}{2}\cdot 2\pi - \sin\theta \Big|_0^{2\pi} + \frac{1}{2}\int_0^{2\pi} \frac{1}{2}(1 + \cos 2\theta)\, d\theta$$

$$= \pi + \frac{1}{4}\left(\theta + \frac{1}{2}\sin 2\theta\right)\Big|_0^{2\pi}$$

$$= \frac{3\pi}{2}.$$

EXAMPLE 7. Compute the area bounded by $r = \cos 3\theta$. This graph is aptly named a **three-leaved rose**.

Solution. By plotting points, we get the graph as in Figure 10.9. All of the leaves have equal areas, so we might as well find the area of one of the leaves and multiply by three. The area of one leaf is

$$\int_{-\pi/6}^{\pi/6} \frac{1}{2}\cos^2 3\theta\, d\theta = \frac{1}{2}\cdot\frac{1}{3}\int_{-\pi/2}^{\pi/2} \cos^2 t\, dt$$

$$= \frac{2}{6}\int_0^{\pi/2} \frac{1}{2}(1 + \cos 2t)\, dt$$

$$= \frac{1}{6}\left(\theta + \frac{1}{2}\sin 2t\right)\Big|_0^{\pi/2}$$

$$= \frac{\pi}{12}.$$

The total area of the three leaves is thus

$$3 \cdot \frac{\pi}{12} = \frac{\pi}{4}.$$

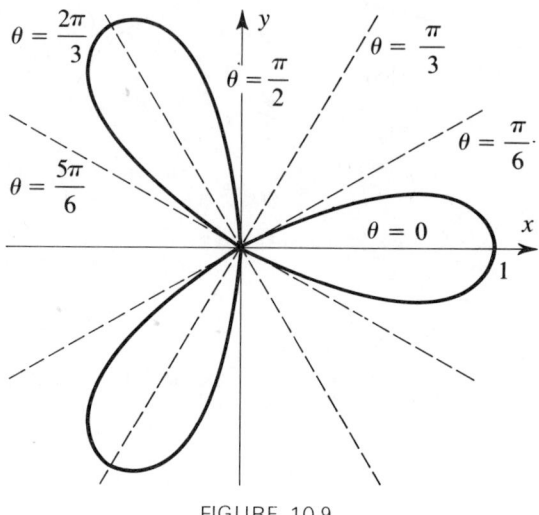

FIGURE 10.9

EXAMPLE 8. Find the area of the region bounded by the *Spiral of Archimedes* $r = \theta$ between $\theta = 0$ and $\theta = \pi$.

Solution. The area is given by

$$\int_0^\pi \frac{1}{2} \theta^2 \, d\theta = \frac{1}{6} \theta^3 \Big|_0^\pi = \frac{\pi^3}{6}.$$

Suppose that an arc is given in polar form $r = f(\theta)$. Then $x = r \cos \theta$ and $y = r \sin \theta$. It follows that

$$\begin{aligned}
ds^2 &= dx^2 + dy^2 \\
&= (dr \cos \theta - r \sin \theta \, d\theta)^2 + (dr \sin \theta + r \cos \theta \, d\theta)^2 \\
&= dr^2 \cos^2 \theta - 2r \, dr \cos \theta \sin \theta \, d\theta + r^2 \sin^2 \theta \, d\theta^2 \\
&\quad + dr^2 \sin^2 \theta + 2r \, dr \sin \theta \cos \theta \, d\theta + r^2 \cos^2 \theta \, d\theta^2 \\
&= r^2 (\sin^2 \theta + \cos^2 \theta) \, d\theta^2 + dr^2 (\cos^2 \theta + \sin^2 \theta) \\
&= r^2 \, d\theta^2 + dr^2;
\end{aligned}$$

that is,

(9) $$ds^2 = r^2 \, d\theta^2 + dr^2.$$

EXAMPLE 9. Find the arc length of the cardioid $r = 1 - \cos\theta$. (See Figure 10.7.)

Solution. The arc length s is given by

$$s = \int_0^{2\pi} ds$$

$$= \int_0^{2\pi} \sqrt{r^2\, d\theta^2 + dr^2}$$

$$= \int_0^{2\pi} \sqrt{(1-\cos\theta)^2\, d\theta^2 + \sin^2\theta\, d\theta^2}$$

$$= \int_0^{2\pi} \sqrt{1 - 2\cos\theta + \cos^2\theta + \sin^2\theta}\, d\theta$$

$$= \int_0^{2\pi} \sqrt{2 - 2\cos\theta}\, d\theta$$

$$= \int_0^{2\pi} \sqrt{2}\sqrt{1 - \cos\theta}\, d\theta.$$

But $\cos\theta = 1 - 2\sin^2\theta/2$, and hence

$$s = \sqrt{2}\int_0^{2\pi} \sqrt{1 - 1 + 2\sin^2\frac{\theta}{2}}\, d\theta$$

$$= 2\int_0^{2\pi} \sin\frac{\theta}{2}\, d\theta$$

$$= -4\cos\frac{\theta}{2}\bigg|_0^{2\pi}$$

$$= -4(-1 - 1)$$

$$= 8.$$

Special care must be exercised when working in polar coordinates. In Example 9, if we had tried to integrate from π to 3π, we would have obtained

$$s = 2\int_\pi^{3\pi} \sin\frac{\theta}{2}\, d\theta$$

$$= -4\cos\frac{\theta}{2}\bigg|_\pi^{3\pi}$$

$$= 0.$$

The reason for the discrepancy is that there is a *cusp* (roughly speaking, a cusp is a sharp corner) at $\theta = 2\pi$ (see Figure 10.7). In order to get the correct answer, it is necessary to integrate from 0 to 2π. This step avoids integrating through the cusp.

PROBLEMS

1. Plot and label each of the following points on one polar graph: $P(3, \pi)$, $P(3, 3\pi)$, $P(-2, 0)$, $P(-1, \pi/2)$, $P(1, \pi/4)$, $P(2, \pi/6)$, $P(1/2, 7\pi/6)$, $P(0, 0)$, $P(0, \pi)$, $P(1, 0)$.
2. Convert each point in Problem 1 to its representation in rectangular coordinates.
3. Give polar coordinates for each of the following points in the xy-plane: $(1, 1)$, $(\sqrt{3}, -1)$, $(3, 4)$, $(0, 2)$, $(-1, 0)$, $(\sqrt{2}, -\sqrt{2})$, $(-1, -1)$, $(0, -1)$, $(0, 0)$, $(-1, 1)$.

In Problems 4 through 12, sketch the polar graph. Names of the figures are given in parentheses.

4. $r = 2 - 2\cos\theta$ (cardioid)
5. $r = 2\cos 3\theta$ (three-leaved rose)
6. $r = 5\cos 2\theta$ (four-leaved rose)
7. $r = 2\sin 5\theta$ (five-leaved rose)
8. $r^2 = \sin 2\theta$ (lemniscate)
9. $r = 1 - 2\cos\theta$ (lemicon)
10. $r = 2\theta$ (Spiral of Archimedes)
11. $r = e^\theta$ (logarithmic spiral)
12. $r = 1/\theta$ (reciprocal spiral)

Find the area enclosed by each of the following figures.

13. $r = 4\sin 4\theta$
14. $r^2 = 4\cos 2\theta$
15. $r = 1 + 2\cos\theta$ (inner loop only)
16. $r = 1/\theta$, $0 < \theta \leq e$

Compute the arc length.

17. $r = a + a\sin\theta$
18. $r = \theta$, $0 \leq \theta \leq \pi$
19. $r = \sin^2(\theta/2)$, $0 \leq \theta \leq \pi$
20. $r = a\theta^2$, $0 \leq \theta \leq \pi$

3. THE CONIC SECTIONS

Several plane figures are of special significance in mathematics. Among them are the *conic sections*. The name conic section is derived from the fact that these figures can be obtained by slicing a right circular cone with a plane, as in Figure 10.10.

SECTION 3: THE CONIC SECTIONS

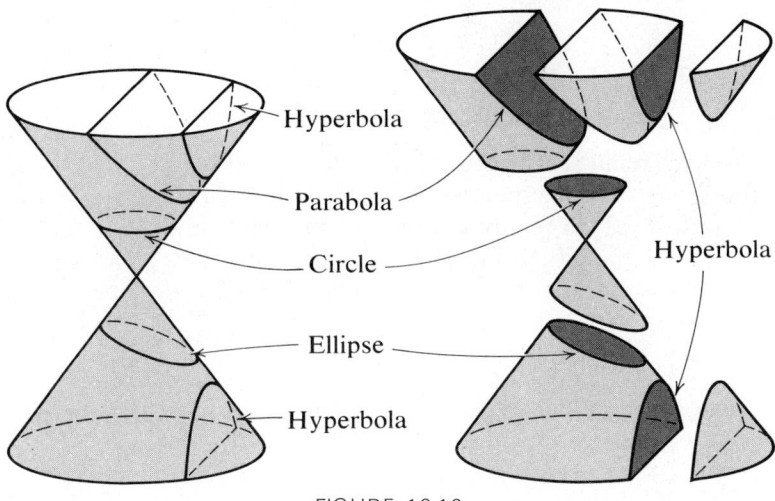

FIGURE 10.10

A *parabola* is the locus of all points (x, y) such that the distance from (x, y) to a fixed point (x_0, y_0) equals the distance from (x, y) to a fixed line L.

The distance from a point (x, y) to a line is always taken to be the perpendicular distance, which is also the shortest distance from the point (x, y) to any point on the line.

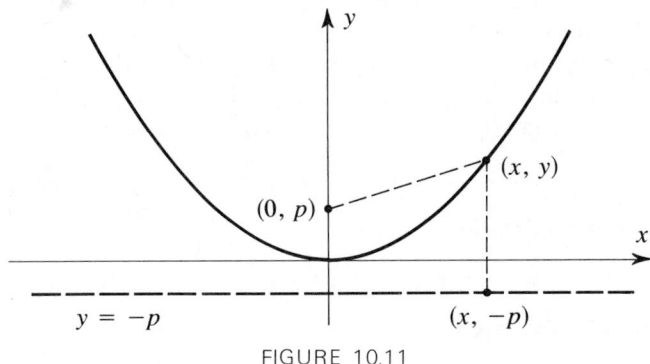

FIGURE 10.11

To simplify the computations, we will take the fixed point at $(0, p)$ and the line L to be $y = -p$, as in Figure 10.11. If (x, y) is any point on the parabola, then by the distance formula,

$$\sqrt{(x - x)^2 + (y + p)^2} = \sqrt{(x - 0)^2 + (y - p)^2}.$$

Squaring both sides, we obtain

$$y^2 + 2yp + p^2 = x^2 + y^2 - 2yp + p^2,$$

which simplifies to

(10) $$x^2 = 4py.$$

The fixed point is called the *focus* and the fixed line is called the *directrix*. The equation $y^2 = 4px$ corresponds to a parabola with directrix $x = -p$ and focus $(p, 0)$. If p is positive, the parabola opens to the right; and if p is negative, the parabola opens to the left. A *standard form* for a parabola is an equation like $x^2 = 4py$ or $y^2 = 4px$. The parabola with equation $x^2 = 4py$ opens up if $p > 0$ and down if $p < 0$.

EXAMPLE 10. Find the equation of the parabola with directrix $x = -2$ and focus at $(2, 0)$.

Solution. Since $p = 2$, the formula $y^2 = 4px$ becomes $y^2 = 8x$.

EXAMPLE 11. Find the equation of the parabola with directrix $y = 3$ that passes through $(0, 0)$.

Solution. This time, $p = -3$ and hence $x^2 = -12y$.

An *ellipse* is the locus of all points (x, y), the sum of whose distances from two fixed points is constant. The fixed points are called *foci* (plural of *focus*).

We simplify the computations by taking the foci at the points $(-c, 0)$ and $(c, 0)$ and by letting the constant be $2a$, as in Figure 10.12. Let (x, y) be any point on the ellipse. Then by the distance formula, we have

$$\sqrt{(x+c)^2 + (y-0)^2} + \sqrt{(x-c)^2 + (y-0)^2} = 2a.$$

This equation is equivalent to

$$\sqrt{(x+c)^2 + y^2} = -\sqrt{(x-c)^2 + y^2} + 2a.$$

Squaring both sides, this becomes

$$x^2 + 2xc + c^2 + y^2 = x^2 - 2xc + c^2 + y^2 + 4a^2$$
$$- 4a\sqrt{(x-c)^2 + y^2},$$

which is equivalent to

$$xc - a^2 = -a\sqrt{(x-c)^2 + y^2}.$$

We again square both sides to eliminate the remaining radical and obtain
$$x^2c^2 - 2a^2xc + a^4 = a^2(x^2 - 2xc + c^2 + y^2),$$
which simplifies to
$$x^2(a^2 - c^2) + y^2a^2 = a^2(a^2 - c^2).$$

But $c < a$, so we have $a^2 - c^2 = b^2$, where $b = \sqrt{a^2 - c^2}$. Dividing both sides by a^2b^2, we get the following formula of an ellipse with foci at $(-c, 0)$ and $(c, 0)$:

(11) $$\frac{x^2}{a^2} + \frac{y^2}{b^2} = 1.$$

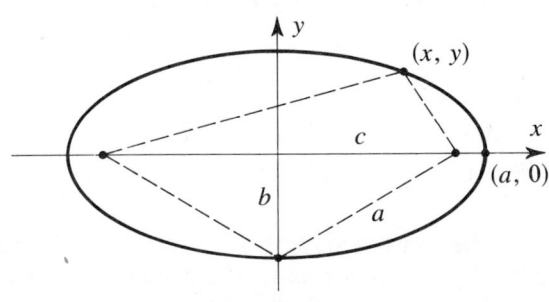

FIGURE 10.12

Because of its simplicity, Formula (11) is called the **standard form** for the formula of an ellipse. Notice that when $y = 0$, we get $x^2 = a^2$ and hence $x = \pm a$. Thus the ellipse passes through the points $(a, 0)$ and $(-a, 0)$. The ellipse also passes through $(0, b)$ and $(0, -b)$. The interval $[-a, a]$ on the x-axis is called the **major axis** and the interval $[-b, b]$ on the y-axis is called the **minor axis**. (If the foci were chosen on the y-axis, then the major axis would lie on the y-axis.) Notice also that we get a circle in the special case where $a = b$—that is, where $c = 0$.

EXAMPLE 12. Find the equation of the ellipse with foci at $(-3, 0)$ and $(3, 0)$ that passes through the point $(0, 4)$.

Solution. Here $c = 3$ and $b = 4$, and hence $a = \sqrt{b^2 + c^2} = 5$. The equation is thus
$$\frac{x^2}{25} + \frac{y^2}{16} = 1.$$

EXAMPLE 13. Find the equation of the ellipse with foci on the y-axis that passes through $(0, 13)$, $(5, 0)$, and $(0, -13)$.

Solution. Note that $b = 13$, $a = 5$, and hence

$$\frac{x^2}{25} + \frac{y^2}{169} = 1.$$

To locate the foci, note that $c^2 = b^2 - a^2 = 144$, so that $c = 12$. Thus the foci are at $(0, 12)$ and $(0, -12)$.

EXAMPLE 14. Rewrite $169x^2 + 144y^2 = 24336$ in standard form and locate the foci.

Solution. Dividing both sides by 24336, we get

$$\frac{x^2}{12^2} + \frac{y^2}{13^2} = 1.$$

The foci are on the y-axis at $(0, 5)$ and $(0, -5)$, since $5^2 = 13^2 - 12^2$. The major axis lies on the y-axis.

A *hyperbola* is the locus of all points (x, y) such that the difference of the distances from two fixed points is constant. The fixed points are again called *foci*.

We again take the foci at $(-c, 0)$ and $(c, 0)$, and we take the constant difference to be $2a$. Let (x, y) be any point on the hyperbola, as in Figure 10.13.

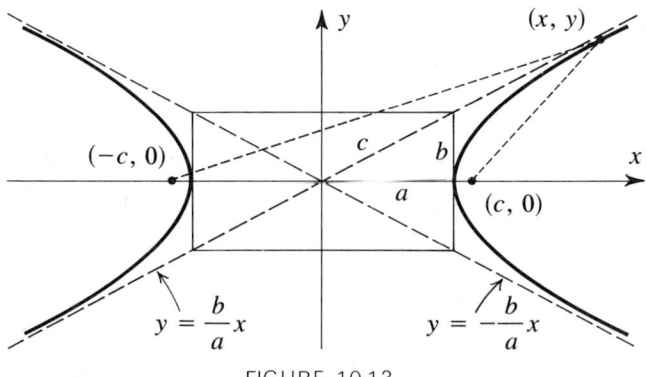

FIGURE 10.13

The distance formula yields

$$\sqrt{(x + c)^2 + y^2} - \sqrt{(x - c)^2 + y^2} = 2a,$$

or, equivalently,
$$\sqrt{(x+c)^2 + y^2} = \sqrt{(x-c)^2 + y^2} + 2a.$$

Squaring both sides, we obtain
$$x^2 + 2xc + c^2 + y^2 = x^2 - 2xc + c^2 + y^2 + 4a^2 + 4a\sqrt{(x-c)^2 + y^2},$$
or
$$xc - a^2 = a\sqrt{(x-c)^2 + y^2}.$$

We again square both sides to get
$$x^2c^2 - 2a^2xc + a^4 = a^2(x^2 - 2xc + c^2 + y^2),$$

which simplifies to
$$x^2(c^2 - a^2) - y^2a^2 = a^2(c^2 - a^2).$$

But $a < c$, so letting $b^2 = c^2 - a^2$ and dividing by a^2b^2, we obtain the following *standard form* for a hyperbola:

(12) $$\frac{x^2}{a^2} - \frac{y^2}{b^2} = 1.$$

We note that if $y = 0$, then $x = \pm a$, so that the graph passes through the points $(a, 0)$ and $(-a, 0)$. Since $c^2 = a^2 + b^2$, the triangle in Figure 10.13 with sides a, b, and c is a right triangle. Solving Formula (12) for y, we get

$$y = \pm \frac{b}{a}\sqrt{x^2 - a^2}.$$

Considering, for convenience, the positive root, we note that the difference $x - \sqrt{x^2 - a^2}$ is small for large x because

$$\lim_{x \to \infty} (x - \sqrt{x^2 - a^2}) = \lim_{x \to \infty} (x - \sqrt{x^2 - a^2}) \frac{x + \sqrt{x^2 - a^2}}{x + \sqrt{x^2 - a^2}}$$

$$= \lim_{x \to \infty} \frac{x^2 - x^2 + a^2}{x + \sqrt{x^2 - a^2}}$$

$$= \lim_{x \to \infty} \frac{a^2}{x(1 + \sqrt{1 - (a/x)^2})}$$

$$= 0.$$

Thus the line $y = (b/a)x$ is an oblique asymptote of the hyperbola. By symmetry, the line $y = -(b/a)x$ is also an oblique asymptote.

If the foci lie on the y-axis instead of the x-axis, then the standard form of the hyperbola is given by

$$\frac{y^2}{b^2} - \frac{x^2}{a^2} = 1.$$

The lines $y = (b/a)x$ and $y = -(b/a)x$ are again asymptotes, as in Figure 10.14.

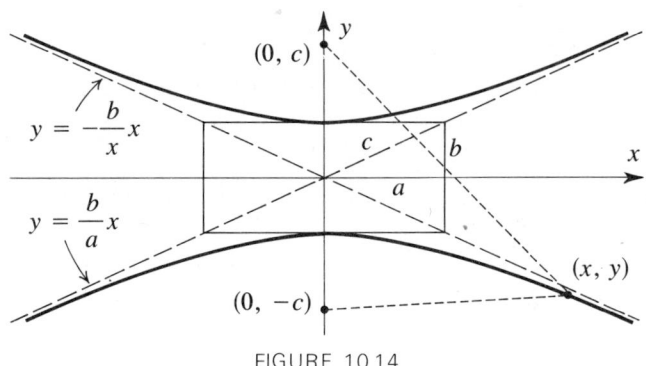

FIGURE 10.14

EXAMPLE 15. Find the equation of the hyperbola with foci at $(-2, 0)$ and $(2, 0)$ if the graph passes through the point $(1, 0)$.

Solution. We have $c = 2$, $a = 1$, and hence $b^2 = 2^2 - 1^2 = 3$. The equation is thus

$$\frac{x^2}{1} - \frac{y^2}{3} = 1.$$

EXAMPLE 16. Rewrite the equation $16y^2 - 9x^2 = 144$ in standard form and sketch the graph.

Solution. We must divide both sides of the equation by 144 to normalize the right-hand side. Thus the equation becomes

$$\frac{y^2}{9} - \frac{x^2}{16} = 1.$$

Here $b = 3$, and $a = 4$, and hence $c = \sqrt{3^2 + 4^2} = 5$. We now have all the necessary information to sketch Figure 10.15.

SECTION 3: THE CONIC SECTIONS

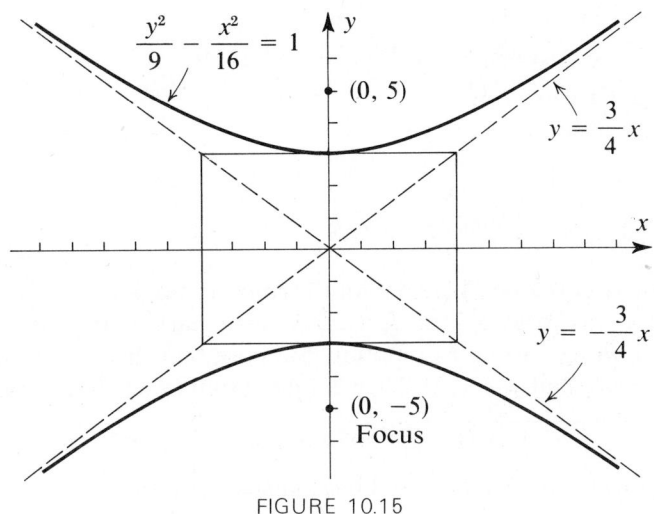

FIGURE 10.15

PROBLEMS

1. Locate the focus and directrix if $x^2 = 10y$.
2. Locate the focus and directrix if $y^2 = -3x$.
3. Find the equation of the parabola that passes through the points $(1, 1)$, $(0, 0)$, and $(1, -1)$.
4. Find the equation of the parabola that passes through $(-8, 2)$, $(0, 0)$, and $(8, 2)$.
5. Put the equation $4y^2 + x = 0$ into standard form and sketch the graph. Locate and label the directrix and the focus.
6. Sketch the graph of $x^2 - 4y = 0$. Label the directrix and the focus.
7. Start with the definition of a parabola and derive the equation of the parabola with directrix $x = 1$ and focus $(5, 2)$.
8. Find the equation of the parabola with directrix $y = k + p$ and focus $(h, k - p)$.
9. Locate the foci if $(x^2/3^2) + (y^2/5^2) = 1$.
10. Locate the foci if $(x^2/a^2) + (y^2/b^2) = 1$ and $a > b$.
11. Find the equation of the ellipse that passes through $(1, 0)$, $(-1, 0)$, $(0, 2)$ and $(0, -2)$.
12. Find the equation of the ellipse that passes through $(4, 0)$, $(0, 4)$, $(-4, 0)$ and $(0, -4)$.
13. Put the equation $25x^2 + 169y^2 = 4225$ into standard form and graph. Locate and label the foci.
14. Put the equation $9x^2 + 4y^2 = 36$ into standard form and graph. Locate and label the foci.
15. Start with the definition and obtain a formula for an ellipse with foci at $(-1, 1)$ and $(5, 1)$ if the sum of the distances from these points is 10.

346 CHAPTER 10: TOPICS IN ANALYTIC GEOMETRY

16. Start with the definition and obtain a formula for an ellipse with foci $(h - c, k)$ and $(h + c, k)$ if the sum of the distances from these points is $2a$, where $a > c \geq 0$.
17. Locate the foci and asymptotes if $y^2 - x^2 = 1$.
18. Locate the foci and asymptotes if $(x^2/25) - (y^2/144) = 1$.
19. Sketch the graph of $9x^2 - 4y^2 = 36$.
20. Sketch the graph of $9y^2 - 16x^2 = 1$.

4. TRANSLATION OF AXES

When we developed formulas for the conic sections, it was always assumed that the graph was located in a very convenient part of the plane. This may appear to be a rather strong assumption. Suppose that the foci of an ellipse are located at $(h - c, k)$ and $(h + c, k)$. Then, for any point (x, y) on the graph, we have

$$\sqrt{(x - h + c)^2 + (y - k)^2} = 2a - \sqrt{(x - h - c)^2 + (y - k)^2}.$$

One of the radicals can be eliminated by squaring both sides:

$$(x - h)^2 + 2(x - h)c + c^2 + (y - k)^2$$
$$= 4a^2 + (x - h)^2 - 2(x - h)c + c^2 + (y - k)^2$$
$$- 4a\sqrt{(x - h - c)^2 + (y - k)^2}.$$

This formula simplifies to

$$(x - h)c - a^2 = -a\sqrt{(x - h - c)^2 + (y - k)^2};$$

and after squaring both sides to eliminate the remaining radical, we obtain

$$(x - h)^2 c^2 - 2(x - h)ca^2 + a^4 = a^2[(x - h)^2 - 2(x - h)c + c^2 + (y - k)^2].$$

Collecting terms gives

$$(x - h)^2(a^2 - c^2) + (y - k)^2 a^2 = a^2(a^2 - c^2).$$

Letting $b^2 = a^2 - c^2$ and dividing both sides by $a^2 b^2$, we get the following formula for an ellipse:

(13) $$\frac{(x - h)^2}{a^2} + \frac{(y - k)^2}{b^2} = 1.$$

Suppose that we now introduce a new coordinate system with origin at the point (h, k), as in Figure 10.16. Then a point with coordinates (x, y) in the old coordinate system will have coordinates $(x - h, y - k)$ in the new coordinate system. Letting $X = x - h$ and $Y = y - k$, Equation (14) becomes

(14) $$\frac{X^2}{a^2} + \frac{Y^2}{b^2} = 1.$$

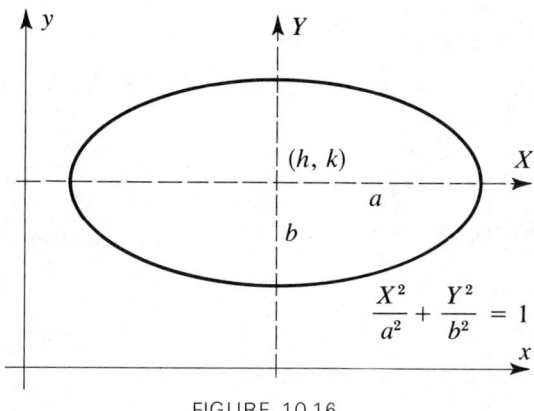

FIGURE 10.16

EXAMPLE 17. Sketch the graph of $x^2 - 10x - 8y + 17 = 0$.

Solution. We need to complete the square for $x^2 - 10x$. Recall that in order to complete the square for $x^2 + bx$, we add and subtract $(b/2)^2$. In this way, we obtain

$$x^2 - 10x + 25 = 25 - 17 + 8y,$$

so that $(x - 5)^2 = 8(y + 1)$. Let $X = x - 5$ and $Y = y + 1$. Then $X^2 = 4pY$ with $p = 2$. This is the equation of a parabola with focus at $(0, p)$ and directrix $Y = -p$ in the X, Y coordinate system. Since the origin in the new coordinate system is the point $(5, -1)$ in the old system, we obtain the graph shown in Figure 10.17.

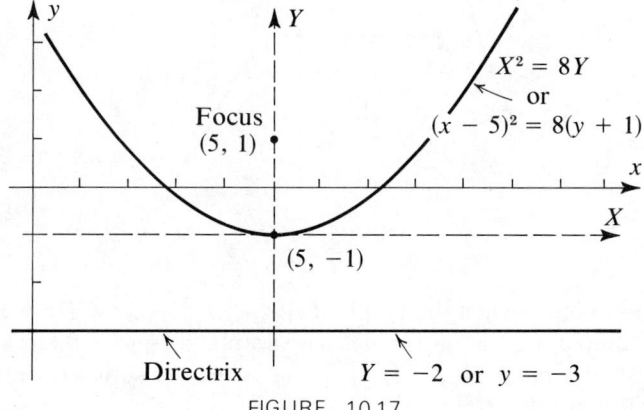

FIGURE 10.17

EXAMPLE 18. Analyze the equation $16x^2 - 9y^2 + 64x + 54y - 161 = 0$.

Solution. We have

$$16(x^2 + 4x + 4) - 9(y^2 - 6y + 9) = 161 + 64 - 81,$$

so that

$$16(x + 2)^2 - 9(y - 3)^2 = 144.$$

Dividing both sides by 144, we obtain

$$\frac{(x+2)^2}{3^2} - \frac{(y-3)^2}{4^2} = 1.$$

If we let $X = x + 2$ and $Y = y - 3$, then we recognize

$$\frac{X^2}{3^2} - \frac{Y^2}{4^2} = 1$$

as a hyperbola with foci on the X-axis. The graph is given in Figure 10.18.

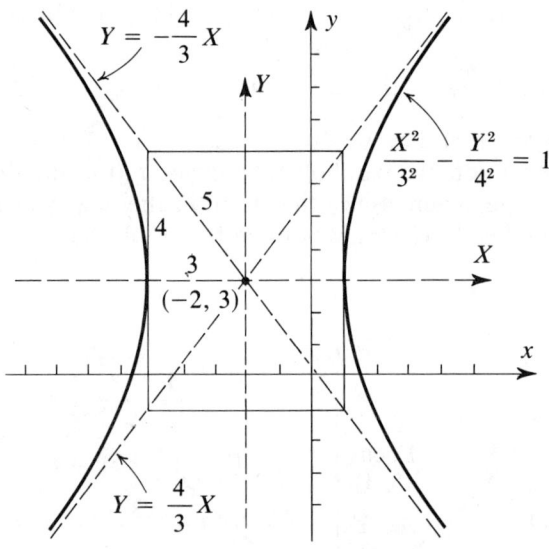

FIGURE 10.18

There are times when the graph of $Ax^2 + By^2 + Cx + Dy + E = 0$ is not a conic. For example, $x^2 + y^2 + 1 = 0$ has no graph, $x^2 + y^2 = 0$ has a graph that consists of just one point, $4(x - 1)^2 + 3(y + 3)^2 = 0$ is satisfied only at $x = 1$, $y = -3$, and the graph of $4(x + 2)^2 - 9(y - 2)^2 = 0$ is a pair of straight lines.

These are considered *degenerate* cases. One point and a pair of intersecting lines are really examples of conic sections, since each can be obtained by slicing a cone with a plane.

PROBLEMS

Identify the following conic sections and sketch the graph, labeling each focus, vertex, directrix, and asymptote.

1. $x^2 + 4y^2 - 2x + 16y - 19 = 0$
2. $x^2 - 9y^2 + 2x - 8 = 0$
3. $y^2 - 6y - 8x + 17 = 0$
4. $x^2 + y^2 + 2x - 4y + 1 = 0$
5. $x^2 - y^2 - 2x = 0$
6. $x^2 + y^2 + 16x - 14y + 77 = 0$
7. $x^2 - 2x - 12y + 25 = 0$
8. $16x^2 + 25y^2 + 32x - 50y - 359 = 0$
9. $9x^2 + 4y^2 + 18x - 24y + 9 = 0$
10. $x^2 - 6x - 8y - 7 = 0$
11. $3y^2 - 2x^2 - 12y - 4x + 4 = 0$
12. $4x^2 + 4y^2 + 16x - 24y + 51 = 0$
13. $25y^2 - 144x^2 + 1000y + 2880x - 800 = 0$
14. $y^2 - 2y - x + 2 = 0$
15. $9x^2 + 25y^2 - 36x + 50y + 60 = 0$
16. Show that an equation of the form $Ax^2 + By^2 + Cx + Dy + E = 0$ is degenerate or else represents

    a. a *circle* if $A = B$.
    b. an *ellipse* if $AB > 0$.
    c. a *parabola* if $AB = 0$.
    d. a *hyperbola* if $AB < 0$.

5. ROTATION OF AXES

Consider the parabola with directrix the line $y = -x - p/\sqrt{2}$ and with focus at the point $(p/\sqrt{2}, p/\sqrt{2})$, as in Figure 10.19. By somewhat tedious computations, the equation of this parabola is

(16) $$x^2 + y^2 - 2xy - 4\sqrt{2}xp - 4\sqrt{2}yp = 0.$$

Starting with Equation (16), how can we recognize it as being the equation of a parabola? We introduce a new set of coordinate axes as in Figure 10.20. The coordinates (x, y) in the old system are related to the coordinates (X, Y) in the new system by the equations

(17) $\quad x = X \cos \theta - Y \sin \theta,$

(18) $\quad y = X \sin \theta + Y \cos \theta,$

or $\quad X = x \cos \theta + y \sin \theta,$

$\quad Y = -x \sin \theta + y \cos \theta,$

where θ is the angle the new X-axis makes with the old x-axis.

350 CHAPTER 10: TOPICS IN ANALYTIC GEOMETRY

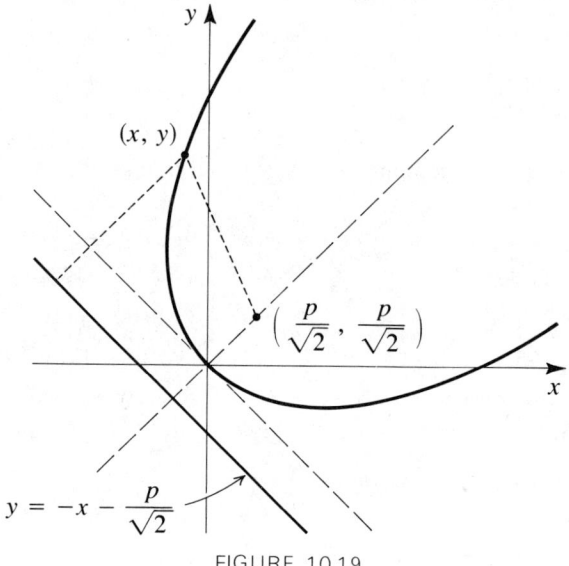

FIGURE 10.19

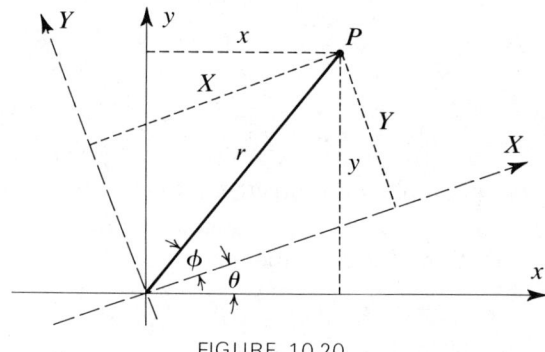

FIGURE 10.20

To verify these equations, we introduce an angle ϕ as in Figure 10.20 and notice that

$$x = r \cos(\theta + \phi)$$
$$= r(\cos\theta \cos\phi - \sin\theta \sin\phi).$$

But $\cos\phi = X/r$ and $\sin\phi = Y/r$, and hence

$$x = r\left(\cos\theta \cdot \frac{X}{r} - \sin\theta \cdot \frac{Y}{r}\right)$$
$$= X \cos\theta - Y \sin\theta.$$

Similarly,
$$y = r \sin(\theta + \phi)$$
$$= r(\sin\theta \cos\phi + \cos\theta \sin\phi)$$
$$= r\left(\sin\theta \cdot \frac{X}{r} + \cos\theta \cdot \frac{Y}{r}\right)$$
$$= X \sin\theta + Y \cos\theta.$$

To get the other pair of equations, we merely interchange x and X, y and Y, and replace θ by $-\theta$.

EXAMPLE 19. Analyze the equation
$$x^2 + y^2 - 2xy - 4\sqrt{2}xp - 4\sqrt{2}yp = 0.$$

Solution. Let $\theta = \pi/4$. Then
$$x = X \cos\frac{\pi}{4} - Y \sin\frac{\pi}{4}$$
$$= \frac{1}{\sqrt{2}}X - \frac{1}{\sqrt{2}}Y,$$

and
$$y = \frac{1}{\sqrt{2}}X + \frac{1}{\sqrt{2}}Y.$$

Substituting into the given equation, we get
$$0 = x^2 + y^2 - 2xy - 4\sqrt{2}xp - 4\sqrt{2}yp$$
$$= (x - y)^2 - 4\sqrt{2}p(x + y)$$
$$= \left(-\frac{2}{\sqrt{2}}Y\right)^2 - 4\sqrt{2}p\frac{2}{\sqrt{2}}X$$
$$= 2Y^2 - 8pX.$$

Simplifying, this becomes
$$Y^2 = 4pX,$$
which we recognize as the equation of a parabola. The graph is given in Figure 10.19.

The foregoing solution is a bit unfair. The angle $\theta = \pi/4$ was pulled out of the air and it happened to work. The next example shows how to pick an angle θ that will simplify the given equation.

CHAPTER 10: TOPICS IN ANALYTIC GEOMETRY

EXAMPLE 20. Analyze the equation
$$x^2 - y^2 + 2\sqrt{3}xy - 2 = 0.$$

Solution. Let $x = X \cos \theta - Y \sin \theta$ and $y = X \sin \theta + Y \cos \theta$. We need to find a value for θ that will yield zero as the coefficient of XY in the new expression. Substituting, we see that

$$\begin{aligned}
2 = x^2 - y^2 &+ 2\sqrt{3}xy \\
= (X \cos \theta &- Y \sin \theta)^2 - (X \sin \theta + Y \cos \theta)^2 \\
&+ 2\sqrt{3}(X \cos \theta - Y \sin \theta)(X \sin \theta + Y \cos \theta) \\
= X^2 \cos^2 \theta &- 2XY \cos \theta \sin \theta + Y^2 \sin^2 \theta - X^2 \sin^2 \theta \\
&- 2XY \cos \theta \sin \theta - Y^2 \cos^2 \theta + 2\sqrt{3}X^2 \cos \theta \sin \theta \\
&+ 2\sqrt{3}XY \cos^2 \theta - 2\sqrt{3}XY \sin^2 \theta - 2\sqrt{3}Y^2 \sin \theta \cos \theta \\
= X^2(\cos^2 \theta &- \sin^2 \theta + 2\sqrt{3} \cos \theta \sin \theta) \\
+ Y^2(\sin^2 \theta &- \cos^2 \theta - 2\sqrt{3} \sin \theta \cos \theta) \\
+ XY(-4 \cos \theta &\sin \theta + 2\sqrt{3} \cos^2 \theta - 2\sqrt{3} \sin^2 \theta).
\end{aligned}$$

We need to find θ such that
$$-4 \cos \theta \sin \theta + 2\sqrt{3} \cos^2 \theta - 2\sqrt{3} \sin^2 \theta = 0.$$

The most direct way to find such a θ is to recall the identities
$$2 \cos \theta \sin \theta = \sin 2\theta$$
and
$$\cos^2 \theta - \sin^2 \theta = \cos 2\theta;$$
then use these identities to rewrite the preceding equation as
$$\sin 2\theta = \sqrt{3} \cos 2\theta,$$
which implies
$$\cot 2\theta = \frac{1}{\sqrt{3}},$$
which implies $2\theta = \pi/3$ and hence $\theta = \pi/6$. With this choice for θ,
$$\begin{aligned}
2 = x^2 - y^2 &+ 2\sqrt{3}xy \\
= X^2 &\left[\cos^2\left(\frac{\pi}{6}\right) - \sin^2\left(\frac{\pi}{6}\right) + 2\sqrt{3} \cos\left(\frac{\pi}{6}\right) \sin\left(\frac{\pi}{6}\right) \right]
\end{aligned}$$

$$+ Y^2 \left[\sin^2\left(\frac{\pi}{6}\right) - \cos^2\left(\frac{\pi}{6}\right) - 2\sqrt{3}\sin\left(\frac{\pi}{6}\right)\cos\left(\frac{\pi}{6}\right) \right]$$

$$+ XY \left[-4\cos\left(\frac{\pi}{6}\right)\sin\left(\frac{\pi}{6}\right) + 2\sqrt{3}\cos^2\left(\frac{\pi}{6}\right) - 2\sqrt{3}\sin^2\left(\frac{\pi}{6}\right) \right]$$

$$= X^2 \left(\frac{3}{4} - \frac{1}{4} + 2\sqrt{3}\frac{\sqrt{3}}{2}\frac{1}{2} \right) + Y^2 \left(\frac{1}{4} - \frac{3}{4} - 2\sqrt{3}\frac{1}{2}\frac{\sqrt{3}}{2} \right)$$

$$+ XY \left(-4\frac{\sqrt{3}}{2}\frac{1}{2} + 2\sqrt{3}\frac{3}{4} - 2\sqrt{3}\frac{1}{4} \right)$$

$$= 2X^2 - 2Y^2 + 0 \cdot XY,$$

which simplifies to

$$X^2 - Y^2 = 1,$$

the equation of a hyperbola. The graph is given in Figure 10.21.

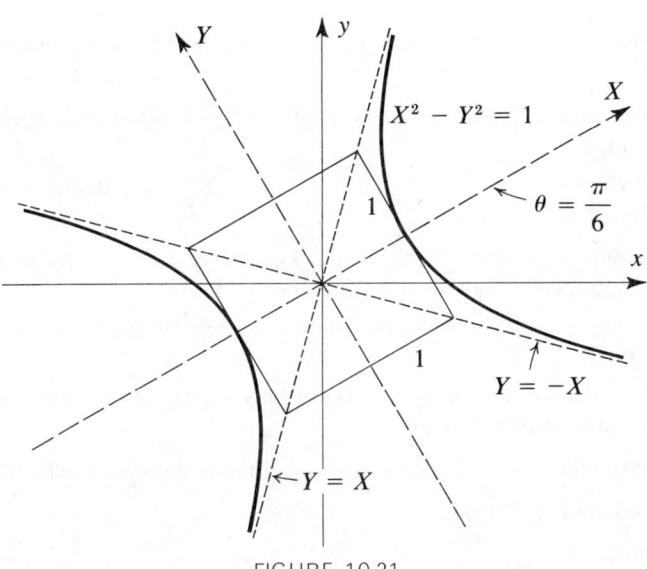

FIGURE 10.21

The method used in Example 20 to find θ can be generalized to an arbitrary equation $Ax^2 + Bxy + Cy^2 + Dx + Ey + F = 0$, $B \neq 0$. The substitutions $x = X\cos\theta - Y\sin\theta$ and $y = X\sin\theta + Y\cos\theta$ lead to a new equation of the form

$$A'X^2 + B'XY + C'Y^2 + D'X + E'Y + F' = 0,$$

where

(19) $$B' = -2(C - A)\cos\theta\sin\theta + B(\cos^2\theta - \sin^2\theta)$$

(see Problem 11). Assuming $B' = 0$, the identities $2\cos\theta\sin\theta = \sin 2\theta$ and $\cos^2\theta - \sin^2\theta = \cos 2\theta$ imply

$$(C - A)\sin 2\theta = B\cos 2\theta.$$

Dividing both sides by $B\sin 2\theta$, this becomes $\cot 2\theta = (C - A)/B$, or

(20) $$\theta = \frac{1}{2}\operatorname{arccot}\frac{C - A}{B}.$$

PROBLEMS

1. Find the new coordinates of the points $(1, 0)$, $(0, 1)$, $(1, \sqrt{3})$, $(\sqrt{3}, 1)$, and $(1, 1)$ if the axes are rotated $\pi/6$.

2. Find the new coordinates of the points $(\sqrt{2}, \sqrt{2})$, $(1, 0)$, $(0, 1)$, $(1, 1)$, and $(4, 1)$ if the axes are rotated $\pi/4$.

3. Find the new equation of the line $(\sqrt{3} - 4)y = (1 + 4\sqrt{3})x + 2$ after the axes are rotated $\pi/6$.

4. Find the new equation of $21x^2 + 31y^2 + 10\sqrt{3}xy - 144 = 0$ after a rotation of $\pi/3$. Sketch the graph.

5. Find the new equation of $x^3 + y^3 + 3x^2y + 3xy^2 + 2x - 2y = 0$ after a rotation of $\pi/4$. Sketch the graph.

6. Analyze the equation $x^2 + y^2 - 2xy - 6\sqrt{2}x - 10\sqrt{2}y + 2 = 0$ by rotating the axes through an appropriate angle. Sketch the graph.

7. Verify that $xy = 1$ is a hyperbola by rotating the axes through an appropriate angle. Sketch the graph.

8. Analyze the equation $16x^2 + 9y^2 - 24xy - 120x - 160y = 0$ by rotating through an appropriate angle. Sketch the graph.

9. Analyze the equation $2xy - 2\sqrt{2}x + 1 = 0$ by rotating the axes. Sketch the graph.

10. Sketch the graph of $5x^2 + 5y^2 + 2\sqrt{3}xy + 4\sqrt{3}x - 4y = 1$.

11. Verify Equation (19).

REVIEW SECTION

We now summarize the basic definitions and results from this chapter. If x and y are both functions of t, then

(a) $$x = x(t) \quad \text{and} \quad y = y(t)$$

are called *parametric equations* with *parameter* t. In parametric equations, *arc length* is given by

(b) $$s = \int_a^b \sqrt{\left(\frac{dx}{dt}\right)^2 + \left(\frac{dy}{dt}\right)^2}\, dt.$$

If $x = x(t)$ and $y = y(t)$, then the *slope of the tangent line* at $t = c$ is

(c) $$\frac{y'(c)}{x'(c)}.$$

If a point P in the plane is connected to the origin with a line L that makes an angle θ with the positive x-axis, and if P is a distance r from the origin, then the numbers r and θ are called *polar coordinates* for the point P and we write

(d) $$P = P(r, \theta).$$

For $r < 0$, we define $P(r, \theta)$ by

(e) $$P(r, \theta) = P(-r, \theta + \pi).$$

Polar coordinates $P(r, \theta)$ are related to rectangular coordinates (x, y) by

(f) $$x = r \cos \theta,$$
(g) $$y = r \sin \theta,$$
(h) $$r^2 = x^2 + y^2,$$

and

(i) $$\tan \theta = \frac{y}{x}.$$

The *area* of the region bounded by the polar graph of $r = f(\theta)$, $a \le \theta \le b$, is given by

(j) $$A = \frac{1}{2} \int_a^b f(\theta)^2 \, d\theta.$$

The *differential of arc length* is

(k) $$ds = \sqrt{r^2 \, d\theta^2 + dr^2}.$$

A *parabola* is the locus of all points P such that the distance from P to a fixed point F (called the *focus*) equals the distance from P to a fixed line L (called the *directrix*). The *standard forms* for a parabola are

(l) $$y^2 = 4px \quad \text{and} \quad x^2 = 4py,$$

where $2p$ is the distance between the focus and the directrix.

An *ellipse* is the locus of all points P such that the sum of the distance from P to a fixed point F_1 with the distance from P to a second fixed point F_2

is a constant. The points F_1 and F_2 are called *foci* (plural of *focus*). The **standard form** for an ellipse is

(m) $$\frac{x^2}{a^2} + \frac{y^2}{b^2} = 1.$$

A *hyperbola* is the locus of all points P such that the difference of the distances from P to two fixed points F_1 and F_2 is constant. The points F_1 and F_2 are called *foci*. The **standard forms** for a hyperbola are

(n) $$\frac{x^2}{a^2} - \frac{y^2}{b^2} = 1 \quad \text{and} \quad \frac{y^2}{b^2} - \frac{x^2}{a^2} = 1.$$

The lines $y = \pm(b/a)x$ are oblique *asymptotes*.

If a new coordinate system is introduced with origin at the point (h, k), then a point with coordinates (x, y) in the old coordinate system has coordinates (X, Y) in the new system, where

(o) $$X = x - h \quad \text{and} \quad Y = y - k.$$

The introduction of such a coordinate system is called *translation of axes*.

If a new coordinate system is introduced with origin at $(0, 0)$ but with axes rotated an angle θ, then a point with coordinates (x, y) in the old system has coordinates (X, Y) in the new system, where

(p) $$X = x \cos \theta + y \sin \theta,$$
(q) $$Y = -x \sin \theta + y \cos \theta;$$

or, equivalently,

(r) $$x = X \cos \theta - Y \sin \theta,$$
(s) $$y = X \sin \theta + Y \cos \theta.$$

The introduction of such a coordinate system is called *rotation of axes*.

REVIEW PROBLEMS

1. Find the area of the region bounded by the polar graph of the given equation.
   a. $r = 2 \cos \theta$
   b. $r = 5 \cos 3\theta$
   c. $r = 2 - \cos \theta$
   d. $r^2 = 2 \sin 2\theta$

2. Find the length of the arc $x = 3 \cos^2 t$, $y = 2 \cos^3 t$, $0 \le t \le \pi/2$.

·3. Find the length of the arc $x = \cos^3 t$, $y = \sin^3 t$, $0 \le t \le \pi/2$.

4. Find the area of the surface generated by rotating the arc $x = \cos^2 t$, $y = \sin t \cos t$, $0 \le t \le \pi/2$, about the x-axis.

5. Find the area of the surface generated by rotating one loop of $r^2 = a^2 \cos 2\theta$ about the line $\theta = 0$.

6. Find the length of the following arcs.

   a. $r = a\cos^2\dfrac{\theta}{2}, 0 \leq \theta \leq \pi$

   b. $r = a\sin^3\dfrac{\theta}{3}, 0 \leq \theta \leq \pi$

   c. $r = a(1 - \cos\theta)$

7. Change each of the following rectangular equations into an equivalent polar equation.

   a. $x^2 + y^2 - 4x = \sqrt{x^2 + y^2}$
   b. $x^2(1 - y) = y^3$

   c. $x \cos A + y \sin A = 2$
   d. $(x^2 + y^2)^2 = y^2 - x^2$

   e. $(x^2 + y^2)^2 + 2ay(x^2 + y^2) - a^2x^2 = 0$

8. Transform each of the following polar equations into an equivalent equation in rectangular coordinates.

   a. $r = 3 \tan\theta \sin\theta$
   b. $r^2 = a^2 \sin 2\theta$

9. Sketch the following conic sections. Label each focus, vertex, center, and asymptote (where appropriate).

   a. $3x^2 - 3xy + y^2 + 6x - 7 = 0$
   b. $x^2 + xy + y^2 - 3 = 0$

   c. $\sqrt{2}x - 2xy = 3$
   d. $3x^2 + 2y^2 - 4y - 7 = 0$

   e. $x^2 + xy + y^2 - x + y - 3 = 0$
   f. $x^2 - 4xy - 2y^2 - 3x + 2y + 6 = 0$

10. Sketch the graph of the given parametric equations. Eliminate the parameter and obtain an equation in x and y.

    a. $x = 2t, y = 3 - t$
    b. $x = e^t, y = e^{-t}$

    c. $x = \sin\theta, y = 2\cos\theta$
    d. $x = \sin 2\theta, y = 2 \sin^2\theta$

    e. $x = \tan^2\theta, y = \sec\theta$
    f. $x = \tan\theta, y = \sec\theta$

    g. $x = 3\cos\theta, y = 5\sin\theta$
    h. $x = 1 + \dfrac{1}{t}, y = t - \dfrac{1}{t}$

11. Find the equation of the tangent line at $t = \pi/3$ if $x(t) = 3\cos t$, $y(t) = 2 \sin t$.

12. Find dy/dx if $x = a(\theta - \sin\theta)$ and $y = a(1 - \cos\theta)$.

13. Sketch the polar graph of $r = a \sin\theta + b \cos\theta$. Rewrite in rectangular coordinates.

14. Find the area of the region common to both $r = \cos\theta$ and $r = \sin\theta$.

15. Find the area of the region common to both $r = 3\cos\theta$ and $r = 1 + \cos\theta$.

16. Find the area of the region outside $r = 1$ and inside $r = 2 \sin\theta$.

17. Find the area of the region outside $r = 1 + \sin\theta$ that is inside $r = 3 \sin\theta$.

18. Find the area of the inner loop of $r = 1 - 2 \sin\theta$.

19. The ellipse $(x^2/a^2) + (y^2/b^2) = 1$ is rotated about the x-axis to generate what is called an *ellipsoid*. Compute the volume of such an ellipsoid.

20. Find the equation of a parabola that passes through the points $(-1, 1)$, $(2, -1)$, and $(1, 2)$ and that has its axis parallel to the y-axis.

358 CHAPTER 10: TOPICS IN ANALYTIC GEOMETRY

21. Show that the parabolas $x^2 = -4a(y - a)$ and $x^2 = 4a(y + a)$
    a. have the same focus, and
    b. intersect at right angles.

22. Find the equation of an ellipse with foci at $(0, 2)$ and $(0, 8)$ and with one vertex at the origin.

23. Find the equation of a hyperbola with foci at $(-1, 0)$ and $(9, 0)$ and with one vertex at the origin.

24. Show that no line which is tangent to the hyperbola $y^2 - x^2 = 1$ passes through the origin.

25. Find the angle between the tangent lines to the graphs of $y = x^2$ and $y = \sqrt{x}$ at their point of intersection.

26. A ball is thrown from a position 2 meters above the ground and strikes the ground 3 seconds later. If the ball is thrown at an intial angle of 30°, find the horizontal distance traveled by the ball.

27. Find all points where the tangent line is horizontal or vertical if $x = \cos 2t$, $y = \cos t$.

28. Neglecting air friction, show that the path of a projectile is given in parametric equations by $x = tv_0 \cos \theta$, $y = -\frac{1}{2}gt^2 + tv_0 \sin \theta$, where θ is the original angle of inclination, v_0 is the magnitude of the initial velocity, and g is the acceleration due to gravity. Find the highest point on the path of the projectile. Show that the path is a parabola.

29. Suppose that under a rotation of axes, the equation $Ax^2 + Bxy + Cy^2 + Dx + Ey + F = 0$ is transformed into $A'X^2 + B'XY + C'Y^2 + D'X + E'Y + F' = 0$. Show that
    a. $B^2 - 4AC = B'^2 - 4A'C'$,
    b. $A + C = A' + C'$, and
    c. $F = F'$.

30. A tire is rolling along the x-axis at a rate of v meters per second. If the radius of the tire is r meters, find the parametric equations of the motion of a point on the tire that starts at $(0, 0)$ when $t = 0$.

31. The expression $B^2 - 4AC$ in Problem 29 is called a **discriminant**. Show that the equation $Ax^2 + Bxy + Cy^2 + Dx + Ey + F = 0$ represents a degenerate conic or else
    a. a parabola if $B^2 - 4AC = 0$,
    b. an ellipse if $B^2 - 4AC < 0$, and
    c. a hyperbola if $B^2 - 4AC > 0$.

32. Sketch the parabola $y = x^2$ on a large piece of graph paper, with the x-axis labeled from -30 to $+30$ and the y-axis labeled from 0 to 1000 (using appropriate scales). Let a and b be two numbers between 0 and 30 and connect the points $(-a, a^2)$ and (b, b^2) with a straight line. Notice that the line crosses the y-axis at a point $(0, c)$ and show that, in general, $c = ab$.

33. On a piece of unlined paper, mark a point about 2 centimeters above the center on the lower edge. Make a sharp crease, as in Figure 10.22. Repeat by making creases in different directions until a parabola appears. Argue that a parabola can indeed be generated in this manner.

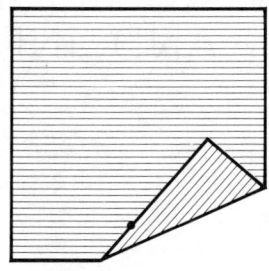

FIGURE 10.22

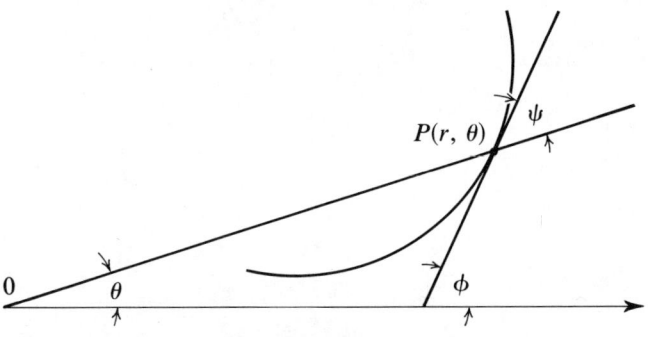

FIGURE 10.23

34. Let ϕ be the angle that the tangent line at the point $P(r, \theta)$ makes with the positive x-axis and let $\psi = \phi - \theta$, as in Figure 10.23. Show that $r = \tan \psi (dr/d\theta)$.

35. If two intersecting lines have slopes m_1 and m_2, respectively, and if α is the angle between them, show that $\tan \alpha = (m_2 - m_1)/(1 + m_1 m_2)$. [*Hint:* Use the addition formula for tangent.]

36. If m_3 is the slope of the line that bisects the angle α in Problem 35, show that $(m_2 - m_3)/(1 + m_2 m_3) = (m_3 - m_1)/(1 + m_1 m_3)$.

37. Let $(x^2/a^2) + (y^2/b^2) = 1$ be the equation of an ellipse, $a > b > 0$. Let L_1 be the line through a point P on the ellipse and the focus $(-c, 0)$, L_2 the line through P and the focus $(c, 0)$, and L_3 the tangent line at P. Show that L_3 bisects the angle between L_1 and L_2. (This has a physical interpretation. Sound travels in a straight line and bounces off a wall at the same angle it hits the wall; that is, "the angle of reflection equals the angle of incidence." In an elliptically shaped room, a whisper originating at one focus is easily audible at the other focus.)

38. Let $x^2 = 4py$ be the equation of a parabola. Let P be a point on the parabola, F the focus. If L_1 is the line through P and F, L_2 is the vertical line through P, and L_3 is the tangent line at P, show that L_3 bisects the angle between L_1 and L_2. (This also has a physical interpretation. Television crews use a parabolic surface with a microphone at the focus to listen in on a quarterback a hundred feet away in the presence of thousands of shouting fans.)

CHAPTER 11

POLYNOMIAL APPROXIMATIONS AND INFINITE SERIES

Since polynomials are easy to work with (easy to evaluate, differentiate, integrate, and so forth), it is sometimes convenient to approximate a given function f by a polynomial function. Often only partial information about the function f is known; perhaps the value of f is given at a finite number of points $x_1, x_2, \ldots, x_n$, or perhaps $f(a)$, $f'(a)$, $f''(a)$, $\ldots$, $f^{(n)}(a)$ are known at one point $x = a$. A polynomial approximation of f will then allow $f(x)$ to be estimated for other choices of x. In this chapter we will start with a function f and then obtain a polynomial of degree n, which in some sense is the "best possible" polynomial approximation of degree n. By taking the limit as n tends to infinity, we get an infinite series that agrees with the original function on some interval. Section 3 deals with the technical question of when a given series converges. Finally, we look at how infinite series can be used to solve a large collection of differential equations.

1. TAYLOR POLYNOMIALS

There are several ways to approximate a given function by a polynomial function. The following example illustrates two of the most common methods.

SECTION 1: TAYLOR POLYNOMIALS 361

EXAMPLE 1. Find a polynomial of degree 2 that approximates the graph of $f(x) = 1/(x^2 + 1)$ on the interval $[-1, 3]$.

Solution 1. Notice that f is a function that goes through the points $(-1, \frac{1}{2})$, $(1, \frac{1}{2})$, and $(3, \frac{1}{10})$. It is possible (see Problem 16) to find a polynomial of degree 2 that also passes through these three points—namely,

$$p(x) = -\frac{1}{20}x^2 + \frac{11}{20}.$$

The graph of f with p superimposed is given in Figure 11.1.

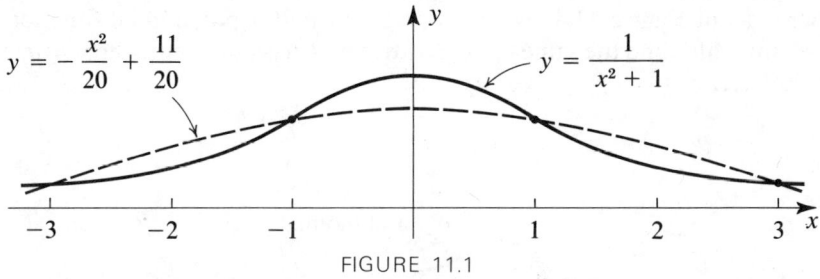

FIGURE 11.1

Solution 2. Notice that $f'(x) = -2x/(x^2 + 1)^2$ and $f''(x) = (6x^2 - 2)/(x^2 + 1)^3$, so that $f(1) = \frac{1}{2}$, $f'(1) = -\frac{1}{2}$, and $f''(1) = \frac{1}{2}$. Notice that the polynomial

$$q(x) = \frac{1}{4}x^2 - x + \frac{5}{4}$$

also has the properties that $q(1) = \frac{1}{2}$, $q'(1) = -\frac{1}{2}$, and $q''(1) = \frac{1}{2}$. A method for finding such polynomial functions will be discussed later in this section. The graph of f with q superimposed is given in Figure 11.2.

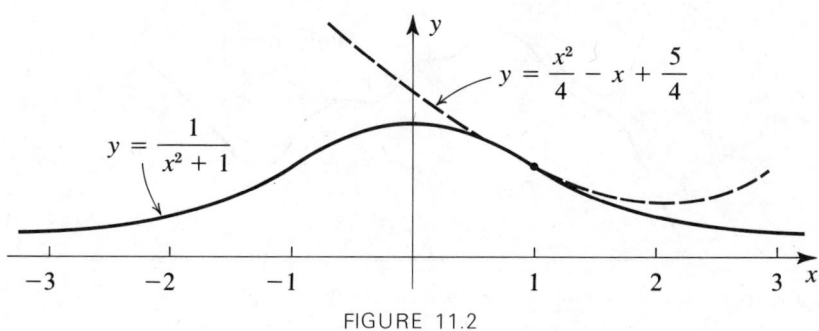

FIGURE 11.2

The preceding example thus describes two ways to obtain polynomial approximations. In the first type of approximations, if the value of f is known at the $n + 1$ points $x_0, x_1, \ldots, x_n$, then exactly one polynomial of degree n at most can be found that passes through the points $(x_i, f(x_i))$. Thus two points determine a line, three points determine a parabola, four points determine a cubic, and so forth. If the numbers x_i are evenly distributed in the interval $[a, b]$, we should hope that the associated polynomial would closely approximate the function f throughout the interval $[a, b]$. Indeed a theorem due to Weierstrass (1815–1897) states that if f is any continuous function on the interval $[a, b]$, then f can be approximated as closely as desired by some polynomial function on $[a, b]$.

The approximations that will be studied in this chapter will be of the second type. Instead of approximating a function fairly closely on an interval, as in Figure 11.3, we will obtain very close approximations to f for x close to a fixed number a, as in Figure 11.4. To do so, we will find a polynomial function that has the same value and the same kth derivative as f has at the number $x = a$, where $k = 1, 2, 3, \ldots, n$.

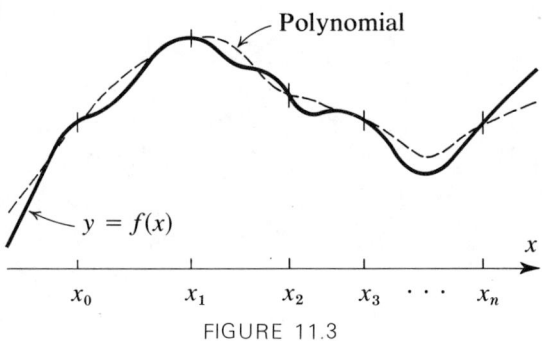

FIGURE 11.3

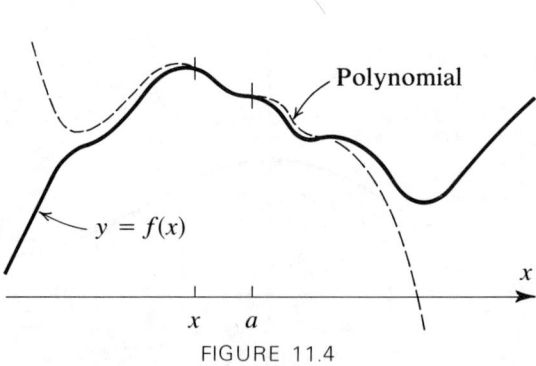

FIGURE 11.4

Suppose that $p(x)$ is a polynomial of degree n such that

$$p(a) = f(a), \quad p'(a) = f'(a), \quad p''(a) = f''(a), \ldots, p^{(n)}(a) = f^{(n)}(a).$$

It is convenient to write $p(x)$ in terms of powers of $(x - a)$:

(1) $$p(x) = b_0 + b_1(x - a) + b_2(x - a)^2 + \cdots + b_n(x - a)^n.$$

It is always possible to rewrite a given polynomial in the preceding form by using long division. For example, if $q(x) = 4x^2 + 2x + 1$ and $a = 2$, then

$$\frac{q(x)}{x - 2} = 4x + 10 + \frac{21}{x - 2},$$

and

$$\frac{4x + 10}{x - 2} = 4 + \frac{18}{x - 2}.$$

Multiplying both equations by $(x - 2)$, we see that

$$q(x) = (4x + 10)(x - 2) + 21$$
$$= [4(x - 2) + 18](x - 2) + 21$$
$$= 21 + 18(x - 2) + 4(x - 2)^2.$$

We will compute the coefficients $b_0, b_1, \ldots, b_n$ in Equation (1) in terms of the function f by repeated differentiation and then evaluating at $x = a$.

$$p'(x) = b_1 + 2b_2(x - a) + 3b_3(x - a)^2 + \cdots + nb_n(x - a)^{n-1}$$
$$p''(x) = 2b_2 + 6b_3(x - a) + \cdots + n(n - 1)b_n(x - a)^{n-2}$$
$$\vdots$$
$$p^{(n)}(x) = n!\, b_n$$

[Recall that $0! = 1$, $1! = 1$, and $n! = n(n - 1) \cdots 2 \cdot 1$ for $n > 1$.] Evaluating at $x = a$ leads to

$$p(a) = b_0 = f(a) \qquad\qquad b_0 = f(a)$$
$$p'(a) = b_1 = f'(a) \qquad\qquad b_1 = f'(a)$$
$$p''(a) = 2b_2 = f''(a) \qquad\qquad b_2 = \frac{f''(a)}{2!}$$
$$p'''(a) = 6b_3 = f'''(a) \qquad\qquad b_3 = \frac{f'''(a)}{3!}$$
$$\vdots \qquad\qquad\qquad\qquad \vdots$$
$$p^{(n)}(a) = n!\, b_n = f^{(n)}(a) \qquad b_n = \frac{f^{(n)}(a)}{n!}$$

We conclude that the polynomial $p(x)$ is given by

$$p(x) = f(a) + f'(a)(x-a) + \frac{f''(a)}{2!}(x-a)^2 + \frac{f'''(a)}{3!}(x-a)^3 + \cdots$$

$$+ \frac{f^{(n)}(a)}{n!}(x-a)^n.$$

Using the standard mathematical conventions $f^{(0)}(x) = f(x)$, $0! = 1$, and $(x-a)^0 = 1$, $p(x)$ can be written as

(2) $$p(x) = \sum_{k=0}^{n} \frac{f^{(k)}(a)}{k!}(x-a)^k = \sum_{k=0}^{n} f^{(k)}(a) \frac{(x-a)^k}{k!}.$$

The polynomial $p(x)$ is called the nth degree **Taylor polynomial** for the function f expanded about the point $x = a$.

EXAMPLE 2. Find the nth Taylor polynomial for the function $f(x) = e^x$ about the point $a = 0$.

Solution. We need to compute $f(0)$, $f'(0)$, $f''(0)$, ..., $f^{(n)}(0)$. But $f'(x) = e^x = f(x)$, and hence

$$f(0) = 1 = f'(0) = f''(0) = \cdots = f^{(n)}(0).$$

It follows that

$$p(x) = f(0) + f'(0)\frac{x-0}{1!} + f''(0)\frac{(x-0)^2}{2!} + \cdots + f^{(n)}(0)\frac{(x-0)^n}{n!}$$

$$= 1 + x + \frac{x^2}{2!} + \frac{x^3}{3!} + \cdots + \frac{x^n}{n!}$$

$$= \sum_{k=0}^{n} \frac{x^k}{k!}.$$

The following theorem tells us how well a Taylor polynomial approximates the function f.

THEOREM 1. (Taylor's Theorem) Suppose that $f^{(n+1)}$ is continuous on some interval containing x and a. Then there exists a number c between x and a such that

$$f(x) = \sum_{k=0}^{n} f^{(k)}(a) \frac{(x-a)^k}{k!} + f^{(n+1)}(c) \frac{(x-a)^{n+1}}{(n+1)!}.$$

Proof. Let
$$R = f(x) - \sum_{k=0}^{n} f^{(k)}(a) \frac{(x-a)^k}{k!}.$$

The theorem will be proved if we can show that
$$R = f^{(n+1)}(c) \frac{(x-a)^{n+1}}{(n+1)!}$$

for some number c between x and a.

Define a new function F by
$$F(t) = f(x) - \sum_{k=0}^{n} f^{(k)}(t) \frac{(x-t)^k}{k!} - R \frac{(x-t)^{n+1}}{(x-a)^{n+1}}.$$

Then $F(a) = F(x) = 0$, so, by the Mean Value Theorem for Derivatives, there exists a number c between x and a such that $F'(c) = 0$. But by the product rule of differentiation,

$$F'(t) = \sum_{k=1}^{n} f^{(k)}(t) \frac{(x-t)^{k-1}}{(k-1)!} - \sum_{k=0}^{n} f^{(k+1)}(t) \frac{(x-t)^k}{k!} + (n+1)R \frac{(x-t)^n}{(x-a)^{n+1}}$$

$$= \sum_{k=0}^{n-1} f^{(k+1)}(t) \frac{(x-t)^k}{k!} - \sum_{k=0}^{n} f^{(k+1)}(t) \frac{(x-t)^k}{k!} + (n+1)R \frac{(x-t)^n}{(x-a)^{n+1}}$$

$$= -f^{(n+1)}(t) \frac{(x-t)^n}{n!} + (n+1)R \frac{(x-t)^n}{(x-a)^{n+1}}.$$

Thus
$$F'(c) = 0 = -f^{(n+1)}(c) \frac{(x-c)^n}{n!} + (n+1)R \frac{(x-c)^n}{(x-a)^{n+1}},$$

and solving for R, we obtain
$$R = f^{(n+1)}(c) \frac{(x-a)^{n+1}}{(n+1)!},$$

which concludes the proof.

Since R here depends on both n and x, we will henceforth write $R_n(x)$ instead of R. The expression
$$R_n(x) = f^{(n+1)}(c) \frac{(x-a)^{n+1}}{(n+1)!}$$

is called the nth **remainder term**, and $|R_n(x)| = |f(x) - p(x)|$ is called the **error**. Notice that
$$|R_n(x)| \leq \frac{M_{n+1}}{(n+1)!} |x-a|^{n+1}$$

gives an **error bound**, where M_k is a number such that $|f^{(k)}(t)| \le M_k$ for all t between x and a.

EXAMPLE 3. Find the fourth-degree Taylor polynomial for the function $f(x) = \sqrt{x}$ about the point $a = 4$. Use it to estimate $\sqrt{3}$ and then show that the error is less than 0.0004.

Solution. We first need to compute derivatives and evaluate these derivatives at $a = 4$.

$$f(x) = \sqrt{x} \qquad f(4) = 2$$

$$f'(x) = \frac{1}{2}x^{-(1/2)} \qquad f'(4) = \frac{1}{4}$$

$$f''(x) = -\frac{1}{4}x^{-(3/2)} \qquad f''(4) = -\frac{1}{32}$$

$$f'''(x) = \frac{3}{8}x^{-(5/2)} \qquad f'''(4) = \frac{3}{256}$$

$$f^{iv}(x) = -\frac{15}{16}x^{-(7/2)} \qquad f^{iv}(4) = -\frac{15}{2048}$$

$$f^{v}(x) = \frac{105}{32}x^{-(9/2)} \qquad f^{v}(c) = \frac{105}{32}c^{-(9/2)}$$

Thus

$$p(x) = f(4) + f'(4)(x-4) + f''(4)\frac{(x-4)^2}{2!} + f'''(4)\frac{(x-4)^3}{3!} + f^{iv}(4)\frac{(x-4)^4}{4!}$$

$$= 2 + \frac{1}{4}(x-4) - \frac{1}{64}(x-4)^2 + \frac{1}{512}(x-4)^3 - \frac{5}{16{,}384}(x-4)^4.$$

Therefore $\qquad \sqrt{3} \approx p(3) = 2 - \frac{1}{4} - \frac{1}{64} - \frac{1}{512} - \frac{5}{16{,}384}$

$$\approx 1.7321$$

with error

$$|R_n(3)| = \left| \frac{105}{32} \cdot c^{-(9/2)} \cdot \frac{(-1)^5}{5!} \right|$$

$$= \frac{105}{32} \cdot c^{-4} \cdot c^{-(1/2)} \cdot \frac{1}{120}.$$

SECTION 1: TAYLOR POLYNOMIALS

But $3 < c < 4$, and hence $c^{-4} < 3^{-4}$ and $c^{-1/2} < 1$. Consequently,

$$|R_n(3)| = \frac{105}{32} \cdot c^{-4} \cdot c^{-(1/2)} \cdot \frac{1}{120}$$

$$< \frac{105}{32} \cdot \frac{1}{3^4} \cdot 1 \cdot \frac{1}{120}$$

$$= \frac{7}{20{,}736}$$

$$< 0.0004.$$

EXAMPLE 4. Find the nth-degree Taylor polynomial for $f(x) = \sin x$ about $a = 0$. Find the remainder term.

Solution.

$$f(x) = \sin x \qquad f(0) = 0$$
$$f'(x) = \cos x \qquad f'(0) = 1$$
$$f''(x) = -\sin x \qquad f''(0) = 0$$
$$f'''(x) = -\cos x \qquad f'''(0) = -1$$
$$f^{iv}(x) = \sin x \qquad f^{iv}(0) = 0$$
$$\vdots \qquad\qquad \vdots$$

It follows that

$$p(x) = 0 + x - \frac{x^3}{3!} + \frac{x^5}{5!} - \frac{x^7}{7!} + \cdots + (-1)^k \frac{x^{2k+1}}{(2k+1)!}$$

$$= \sum_{i=0}^{k} (-1)^i \frac{x^{2i+1}}{(2i+1)!},$$

where $n = 2k + 1$ or $n = 2k + 2$.
If $n = 2k + 1$, then the remainder term is

$$R_n(x) = (-1)^{k+1} \sin(c) \frac{x^{2k+2}}{(2k+2)!},$$

where c is some number between x and 0.

EXAMPLE 5. Find the nth-degree Taylor polynomial for $f(x) = \ln x$ about the point $a = 1$.

Solution.

$$f(x) = \ln x \qquad\qquad f(1) = 0$$
$$f'(x) = x^{-1} \qquad\qquad f'(1) = 1$$
$$f''(x) = -x^{-2} \qquad\qquad f''(1) = -1$$
$$f'''(x) = 2x^{-3} \qquad\qquad f'''(1) = 2$$
$$f^{\text{iv}}(x) = -3!\,x^{-4} \qquad\qquad f^{\text{iv}}(1) = -3!$$

$$\vdots \qquad\qquad\qquad\qquad \vdots$$

$$f^{(k)}(x) = (-1)^{k-1}(k-1)!\,x^{-k} \qquad f^{(k)}(1) = (-1)^{k-1}(k-1)!$$

The Taylor polynomial is thus

$$p(x) = (x-1) - \frac{(x-1)^2}{2!} + 2\frac{(x-1)^3}{3!} - 3!\frac{(x-1)^4}{4!} + \cdots$$

$$+ (-1)^{n-1}(n-1)!\,\frac{(x-1)^n}{n!}$$

$$= (x-1) - \frac{(x-1)^2}{2} + \frac{(x-1)^3}{3} - \frac{(x-1)^4}{4} + \cdots$$

$$+ (-1)^{n-1}\frac{(x-1)^n}{n}$$

$$= \sum_{k=1}^{n} (-1)^{k-1}\frac{(x-1)^k}{k}.$$

PROBLEMS

Find the nth-degree Taylor polynomial for the following functions.

1. $f(x) = \cos x$, $a = 0$
2. $f(x) = \sin x$, $a = \dfrac{\pi}{2}$
3. $f(x) = \dfrac{1}{x}$, $a = 1$
4. $f(x) = \tan x$, $a = 0$, $n = 3$
5. $f(x) = 3x^3 + 5x^2 + 4x + 1$, $a = 0$
6. $f(x) = \sin x \cos x$, $a = 0$, $n = 3$

Estimate the following numbers by using an appropriate Taylor polynomial of degree 3. Estimate the error.

7. $\sqrt{50}$
8. e
9. $\sin(1°)\left(1° = \dfrac{\pi}{180}\right)$
10. $(1.02)^{10}$.

11. $\cos(28°)$ 12. $\ln 3$

13. a. Find the fourth-degree Taylor polynomial for $\cos^2 x$, $a = 0$.

    b. Find the fourth-degree Taylor polynomial for $\cos x$, $a = 0$; then compute the square of this polynomial.

    c. Find the eighth-degree Taylor polynomial for $f(x) = \cos^2 x$ expanded about 0, without differentiating f.

14. Show that the derivative of the nth-degree Taylor polynomial for $\sin x$ is the $(n-1)$th-degree Taylor polynomial for $\cos x$, both expanded about $a = 0$.

15. Obtain the nth-degree Taylor polynomial for $\ln x$ from the $(n-1)$th-degree Taylor polynomial for $1/x$ where $a = 1$.

16. (*Lagrange Interpolation Formula.*) Let $x_1, x_2, \ldots, x_n$ be n distinct numbers and let $y_1, y_2, \ldots, y_n$ be any collection of n numbers. Show that the polynomial

$$p(x) = \sum_{i=1}^{n} y_i \frac{(x - x_1)(x - x_2)\cdots(x - x_{i-1})(x - x_{i+1})\cdots(x - x_n)}{(x_i - x_1)\cdots(x_i - x_{i-1})(x_i - x_{i+1})\cdots(x_i - x_n)}$$

goes through the n points (x_i, y_i), $i = 1, 2, 3, \ldots, n$. What is the degree of $p(x)$?

17. Use Problem 16 to find a polynomial $p(x)$ of degree 2 that passes through the three points $(-1, \frac{1}{2})$, $(1, \frac{1}{2})$, and $(3, \frac{1}{10})$. Use $p(x)$ to estimate the value of the function $f(x) = 1/(x^2 + 1)$ at $x = 0$, $x = 2$, and $x = \frac{1}{2}$.

18. Use a second-degree Taylor polynomial about $a = 1$ to estimate the value of $f(x) = 1/(x^2 + 1)$ at $x = 0$, $x = 2$, and $x = \frac{1}{2}$. Compare these estimates with those from Problem 17.

THE USE OF TAYLOR POLYNOMIALS FOR COMPUTING (OPTIONAL)

Taylor polynomials of degree n provide useful tools for computing on a hand-held calculator. However, some algebra is required first before the polynomials can be easily evaluated. The approximation

$$\sin x \approx x - \frac{x^3}{3!} + \frac{x^5}{5!}$$

requires 9 multiplications, 6 divisions, and 2 additions (or subtractions). Also, some memory is required in order to store intermediate answers. Rewritten in the equivalent form

$$\sin x \approx ((x^2/5/4 - 1)x^2/3/2 + 1)x,$$

only 5 multiplications, 4 divisions, and 2 additions are required, there is no need for memory, and there is less roundoff error. It is therefore worthwhile learning to rewrite polynomials in the latter form.

CHAPTER 11: POLYNOMIAL APPROXIMATIONS AND INFINITE SERIES

EXAMPLE 6. Rewrite the approximation

$$\arctan x \approx x - \frac{x^3}{3} + \frac{x^5}{5}$$

by regrouping the terms of the right-hand side.

Solution.

$$x - \frac{x^3}{3} + \frac{x^5}{5} = x\left(1 - \frac{x^2}{3} + \frac{x^4}{5}\right)$$

$$= x\left(1 - \frac{x^2}{3}\left(1 - \frac{3x^2}{5}\right)\right)$$

$$= x\left(1 + \frac{x^2}{3}\left(-1 + \frac{3x^2}{5}\right)\right)$$

$$= ((3x^2/5 - 1)x^2/3 + 1)x.$$

Notice that this result agrees with Formula (64) of Chapter 7. Formulas (60) to (64) of Chapter 7 are all based on Taylor polynomials. For better approximations, Taylor polynomials of higher degree should be used. For example, the seventh-degree Taylor polynomial for arctan x can be written

$$\arctan x \approx (((5x^2/(-7) + 1)3x^2/5 - 1)x^2/3 + 1)x.$$

In general, a polynomial can be regrouped as follows:

$$a_0 + a_1 x + a_2 x^2 + a_3 x^3 + \cdots + a_n x^n$$

$$= a_0 + a_1 x (1 + (a_2/a_1)x + (a_3/a_1)x^2 + \cdots + (a_n/a_1)x^{n-1})$$

$$= a_0 + a_1 x (1 + (a_2/a_1)x(1 + (a_3/a_2)x + \cdots + (a_n/a_{n-1})x^{n-2}))$$

$$= a_0 + a_1 x (1 + (a_2/a_1)x(1 + (a_3/a_2)x(1 + \cdots$$

$$+ (a_{n-1}/a_{n-2})x(1 + (a_n/a_{n-1})x)\cdots)))$$

$$= (((\cdots((a_n x/a_{n-1} + 1)a_{n-1}x + 1)\cdots$$

$$+ 1)a_3 x/a_2 + 1)a_2 x/a_1 + 1)a_1 x + a_0$$

if $a_i \neq 0$ for any i. If $a_i = 0$ for some i, then factor out x^k instead of x at the appropriate spot, as in Example 6.

PROBLEMS

1. Rewrite $\sin x \approx x - \frac{x^3}{3!} + \frac{x^5}{5!} - \frac{x^7}{7!}$ by regrouping the terms on the right-hand side.

2. Use the result of Problem 1 to compute $\sin x$ for $x = 0.1, 0.2, \pi/6$, and $\pi/4$.

3. Rewrite $e^x \approx 1 + x + \frac{x^2}{2!} + \frac{x^3}{3!} + \frac{x^4}{4!} + \frac{x^5}{5!} + \frac{x^6}{6!} + \frac{x^7}{7!}$ by regrouping.

4. Use the result of Problem 3 to compute $e = e^1$. How does this compare with the estimate $e \approx \left(1 + \frac{1}{256}\right)^{256} = \left(\left(\left(\left(\left(\left(1 + \frac{1}{256}\right)^2\right)^2\right)^2\right)^2\right)^2\right)^2\right)^2$?

5. Find the sixth-degree Taylor polynomial for $\ln(1-x)$ expanded about $x = 0$.

6. Find the sixth-degree Taylor polynomial for $\ln(1+x)$, $a = 0$.

7. Use $\ln[(1+x)/(1-x)] = \ln(1+x) - \ln(1-x)$, together with the results of Problems 5 and 6, to get a Taylor polynomial for $\ln[(1+x)/(1-x)]$.

8. Regroup the terms of the polynomial in Problem 7.

9. Use Problem 8 to compute $\ln(1.2)$. [Solve $1.2 = (1+x)/(1-x)$ for x.]

10. Estimate $\int_0^{1/2} e^{x^2} dx$ by integrating the sixth-degree Taylor polynomial of $y = e^{x^2}$ term by term and evaluating by first regrouping the terms of the antiderivative.

2. TAYLOR SERIES

For each n, the function $f(x) = e^x$ has an nth Taylor polynomial $p_n(x) = \sum_{k=0}^{n} x^k/k!$. Thus we get a sequence $\{p_n(x)\}_{n=0}^{\infty}$ of polynomials that may or may not converge for $x = b$. There are two questions that need to be answered: (a) For what values of x does $\{p_n(x)\}_{n=0}^{\infty}$ converge? (b) For what values of x does $\lim_{n \to \infty} p_n(x) = e^x$?

For $x = 1$, we get the following sequence $\{p_n(1)\}_{n=0}^{\infty}$:

$$p_0(1) = 1$$
$$p_1(1) = 2$$
$$p_2(1) = 2.5$$
$$p_3(1) \approx 2.6666667$$
$$p_4(1) \approx 2.7083333$$
$$p_5(1) \approx 2.7166667$$
$$p_6(1) \approx 2.7180555$$
$$p_7(1) \approx 2.7182539$$
$$p_8(1) \approx 2.7182787$$
$$p_9(1) \approx 2.7182815$$
$$p_{10}(1) \approx 2.7182818$$
$$\vdots \qquad \vdots$$

Since $e^1 = e = 2.718281828459\ldots$, it appears that $\lim_{n \to \infty} p_n(x) = e^x$ at least for $x = 1$.

Before we answer the preceding questions for other choices of x, we need some definitions and terminology. Given a function f that has derivatives of all orders at $x = a$, the **Taylor series** for f expanded about the point $x = a$ is the expression

(3) $$\sum_{n=0}^{\infty} f^{(n)}(a) \frac{(x-a)^n}{n!}.$$

We say that the Taylor series *converges* at $x = b$ if the sequence $\{p_n(b)\}_{n=0}^{\infty}$ converges, where

$$p_n(x) = \sum_{k=0}^{n} f^{(k)}(a) \frac{(x-a)^k}{k!}.$$

The sequence $\{p_n(b)\}_{n=0}^{\infty}$ is called a *sequence of partial sums*. If the Taylor series converges at $x = b$, then $\lim_{n \to \infty} p_n(b)$ is called the *sum* of the Taylor series at $x = b$, and we write

(4) $$\sum_{n=0}^{\infty} f^{(n)}(a) \frac{(b-a)^n}{n!} = \lim_{n \to \infty} p_n(b).$$

Strictly speaking, the sum of a Taylor series is not a sum in the usual sense. Rather, it is the limit of a sequence of partial sums. The word sum is very suggestive, however, just as the symbol dy/dx is suggestive of a quotient and $\int$ is suggestive of "S" for "sum."

Notice also that Expression (3) has two distinct meanings. On the one hand, it is merely a formal expression called a Taylor series. On the other hand, if the sequence of partial sums converges, then the expression also represents the sum of the Taylor series.

The following theorem answers the question of convergence for many Taylor series.

THEOREM 2. (Ratio Test for Taylor Series) Let f be any function that has derivatives of all orders at $x = a$ and let $c_n = f^{(n)}(a)/n!$. If

$$\lim_{n \to \infty} \left| \frac{(b-a)c_{n+1}}{c_n} \right| = r < 1,$$

the Taylor series converges and

$$\sum_{n=0}^{\infty} f^{(n)}(a) \frac{(b-a)^n}{n!} = f(b).$$

In particular, if $f(x) = e^x$ and $a = 0$, then $c_n = 1/n!$, and for any real number b,

$$\lim_{n \to \infty} \left| \frac{bc_{n+1}}{c_n} \right| = \lim_{n \to \infty} \left| \frac{b/(n+1)!}{1/n!} \right|$$

$$= \lim_{n \to \infty} \left| \frac{b}{n+1} \right|$$

$$= |b| \lim_{n \to \infty} \frac{1}{n+1}$$

$$= 0.$$

Since $0 < 1$, the Taylor series $\sum_{n=0}^{\infty} x^n/n!$ converges at $x = b$ for every real number b. For the function $f(x) = e^x$, the answer to both questions at the beginning of this section is "For all x."

The question of convergence will be investigated in more detail in the next section.

We list some other Taylor series that converge for the given values of x.

$$e^x = \sum_{n=0}^{\infty} \frac{x^n}{n!}, \qquad x \text{ any real number}$$

$$\sin x = \sum_{n=0}^{\infty} (-1)^n \frac{x^{2n+1}}{(2n+1)!}, \qquad x \text{ any real number}$$

$$\cos x = \sum_{n=0}^{\infty} (-1)^n \frac{x^{2n}}{(2n)!}, \qquad x \text{ any real number}$$

$$\ln x = \sum_{n=1}^{\infty} (-1)^{n-1} \frac{(x-1)^n}{n}, \qquad 0 < x < 2$$

$$\frac{1}{1-x} = \sum_{n=0}^{\infty} x^n, \qquad -1 < x < 1$$

$$\cosh x = \sum_{n=0}^{\infty} \frac{x^{2n}}{(2n)!}, \qquad x \text{ any real number (See Chapter 7.)}$$

$$\sinh x = \sum_{n=0}^{\infty} \frac{x^{2n+1}}{(2n+1)!}, \qquad x \text{ any real number}$$

$$\operatorname{arctanh} x = \sum_{n=0}^{\infty} \frac{x^{2n+1}}{2n+1}, \qquad -1 < x < 1$$

$$\arctan x = \sum_{n=0}^{\infty} (-1)^n \frac{x^{2n+1}}{2n+1}, \qquad -1 < x < 1$$

$$\frac{1}{x^2+1} = \sum_{n=0}^{\infty} (-1)^n x^{2n}, \qquad -1 < x < 1$$

Most of the preceding series are actually *Maclaurin* series—that is, Taylor series expanded about $x = 0$.

The foregoing series can be used to generate new ones. For example, if the x in $1/(1 - x) = \sum_{n=0}^{\infty} x^n$ is replaced by $-x^3$, then we get the series

$$\frac{1}{1 + x^3} = \sum_{n=0}^{\infty} (-x^3)^n$$

$$= \sum_{n=0}^{\infty} (-1)^n x^{3n}.$$

Since the original series converges for $-1 < x < 1$, it follows that the new series converges for $-1 < -x^3 < 1$—that is, for $-1 < x < 1$.

Taylor series may be integrated and differentiated term by term. In fact, the new series converges for those values of x for which the old series converges. (This is actually a fairly difficult theorem and will not be proven here.)

EXAMPLE 7. Compute $\int \frac{\sin x}{x} dx$.

Solution. Guessing, substitution, and integration by parts do not seem to work. However,

$$\int \frac{\sin x}{x} dx = \int \frac{1}{x} \sum_{n=0}^{\infty} (-1)^n \frac{x^{2n+1}}{(2n+1)!} dx$$

$$= \int \sum_{n=0}^{\infty} (-1)^n \frac{x^{2n}}{(2n+1)!} dx$$

$$= \sum_{n=0}^{\infty} (-1)^n \int \frac{x^{2n}}{(2n+1)!} dx$$

$$= \sum_{n=0}^{\infty} (-1)^n \frac{1}{2n+1} \cdot \frac{x^{2n+1}}{(2n+1)!} + C.$$

EXAMPLE 8. Compute $\int e^{x^2} dx$.

Solution. Let $u = x^2$. Then

$$e^{x^2} = e^u = \sum_{n=0}^{\infty} \frac{u^n}{n!} = \sum_{n=0}^{\infty} \frac{(x^2)^n}{n!} = \sum_{n=0}^{\infty} \frac{x^{2n}}{n!},$$

so that

$$\int e^{x^2} dx = \int \sum_{n=0}^{\infty} \frac{x^{2n}}{n!} dx$$

$$= \sum_{n=0}^{\infty} \int \frac{x^{2n}}{n!} dx$$

$$= \sum_{n=0}^{\infty} \frac{1}{2n+1} \frac{x^{2n+1}}{n!} + C$$

$$= \sum_{n=0}^{\infty} \frac{x^{2n+1}}{n!(2n+1)} + C.$$

It was stated earlier that $f(x) = e^{x^2}$ has no elementary antiderivatives. The series in Example 8 is an antiderivative of e^{x^2}, but this function cannot be written in terms of *finitely* many sums, products, quotients, roots, and differences of polynomials, logarithms, exponentials, or circular functions.

EXAMPLE 9.

$$\frac{d}{dx} \sin x = \frac{d}{dx} \sum_{n=0}^{\infty} (-1)^n \frac{x^{2n+1}}{(2n+1)!}$$

$$= \sum_{n=0}^{\infty} (-1)^n \frac{d}{dx} \frac{x^{2n+1}}{(2n+1)!}$$

$$= \sum_{n=0}^{\infty} (-1)^n (2n+1) \frac{x^{2n}}{(2n+1)!}$$

$$= \sum_{n=0}^{\infty} (-1)^n \frac{x^{2n}}{(2n)!}$$

$$= \cos x.$$

EXAMPLE 10.

$$\int \frac{dx}{x^2+1} = \int \sum_{n=0}^{\infty} (-1)^n x^{2n} \, dx$$

$$= \sum_{n=0}^{\infty} (-1)^n \int x^{2n} \, dx$$

$$= \sum_{n=0}^{\infty} (-1)^n \frac{x^{2n+1}}{2n+1} + C$$

$$= \arctan x + C \qquad (-1 < x < 1).$$

PROBLEMS

Find Taylor series for the following functions.

1. $f(x) = \sin x + \cos x$, $a = 0$. (Series can be added term by term.)
2. $f(x) = \sin x \cos x \left(= \frac{1}{2} \sin 2x \right)$, $a = 0$.
3. $f(x) = \dfrac{1}{1 - x^2}$, $a = 0$, $-1 < x < 1$.
4. $f(x) = \dfrac{\ln x}{x - 1}$, $a = 1$, $0 < x < 2$.
5. $f(x) = \dfrac{x}{1 - x}$, $a = 0$, $-1 < x < 1$.
6. $F(x) = b^x$, $a = 0$, $b > 1$.

Use Taylor series to compute the following integrals.

7. $\displaystyle\int x \sin x \, dx$
8. $\displaystyle\int \frac{\tan x}{x} \, dx$
9. $\displaystyle\int \arctan x \, dx$
10. $\displaystyle\int \cos\sqrt{x} \, dx$
11. $\displaystyle\int e^{x^3} \, dx$
12. $\displaystyle\int \sin(x^2) \, dx$

Verify that term-by-term differentiation gives correct results for the following Taylor series.

13. $e^x = \displaystyle\sum_{n=0}^{\infty} \frac{x^n}{n!}$
14. $\ln x = \displaystyle\sum_{n=1}^{\infty} (-1)^{n-1} \frac{(x-1)^n}{n}$
15. $\dfrac{1}{1-x} = \displaystyle\sum_{n=0}^{\infty} x^n$
16. $\cos x = \displaystyle\sum_{n=0}^{\infty} (-1)^n \frac{x^{2n}}{(2n)!}$

17. Suppose that f satisfies $f''(x) = -f(x)$, $f(0) = 0$, $f'(0) = 1$. Find f by computing its Taylor series at 0.
18. If f satisfies $f'(x) = f(x)$ for all x, and $f(0) = 1$, what is $f(x)$?
19. If $f''(x) = f(x)$ for all x, $f(0) = 0$, and $f'(0) = 1$, find $f(x)$.
20. If $f(1) = 0$ and $f'(x) = 1/x$ for all $x > 0$, what is $f(x)$?

3. TESTS FOR CONVERGENCE

An *infinite series* is an expression of the form $\sum_{n=1}^{\infty} a_n$. A series $\sum_{n=1}^{\infty} a_n$ *converges* if

$$\lim_{k \to \infty} \left(\sum_{n=1}^{k} a_n \right)$$

exists, and in this case the *sum* of the series is

$$\lim_{k \to \infty} \left(\sum_{n=1}^{k} a_n \right) = L.$$

The sequence $\{\sum_{n=1}^{k} a_n\}_{k=1}^{\infty}$ is called the *sequence of partial sums*. Hence a series converges if its sequence of partial sums converges. As is common practice, if a series converges, then the expression $\sum_{n=1}^{\infty} a_n$ will represent both the series and the sum of the series.

EXAMPLE 11. Show that $\sum_{n=1}^{\infty} \left(\frac{1}{2}\right)^n$ converges and find the sum.

Solution. Note that

$$\frac{1}{2} + \frac{1}{2^2} = \frac{3}{2^2} = \frac{2^2 - 1}{2^2},$$

$$\frac{1}{2} + \frac{1}{2^2} + \frac{1}{2^3} = \frac{7}{2^3} = \frac{2^3 - 1}{2^3},$$

and, in general,

$$\sum_{n=1}^{k} \left(\frac{1}{2}\right)^n = \frac{2^k - 1}{2^k}.$$

Hence

$$\lim_{k \to \infty} \sum_{n=1}^{k} \left(\frac{1}{2}\right)^n = \lim_{k \to \infty} \frac{2^k - 1}{2^k}$$
$$= 1.$$

Thus $\sum_{n=1}^{\infty} \left(\frac{1}{2}\right)^n$ converges and has sum equal to 1.

The preceding series is a special case of series of the form $\sum_{n=1}^{\infty} ar^n$, called *geometric series*. In studying the convergence of such series, it is helpful to verify the following identity:

(5) $\qquad 1 + r + r^2 + r^3 + \cdots + r^n = \dfrac{1 - r^{n+1}}{1 - r} \qquad (r \neq 1).$

This identity can be proved by mathematical induction (on n) or by multiplying the left-hand side by $1 - r$ to get the following collapsing sum:

$$(1 - r)(1 + r + r^2 + \cdots + r^n) = 1 - r + r - r^2 + r^2 - \cdots + r^n - r^{n+1}$$
$$= 1 - r^{n+1}.$$

378   CHAPTER 11: POLYNOMIAL APPROXIMATIONS AND INFINITE SERIES

If $r \neq 1$, we can divide both sides by $1 - r$ to get Formula (5).

The proof of the following theorem is based on taking the limit as n tends to infinity of both sides of Formula (5).

THEOREM 3. Suppose that $a \neq 0$. The geometric series $\sum_{n=0}^{\infty} ar^n$ converges if $|r| < 1$ and diverges if $|r| \geq 1$. If the series converges, then the sum is equal to $a/(1 - r)$.

EXAMPLE 12. Show that $\sum_{n=0}^{\infty} \dfrac{2^n}{5 \cdot 3^{n+1}}$ converges and find the sum.

Solution. This is a geometric series with $a = 1/15$ and $r = 2/3$. Hence $|r| < 1$ and the series converges and has sum

$$\frac{1/15}{1 - (2/3)} = \frac{1}{5}.$$

The following **ratio test** is very closely related to the foregoing discussion of geometric series. It states, roughly speaking, that a series converges if it "acts like" a geometric series with $|r| < 1$. To "act like" a geometric series means that the ratio of successive terms is nearly a constant; that is,

$$\lim_{n \to \infty} \left| \frac{a_{n+1}}{a_n} \right|$$

exists. $\Bigg($ For a geometric series,

$$\lim_{n \to \infty} \left| \frac{ar^{n+1}}{ar^n} \right| = \lim_{n \to \infty} |r| = |r|.\Bigg)$$

THEOREM 4. (Ratio Test)   If

$$\lim_{n \to \infty} \left| \frac{a_{n+1}}{a_n} \right| = r,$$

then the series $\sum_{n=1}^{\infty} a_n$ converges if $r < 1$ and diverges if $r > 1$.

Notice that the ratio test fails to give a method for computing the sum of a series.

EXAMPLE 13. Test the series $\sum_{n=1}^{\infty} \dfrac{4^n}{5 + 7^n}$ for convergence.

Solution.
$$\lim_{n \to \infty} \left| \frac{a_{n+1}}{a_n} \right| = \lim_{n \to \infty} \left| \frac{4^{n+1}}{5 + 7^{n+1}} \div \frac{4^n}{5 + 7^n} \right|$$
$$= \frac{4}{5};$$

hence the series converges by the ratio test.

EXAMPLE 14. For what values of x does $\sum_{n=1}^{\infty} \frac{(x-1)^n}{n}$ converge?

Solution. Let $a_n = (x-1)^n/n$. Then
$$\lim_{n \to \infty} \left| \frac{a_{n+1}}{a_n} \right| = \lim_{n \to \infty} \left| \frac{(x-1)^{n+1}}{n+1} \div \frac{(x-1)^n}{n} \right|$$
$$= \lim_{n \to \infty} \frac{n}{n+1} |x-1|$$
$$= |x-1|,$$

since $\lim_{n \to \infty} n/(n+1) = 1$. Now $|x-1| < 1$ if $-1 < x - 1 < 1$, which is true if $0 < x < 2$. Thus the series converges for $0 < x < 2$. The ratio test fails to determine whether or not the series converges at $x = 0$ and $x = 2$.

This example is a special case of the following theorem.

THEOREM 5. Every power series $\sum_{n=0}^{\infty} a_n(x-c)^n$ has a radius of convergence r such that the series converges if $|x - c| < r$ and diverges if $|x - c| > r$. If $\lim_{n \to \infty} |a_{n+1}/a_n| = K$, then $r = 1/K$. The radius of convergence can be zero, positive, or infinite.

Theorem 5 is illustrated in Figure 11.5.

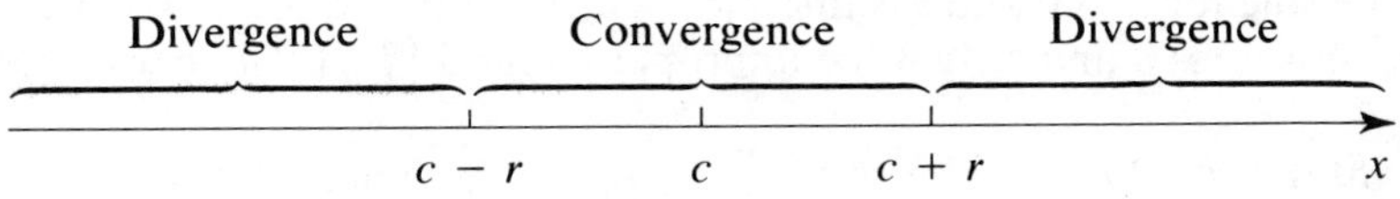

Divergence	Convergence	Divergence
	$c - r \quad c \quad c + r$	x

FIGURE 11.5

CHAPTER 11: POLYNOMIAL APPROXIMATIONS AND INFINITE SERIES

EXAMPLE 15. Find the radius of convergence of the series

$$\sum_{n=0}^{\infty} \frac{n^2}{2^n} (x-5)^n.$$

Solution. Let $a_n = n^2/2^n$. Then

$$\lim_{n \to \infty} \frac{(n+1)^2}{2^{n+1}} \frac{2^n}{n^2} = \lim_{n \to \infty} \frac{1}{2} \left(\frac{n+1}{n} \right)^2$$

$$= \frac{1}{2},$$

and hence the radius of convergence is $r = 1/\frac{1}{2} = 2$. Thus the series converges if $|x - 5| < 2$—that is, if $3 < x < 7$.

EXAMPLE 16. Find the radius of convergence of $\sum_{n=0}^{\infty} \frac{n!(x-5)^n}{n^n}$.

Solution. Let $a_n = n!/n^n$. Then the radius of convergence is given by $\lim_{n \to \infty} a_n/a_{n+1}$ if this limit exists. (Note that this is the reciprocal of the limit in Theorem 5.) Computing gives

$$\lim_{n \to \infty} \frac{a_n}{a_{n+1}} = \lim_{n \to \infty} \frac{n! \, (n+1)^{n+1}}{n^n \, (n+1)!}$$

$$= \lim_{n \to \infty} \left(\frac{n+1}{n} \right)^n$$

$$= \lim_{n \to \infty} \left(1 + \frac{1}{n} \right)^n$$

$$= e.$$

Thus the radius of convergence is $e \approx 2.718281828$.

The following theorem shows that the convergence of some series is related to the convergence of certain improper integrals.

THEOREM 6. (Integral Test) Suppose that $f(x)$ is nonnegative valued and decreasing for $x > 0$ and assume $f(n) = a_n$ for $n = 1, 2, 3, \ldots$. Then the series $\sum_{n=1}^{\infty} a_n$ converges if and only if the improper integral $\int_1^{\infty} f(x)\,dx$ converges.

Sketch of Proof. The series $\sum_{n=2}^{\infty} a_n$ can be interpreted as the area of the region in Figure 11.7. Similarly, the improper integral $\int_1^{\infty} f(x)\,dx$ can be interpreted as the area of the region in Figure 11.6, and hence $\sum_{n=2}^{\infty} a_n \leq \int_1^{\infty} f(x)\,dx$. Now $\sum_{n=1}^{\infty} a_n$ is finite if and only if $\sum_{n=2}^{\infty} a_n$ is finite. It follows that if the integral is finite, then the series must converge.

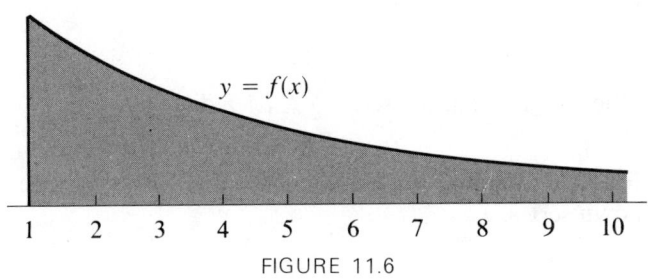
FIGURE 11.6

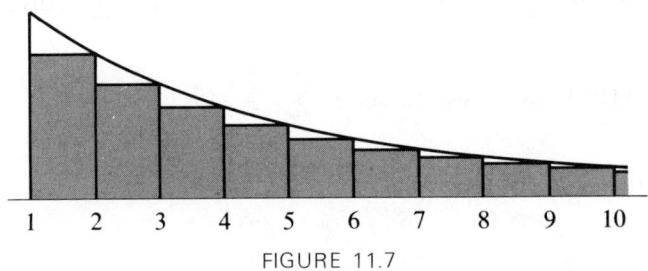
FIGURE 11.7

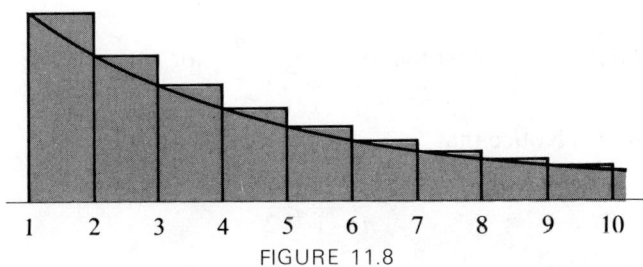
FIGURE 11.8

On the other hand, by sliding the rectangles to the right, as in Figure 11.8, we see that $\int_1^{\infty} f(x)\,dx \leq \sum_{n=1}^{\infty} a_n$. It follows that if $\int_1^{\infty} f(x)\,dx$ is infinite, then $\sum_{n=1}^{\infty} a_n$ must diverge.

EXAMPLE 17. Test the series $\sum_{n=1}^{\infty} \dfrac{1}{n}$ for convergence.

CHAPTER 11: POLYNOMIAL APPROXIMATIONS AND INFINITE SERIES

Solution. Let $f(x) = 1/x$. Then $f(x) \geq 0$, $f(x)$ is decreasing for $x > 0$, and $f(n) = 1/n$ for $n = 1, 2, 3, \ldots$. Now

$$\int_1^\infty \frac{1}{x} dx = \lim_{c \to \infty} \int_1^c \frac{1}{x} dx$$

$$= \lim_{c \to \infty} \ln c$$

$$= \infty,$$

and hence by the integral test, the series $\sum_{n=1}^\infty 1/n$ diverges.

The following theorem yields a useful test of convergence of new series in terms of known series.

THEOREM 7. *(Comparison Test)* Suppose that $\sum a_n$ and $\sum b_n$ are series with nonnegative terms and assume that $a_n \leq b_n$ for $n = 1, 2, 3, \ldots$. If $\sum b_n$ converges, then $\sum a_n$ converges. If $\sum a_n$ diverges, then $\sum b_n$ diverges.

EXAMPLE 18. Test the series $\sum_{n=1}^\infty \dfrac{2^n}{3^{n+1}}$ for convergence.

Solution. For each n,

$$\frac{2^n}{3^{n+1}} < \frac{2^n}{3^n} = \left(\frac{2}{3}\right)^n.$$

The series $\sum_{n=1}^\infty (2/3)^n$ is known to converge, hence $\sum_{n=1}^\infty 2^n/3^{n+1}$ converges also.

EXAMPLE 19. Test the series $\sum_{n=1}^\infty \dfrac{n}{n+1}$ for convergence.

Solution. Notice that

$$\frac{n}{n+1} \geq \frac{1}{2}$$

for each n. The series

$$\sum_{n=1}^\infty \frac{1}{2} = \frac{1}{2} + \frac{1}{2} + \frac{1}{2} + \frac{1}{2} + \cdots$$

diverges, and hence the series $\sum n/(n+1)$ also diverges.

EXAMPLE 20. For what numbers c does $\sum_{n=1}^\infty n^c$ converge?

Solution. If $c \geq -1$, then $n^c \geq 1/n$, and hence the series diverges by the comparison test. If $c < -1$, then $f(x) = x^c$ is nonnegative valued and decreasing for $x > 0$. Also,

$$\int_1^\infty x^c \, dx = \lim_{L \to \infty} \int_1^L x^c \, dx$$

$$= \lim_{L \to \infty} \frac{x^{c+1}}{c+1} \bigg|_1^L$$

$$= \lim_{L \to \infty} \frac{1}{c+1}(L^{c+1} - 1)$$

$$= \frac{-1}{c+1}$$

because $\lim_{L \to \infty} L^{c+1} = 0$ for $c < -1$. By the integral test, the series $\sum n^c$ converges for $c < -1$.

PROBLEMS

Find the radius of convergence.

1. $\sum_{n=0}^\infty x^n$

2. $\sum_{n=1}^\infty nx^n$

3. $\sum_{n=1}^\infty \frac{x^n}{n^2}$

4. $\sum_{n=1}^\infty \frac{n^2}{2^n}(x-3)^n$

5. $\sum_{n=1}^\infty \frac{n^n}{2^n}(x+2)^n$

6. $\sum_{n=1}^\infty \frac{n^n 2^n}{n!}(x-53)^n$

7. $\sum_{n=0}^\infty \frac{2^n}{n!}(x-2)^n$

8. $\sum_{n=1}^\infty \frac{n^3 3^n n!}{n^n}(x+5)^n$

Test for convergence.

9. $\sum_{n=1}^\infty \frac{1}{\sqrt{n}}$

10. $\sum_{n=1}^\infty n^{-3/2} 2^n$

11. $\sum_{n=1}^\infty ne^{-n^2}$

12. $\sum_{n=1}^\infty \frac{n 2^n}{(n+1)^2 3^n}$

13. $\sum_{n=1}^\infty \frac{\sqrt{n}}{(n+1)!}$

14. $\sum_{n=1}^\infty \frac{n^{1000} n!}{n^n}$

15. $\sum_{n=1}^\infty \frac{n+1}{n^{5/2}}$

16. $\sum_{n=1}^\infty \frac{1}{(1.001)^n}$

17. If f has derivatives of all orders at $x = a$, and if $f(x) = \sum_{n=0}^\infty c_n(x-a)^n$ for all x in some interval containing a, show that $c_n = f^{(n)}(a)/n!$. Comment on what you think the

statement "best possible," which appears in the introduction to Chapter 11 (page 360), might mean. [*Hint:* Compute the nth derivative of each side of $f(x) = \sum_{n=0}^{\infty} c_n(x-a)^n$ and evaluate at $x = a$.]

4. SERIES SOLUTIONS TO DIFFERENTIAL EQUATIONS

The use of power series affords a technique for finding solutions to a wide variety of differential equations. The method is illustrated in the following example.

EXAMPLE 21. Solve $y'' + y' + y = 0$ if $y(0) = 1$ and $y'(0) = 0$.

Solution. Assume a solution of the form $y = \sum_{n=0}^{\infty} a_n x^n$. We can then plug the series into the differential equation and evaluate the coefficients:

$$0 = \frac{d^2}{dx^2}\left(\sum_{n=0}^{\infty} a_n x^n\right) + \frac{d}{dx}\left(\sum_{n=0}^{\infty} a_n x^n\right) + \sum_{n=0}^{\infty} a_n x^n$$

$$= \sum_{n=2}^{\infty} n(n-1)a_n x^{n-2} + \sum_{n=1}^{\infty} na_n x^{n-1} + \sum_{n=0}^{\infty} a_n x^n$$

$$= \sum_{n=0}^{\infty} (n+2)(n+1)a_{n+2} x^n + \sum_{n=0}^{\infty} (n+1)a_{n+1} x^n + \sum_{n=0}^{\infty} a_n x^n$$

$$= \sum_{n=0}^{\infty} [(n+2)(n+1)a_{n+2} + (n+1)a_{n+1} + a_n] x^n,$$

and hence the coefficients $[(n+2)(n+1)a_{n+2} + (n+1)a_{n+1} + a_n]$ of x^n must all be equal to zero. Since $y(0) = 1$, it follows that $1 = y(0) = a_0 + a_1 \cdot 0 + a_2 \cdot 0^2 + \cdots + = a_0$. Since $y'(0) = 0$, it follows that the constant term a_1 of $\sum_{n=1}^{\infty} na_n x^{n-1} = y'(x)$ must be zero. So far, $a_0 = 1$ and $a_1 = 0$. To compute the remaining coefficients, we need to solve for a_{n+2} in terms of a_{n+1} and a_n. Now

$$(n+2)(n+1)a_{n+2} + (n+1)a_{n+1} + a_n = 0$$

leads to

(6) $$a_{n+2} = -\frac{a_n + (n+1)a_{n+1}}{(n+2)(n+1)}.$$

In particular,

$$a_2 = -\frac{a_0 + a_1}{2} = -\frac{1}{2},$$

and

$$a_3 = -\frac{a_1 + 2a_2}{6} = \frac{1}{6},$$

$$a_4 = -\frac{a_2 + 3a_3}{12} = 0,$$

$$a_5 = -\frac{a_3 + 4a_4}{20} = -\frac{1}{120}.$$

Equation (6) gives an algorithm for computing as many additional coefficients as we wish. If we stop with the sixth coefficient, $a_5 = -1/120$, then the polynomial

$$f(x) = 1 - \frac{x^2}{2} + \frac{x^3}{6} - \frac{x^5}{120}$$

should approximate a solution to the original differential equation $y + y' + y'' = 0$. Checking, we have

$$f(x) + f'(x) + f''(x) = 1 - \frac{1}{2}x^2 + \frac{1}{6}x^3 - \frac{1}{120}x^5 - x + \frac{1}{2}x^2$$

$$- \frac{1}{24}x^4 - 1 + x - \frac{1}{6}x^3$$

$$= -\frac{1}{24}x^4 - \frac{1}{120}x^5,$$

which is close to zero for $-1 < x < 1$.

EXAMPLE 22. Solve $y'' + 4y = 0$ if $y(0) = 0$ and $y'(0) = 1$.

Solution. Again, we assume a solution $y = \sum_{n=0}^{\infty} a_n x^n$. Thus

$$0 = D_x^2 \left(\sum_{n=0}^{\infty} a_n x^n \right) + 4 \sum_{n=0}^{\infty} a_n x^n$$

$$= \sum_{n=2}^{\infty} n(n-1) a_n x^{n-2} + \sum_{n=0}^{\infty} 4 a_n x^n$$

$$= \sum_{n=0}^{\infty} (n+2)(n+1) a_{n+2} x^n + \sum_{n=0}^{\infty} 4 a_n x^n$$

$$= \sum_{n=0}^{\infty} [(n+2)(n+1) a_{n+2} + 4 a_n] x^n.$$

Since this series is identically zero, each coefficient must be zero. The conditions $y(0) = 0$ and $y'(0) = 1$ yield $a_0 = 0$ and $a_1 = 1$. Now

$$a_{n+2} = -\frac{4a_n}{(n+2)(n+1)},$$

so, in particular, $a_2 = 0$, $a_3 = -2^2/3!$, $a_4 = 0$, and $a_5 = 2^4/5!$. In general,

$$a_{2n} = 0 \quad \text{and} \quad a_{2n+1} = (-1)^n \frac{2^{2n}}{(2n+1)!}.$$

Thus the solution is

$$y = \sum_{n=0}^{\infty} (-1)^n \frac{2^{2n} x^{2n+1}}{(2n+1)!} = \frac{1}{2} \sum_{n=0}^{\infty} (-1)^n \frac{(2x)^{2n+1}}{(2n+1)!},$$

which happens to be the series for $\frac{1}{2} \sin 2x$. Usually we are not so lucky as to recognize the series as being the series of a familiar function.

PROBLEMS

Find power series solutions to each of the following differential equations.
1. $y + y' = 0$, $y(0) = 1$
2. $y - y' = 0$, $y(0) = 2$
3. $y''' - y = 0$, $y(0) = 1$, $y'(0) = 0$, $y''(0) = 0$
4. $xy' - 1 = 0$, $y(1) = 0$ (Expand about $x = 1$.)
5. $2xy - y' = 0$, $y(0) = 1$
6. $y' - 2x = 0$, $y(0) = 5$
7. $2y + 3y' = 0$, $y(0) = 0$
8. $x^2 y' + y' - 1 = 0$, $y(0) = 0$
9. $x^2 y' - y' + 1 = 0$, $y(0) = 0$
10. Use separation of variables (see Chapter 5) to get a solution to Problem 5.
11. Suppose that $y = f(x)$ satisfies the differential equation $y' + x + xy - 1 = 0$ with $f(0) = 1$. Use the first few terms of the series solution of the differential equation in order to estimate the integral $\int_0^{1/2} f(x)\, dx$.

REVIEW SECTION

We conclude with a summary of definitions and basic results.

The nth-degree **Taylor polynomial** for the function f expanded about the point $x = a$ is given by

(a)
$$p_n(x) = \sum_{k=0}^{n} f^{(k)}(a) \frac{(x-a)^k}{k!}.$$

Taylor's theorem states that if $f^{(n+1)}$ is continuous on some interval containing x and a, then

(b) $$f(x) = p_n(x) + f^{(n+1)}(c)\frac{(x-a)^{n+1}}{(n+1)!}$$

for some number c between a and x. The term

(c) $$R_n(x) = f^{(n+1)}(c)\frac{(x-a)^{n+1}}{(n+1)!}$$

is called the nth **remainder term**, and the **error** is

(d) $$|R_n(x)|.$$

The expression

(e) $$\sum_{n=0}^{\infty} f^{(n)}(a)\frac{(x-a)^n}{n!}$$

is called the **Taylor series** for f expanded about a. The Taylor series converges if

(f) $$\lim_{n \to \infty} R_n(x) = 0.$$

A **Maclaurin series** is a Taylor series expanded about $x = 0$.
An **infinite series** is an expression of the form

(g) $$\sum_{n=1}^{\infty} a_n,$$

which **converges** if

(h) $$\lim_{k \to \infty} \left(\sum_{n=1}^{k} a_n \right)$$

exists, in which case the **sum** of the series is the limit

(i) $$\lim_{k \to \infty} \left(\sum_{n=1}^{k} a_n \right) = L.$$

The sequence

(j) $$\left\{ \sum_{n=1}^{k} a_n \right\}_{k=1}^{\infty}$$

is called the **sequence of partial sums**. A series converges if and only if the corresponding sequence of partial sums converges. Otherwise the series is said to **diverge**.

A series of the form

(k) $$\sum_{n=0}^{\infty} ar^n$$

is called a **geometric series**, and it **converges** if $|r| < 1$ and **diverges** if $|r| > 1$. If $|r| < 1$, then the **sum** is

(1) $$\sum_{n=0}^{\infty} ar^n = \frac{a}{1-r}.$$

The **ratio test** states that if $\lim_{n \to \infty} |a_{n+1}/a_n| = c$, then $\sum a_n$ converges if $c < 1$ and diverges if $c > 1$.

Every power series $\sum a_n(x - c)^n$ has a **radius of convergence** r such that the series converges if $|x - c| < r$ and diverges if $|x - c| > r$. (The radius of convergence can be zero, positive, or infinite.)

The **integral test** states that if $f(x)$ is nonnegative valued and decreasing for $x > 0$, then the series $\sum_{n=1}^{\infty} f(n)$ converges if and only if the improper integral $\int_1^\infty f(x)\, dx$ converges.

The **comparison test** states that if $0 \le a_n \le b_n$ for $n = 1, 2, 3, \ldots$, then $\sum a_n$ converges if $\sum b_n$ converges, and $\sum b_n$ diverges if $\sum a_n$ diverges.

REVIEW PROBLEMS

1. Find the Taylor series for the following functions expanded about the given points.

a. $y = \dfrac{1}{(1 - 2x)^2}$, $a = 0$

b. $x \to xe^x$, $a = 0$

c. $f(x) = \dfrac{1}{x}$, $a = 10$

d. $f(x) = \ln x$, $a = 10$

e. $y = \sqrt{1 - x}$, $a = 0$

f. $f(x) = \sqrt[3]{1 + x^2}$, $a = 0$

g. $x \to \dfrac{1}{(1 + x)^2}$, $a = 0$

h. $f(x) = \dfrac{1}{\sqrt{1 - x^2}}$, $a = 0$

i. $f(x) = \arcsin x$, $a = 0$

j. $y = \text{arcsec } x$, $a = \sqrt{2}$

k. $f(x) = \dfrac{1}{x + 1}$, $a = 3$

l. $y = \cos x$, $a = \dfrac{\pi}{3}$

2. Test the following series for convergence.

a. $\sum \dfrac{1}{\sqrt{n}}$

b. $\sum \dfrac{1000}{3^n}$

c. $\sum \dfrac{n}{2^n}$

d. $\sum \dfrac{n!}{2^n}$

e. $\sum n^{1/n}$

f. $\sum \dfrac{(n!)^2}{(2n)!}$

g. $\sum \dfrac{1}{n \ln n}$

h. $\sum \dfrac{1}{n \ln^2 n}$

i. $\sum \text{arccot } n$

j. $\sum_{n=2}^{\infty} \ln\left(1 + \frac{1}{n^2}\right)$

k. $\sum n^{-(n+1)/n}$

l. $\sum \frac{1}{n\sqrt{n^2+1}}$

m. $\sum \frac{1}{[1+(1/n)]^n}$

3. Use appropriate Taylor polynomials to approximate each of the following integrals with error less than 0.001.

a. $\int_0^1 e^{x^3} dx$

b. $\int_0^1 \sin x^2 \, dx$

c. $\int_0^1 \cos(-x^2) \, dx$

d. $\int_0^1 x^3 e^{-x} \, dx$

e. $\int_0^1 \cos\sqrt{x} \, dx$

f. $\int_0^1 \tan x^2 \, dx$

4. Find the interval of convergence.

a. $\sum \frac{x^n}{n^n}$

b. $\sum \frac{n! x^n}{n^n}$

c. $\sum \frac{(-1)^n (x-1)^n}{n^2}$

d. $\sum \frac{2n+1}{n} \frac{(x-3)^n}{3^n}$

e. $\sum \frac{(x-2)^{2n}}{n!}$

f. $\sum n x^n$

5. Evaluate the sum of each series.

a. $\sum_{n=2}^{\infty} \frac{1}{n^2 - 1}$

b. $\sum_{n=0}^{\infty} \frac{1}{(n+1)(n+2)}$

6. Suppose that $\sum_{i=1}^{n} a_i = \left(1 - \frac{1}{3^n}\right)$, $(n = 1, 2, 3, \ldots)$. Find a_i.

7. Let A be a fixed positive real number.
   a. Find the Maclaurin series for $f(x) = (1+x)^A$.
   b. Find the interval of convergence of this series.
   c. Use part (a) to estimate $(1.2)^{5/9}$.
   d. Show that the Maclaurin series reduces to $\sum_{k=0}^{n} \frac{n!}{k!(n-k)!} x^k$ if $A = n$ is a positive integer.

8. Use the identity $\sin^2 x = \frac{1 - \cos 2x}{2}$ to get a Maclaurin series for $\sin^2 x$.

9. Use a fifth-degree Taylor polynomial to estimate $\sqrt{1.2}$.

10. Suppose that $\sum a_n$ converges and $a_n > 0$ for all n.
    a. Show that $\sum a_n^2$ converges.
    b. Does $\sum \ln(1 + a_n)$ converge? Justify your answer.
11. Suppose that $f(x) = \sum_{n=0}^{\infty} a_n x^n$. Show that
    a. if f is an odd function, then $a_0 = a_2 = a_4 = \cdots = 0$, and
    b. if f is an even function, then $a_1 = a_3 = a_5 = \cdots = 0$.

APPENDIX

MISCELLANEOUS FACTS AND FORMULAS

Here are a few standard tidbits of facts and formulas.
If x is a number such that $Ax^2 + Bx + C = 0$, then

(1) $$x = \frac{-B \pm \sqrt{B^2 - 4AC}}{2A}.$$

Equation (1) is called the **quadratic formula**, and the number x is called a **root** of the equation $Ax^2 + Bx + C = 0$.

EXAMPLE 1. Find a number x such that $x^2 + 3x + 1 = 0$.

Solution. By the quadratic formula, every root must be of the form

$$x = \frac{-3 \pm \sqrt{9 - 4}}{2}$$

$$= \frac{-3 \pm \sqrt{5}}{2}.$$

There are two roots

$$\frac{-3+\sqrt{5}}{2} \quad \text{and} \quad \frac{-3-\sqrt{5}}{2}.$$

As a check to make sure that $(-3+\sqrt{5})/2$ is indeed a root, notice that

$$\left(\frac{-3+\sqrt{5}}{2}\right)^2 + 3\left(\frac{-3+\sqrt{5}}{2}\right) + 1 = \frac{9-6\sqrt{5}+5}{4} + \frac{-9+3\sqrt{5}}{2} + 1$$

$$= \frac{14-18-6\sqrt{5}+6\sqrt{5}}{4} + 1$$

$$= -1 + 1$$

$$= 0.$$

In the expression

$$\frac{-B \pm \sqrt{B^2 - 4AC}}{2A},$$

if $B^2 - 4AC < 0$, then the equation $Ax^2 + Bx + C = 0$ has no **real** roots. The roots in this case are called **complex**. For example, $\sqrt{-1}$ is a complex root of $x^2 + 1 = 0$. Such roots are useful but are beyond the scope of this book.

The **area of a circle** of radius r is given by

(2) $$A = \pi r^2.$$

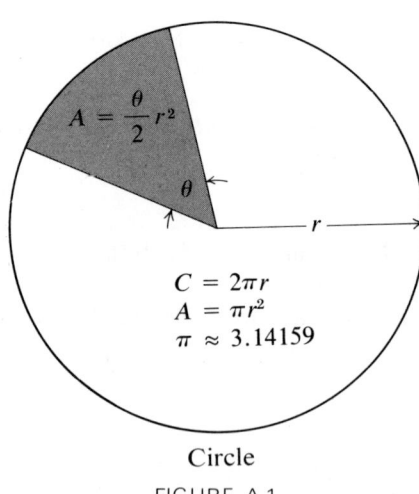

Circle

FIGURE A.1

APPENDIX: MISCELLANEOUS FACTS AND FORMULAS   393

The **area of a sector** (pie-shaped region) of a circle of radius r and with central angle θ (in radians) is

(3) $$A = \frac{\theta}{2}r^2.$$

The **circumference** (length around the outside) **of a circle** of radius r equals

(4) $$C = 2\pi r.$$

The **area of a triangle** of base b and height h is

(5) $$A = \frac{1}{2}bh.$$

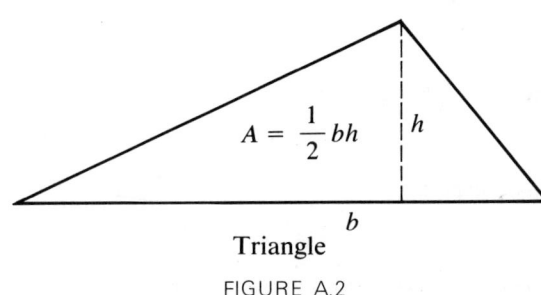

Triangle

FIGURE A.2

If c is the length of the **hypotenuse** (side opposite the right angle) of a right triangle whose other sides have lengths a and b, then the **Pythagorean Theorem** states that

(6) $$c^2 = a^2 + b^2.$$

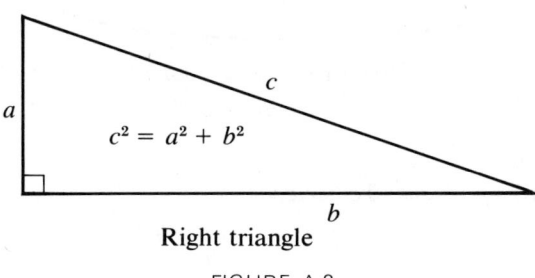

Right triangle

FIGURE A.3

The **area of a trapezoid** is given by

(7) $$A = \frac{1}{2}(a+b)h,$$

where a and b are the lengths of the parallel sides and h is the distance between them.

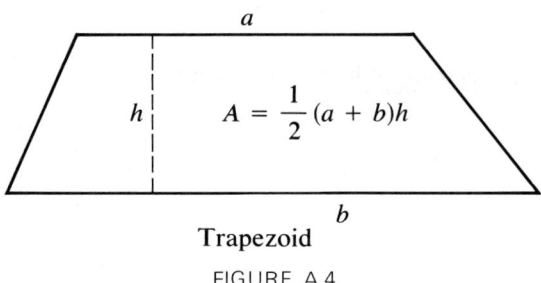

Trapezoid

FIGURE A.4

The **volume of a right circular cylinder** is

(8) $$V = \pi r^2 h$$

and the **lateral surface area** is

(9) $$S = 2\pi r h,$$

where r is the radius of the base and h is the height.

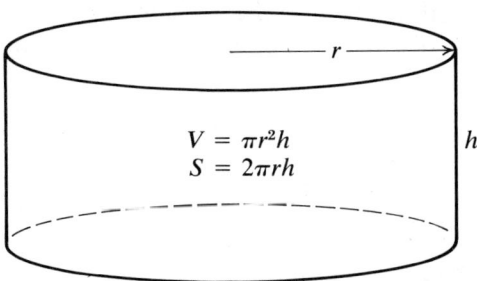

Right circular cylinder

FIGURE A.5

The **volume of a right circular cone** is

(10) $$V = \frac{1}{3}\pi r^2 h$$

and the *lateral surface area* is

(11) $$S = \pi r \sqrt{r^2 + h^2}.$$

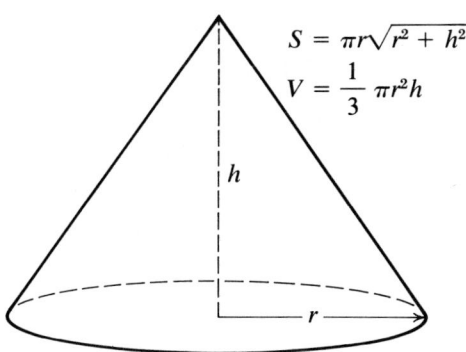

Right circular cone

FIGURE A.6

The *volume of a sphere* is

(12) $$V = \frac{4}{3} \pi r^3$$

and the *surface area* is

(13) $$S = 4\pi r^2.$$

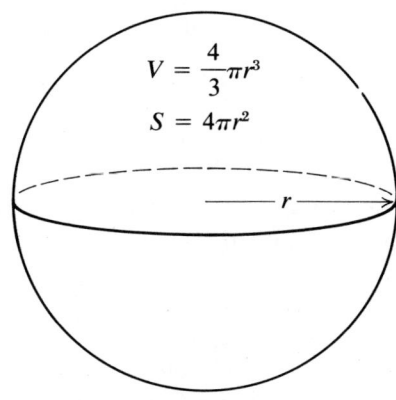

Sphere

FIGURE A.7

ANSWERS TO SELECTED PROBLEMS

Chapter 0, Section 1

1. 5, 2
3. a. $-\sqrt{3} < x < \sqrt{3}$
   c. $-11/3 < x < 1$
   b. $-4 < x < -2$
   d. $x < -3/5$ or $x > 7$

Chapter 0, Section 2

1. $g(x) = \sqrt{x^2 + 1}$
3. $(f+g)(x) = x^2 + \sqrt{x} + 1$, $(fg)(x) = x^2\sqrt{x} + x^2$, $(f-g)(x) = x^2 - \sqrt{x} - 1$, $(f/g)(x) = x^2/(\sqrt{x} + 1)$, $f(g(x)) = x + 2\sqrt{x} + 1$, $g(f(x)) = |x| + 1$, $f(f(x)) = x^4$, $g(g(x)) = \sqrt{\sqrt{x} + 1} + 1$
5. $(f+g)(x) = 3x^2 + 9 + 1/x$, $(fg)(x) = 15x^2 + 3x + 20 + 4/x$, $(f-g)(x) = 3x^2 - 1 - 1/x$, $(f/g)(x) = (3x^3 + 4x)/(5x + 1)$ with $x \neq 0$, $f(g(x)) = 3(5x + 1)^2/x^2 + 4$, $g(f(x)) = 5 + 1/(3x^2 + 4)$, $f(f(x)) = 27x^4 + 72x^2 + 52$, $g(g(x)) = 5 + x(5x + 1)$ with $x \neq 0$
7. $(f+g)(x) = x + 1/x$, $(fg)(x) = 1$ with $x \neq 0$, $(f-g)(x) = x - 1/x$, $(f/g)(x) = x^2$ with $x \neq 0$, $f(g(x)) = 1/x$, $g(f(x)) = 1/x$, $f(f(x)) = x$, $g(g(x)) = x$ for $x \neq 0$
9. $(f+g)(x) = 2x$ for $x \neq 0$, $(fg)(x) = x^2 - 1/x^2$, $(f-g)(x) = 2/x$, $(f/g)(x) = (x^2 + 1)/(x^2 - 1)$

for $x \neq 0$, $f(g(x)) = (x^4 - x^2 + 1)/(x^3 - x)$, $g(f(x)) = (x^4 + x^2 + 1)/(x^3 + x)$, $f(f(x)) = (x^4 + 3x^2 + 1)/(x^3 + x)$, $g(g(x)) = (x^4 - 3x^2 + 1)/(x^3 - x)$

11. $h(x) = x + 1/x$, $g(x) = x^2$
13. $h(x) = \sqrt{x}$, $g(x) = 3x^2 + 4$
15. $h(x) = x + 1/x$, $g(x) = \sqrt{x}$
17. $h(x) = x + x^2 + x^3$, $g(x) = 1/x$
19.
21.

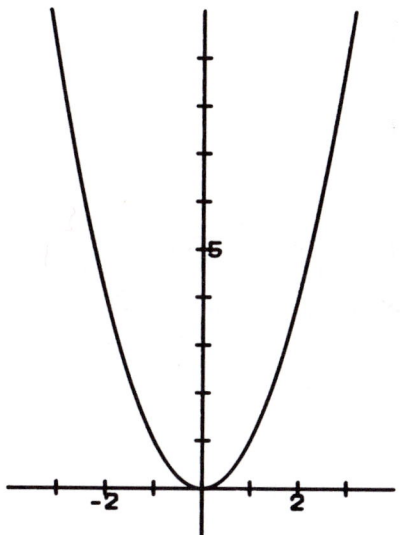
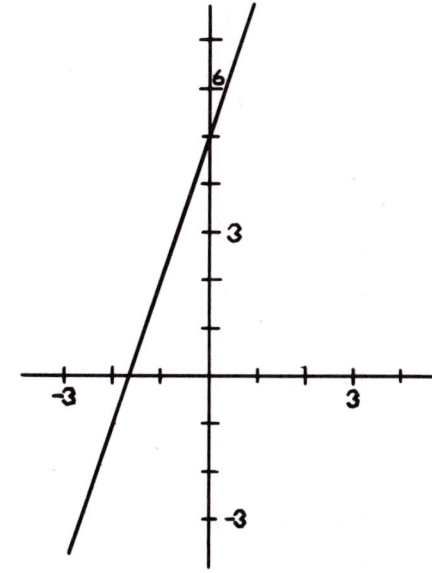

Chapter 1, Section 1

1. 10

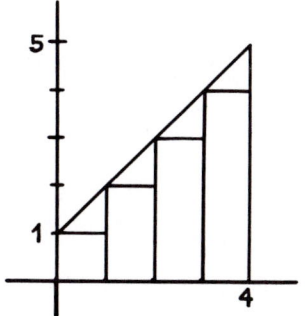

398 ANSWERS TO SELECTED PROBLEMS

3. 3.5

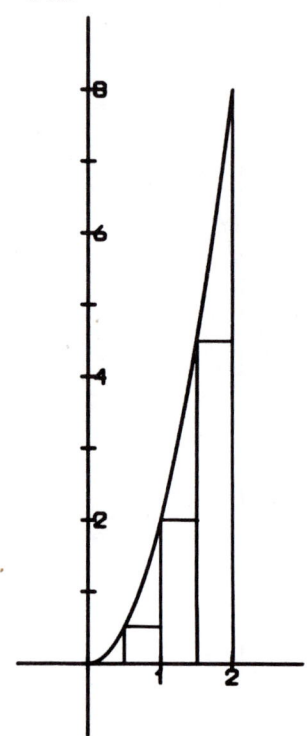

5. 5/32 ≈ 0.156

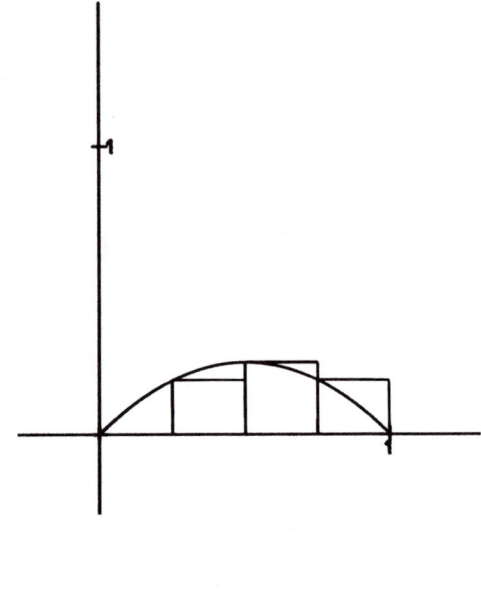

7. 6.146

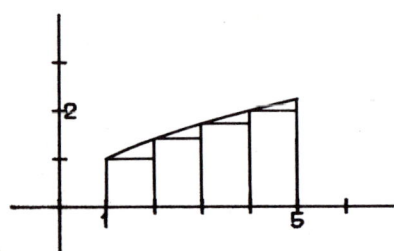

9. 12   11. 16/3   13. 1/6
15. *Hint:* Use the identity $(x + y)^2 = x^2 + 2xy + y^2$

ANSWERS TO SELECTED PROBLEMS

Chapter 1, Section 2

1. 33

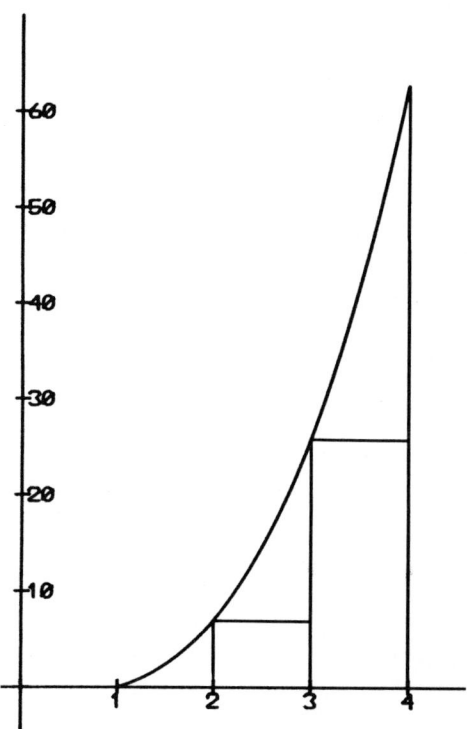

3. 6.146

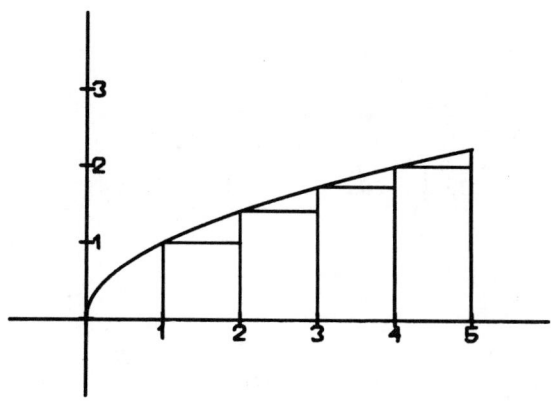

5. 0.927

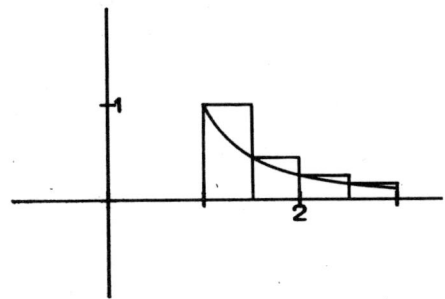

7. 0.950   9. 2.269   11. 58.875   13. 7.505
15. 0.648

Chapter 1, Section 3

1. 15   3. 3600   5. 650c
7. 5050   9. 71/28   11. 1770

13. $\frac{1}{6}[m(m+1)(2m+1) - (n-1)n(2n-1)]$

15. $n^2(n^2+1)(2n^2+1)/6 + n^2(n^2+1)/2 + n^2$
19. 42625   21. 25333

Chapter 1, Section 4

1. Converges to 0   3. Converges to 5
5. Converges to -2 7. Converges to 1
9. Converges to 0   11. Converges to 0
13. Converges to 2   15. Converges to 4/3

Chapter 1, Section 5

1. 52/3   3. $-9/2$ 5. 0   7. 0

9. $(b^2 - a^2)/2$ 11. $(b^4 - a^4)/4$ 13. $\int_1^4 4x\,dx$ 15. $\int_2^4 5\sqrt{x}\,dx$

Chapter 1, Section 6

1. 26   3. $9t^2/2 + 27t^3 - 3/2$ 5. $64t^3 + 64t^2 + 20t$
7. 366   9. $f(t) = t^3 + 5t^2/2 + t$ 13. 315

Chapter 1, Review Section

1. 8/3
3. 1/4
5. 2/3
7. 1/5
9. 2,358,350
11. 0.5
13. 0
15. 2
17. 1
19. $4478.33
21. $(\pi PR^4)/(8cL)$

Chapter 2, Section 1

1. 2
3. -1
5. 0

Chapter 2, Section 2

1. $y = -2x + 1$

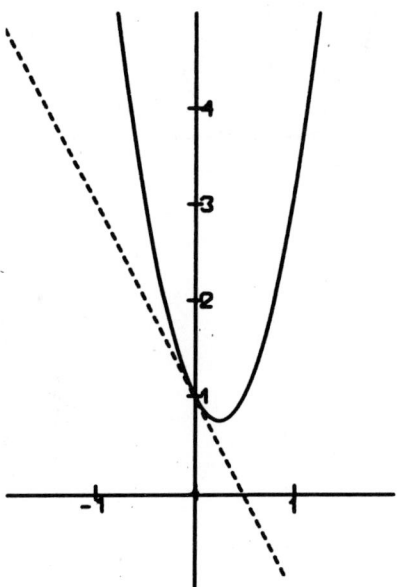

3. $y = x/4 + 1$

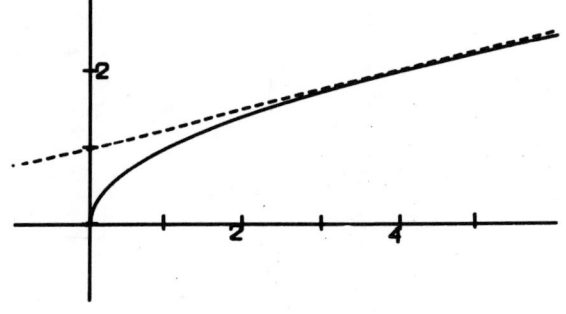

5. $y = 4x + 3$

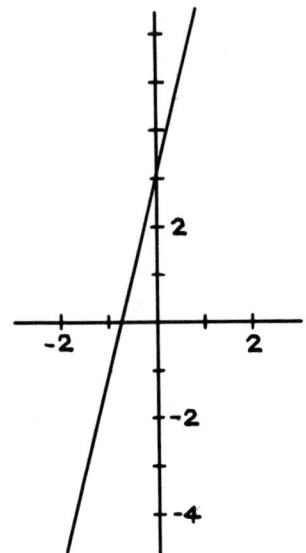

7. $y = -4x - 3$

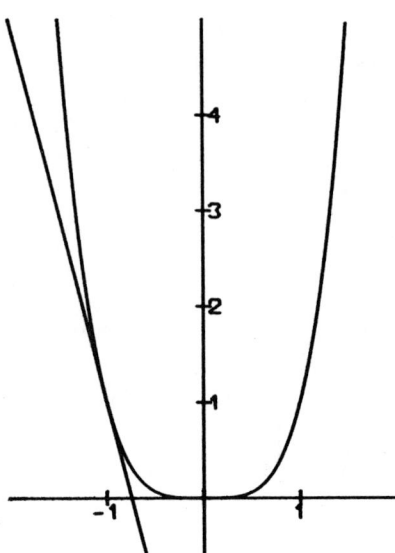

9. $y = 8x - 4$

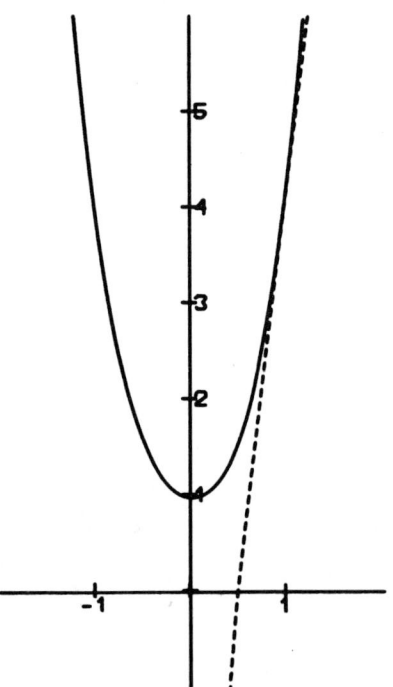

Chapter 2, Section 3

1. 8
3. $3a^2$
5. $2a$
7. $x/\sqrt{x^2+1}$
9. 4
11. 0.5
13. $|x-1| < \frac{1}{7}10^{-3}$ if $0 < x < 2$
15. $|x-1| < \frac{1}{7}E$ if $0 < x < 2$

Chapter 2, The Binomial Theorem

1. 720
3. 1
5. 100
7. $5^3 + 3 \cdot 5^2 \cdot 2 + 3 \cdot 5 \cdot 2^2 + 2^3$
9. $x^5 + 5x^4h + 10x^3h^2 + 10x^2h^3 + 5xh^4 + h^5$
11. $x^n + nx^{n-1}h + \frac{n(n-1)}{2}x^{n-2}h^2$
13. *Hint:* Show that $\frac{n!}{(i-1)!(n-i+1)!} + \frac{n!}{i!(n-i)!} = \frac{(n+1)!}{i!(n+1-i)!}$.

Chapter 2, Section 4

1. $f'(x) = 1$
3. $s'(t) = 2t + 2$
5. $\frac{dy}{dx} = 1/(x+1)^2$
7. $f'(w) = -2w/(w^2+1)^2$
9. $x \to 1 - 1/x^2$
11. $y' = 1/\sqrt{x}$
13. $y' = 1/(2\sqrt{x+2})$
15. $f'(w) = -1/[2(w+2)^{3/2}]$
17. $f'(z) = 2z + 2$
19. $f'(x) = -2/x^3$
21. $f'(x) = 3(x+2)^2$
23. $f'(x) = x^2/(x^3+1)^{2/3}$
25. The temperature is falling at 6 degrees per hour.
27. 35 pounds/month

Chapter 2, Section 5

1. $c = 2$
3. $c = (\sqrt{129} - 3)/6$
5. $c = 2/\sqrt{3}$
9. 22/3
11. 85.6, 73.6

Chapter 2, Section 6

1. $f(x) = 1$
3. $f(x) = 1 + 2x$
5. $f(y) = 3y^2 + 2y$
7. $f(t) = 4t^3$
9. $f(t) = -1/(t+1)^2$
11. $f(x) = (x^2 - 1)/x^2$

Chapter 2, Section 7

1. 28
3. 1/11
5. 74/3
7. $x^3/3$
9. 2/3
11. 10
13. -4
15. 39.225
17. The Fundamental Theorem does not apply.
19. $2\sqrt{x^2+1} - 2\sqrt{x}$
21. 100 friends, 360 friends

Chapter 2, Section 8

1. $-1/x + C$
3. $6\sqrt{x} + C$
5. $-1/(2t^2) + C$
7. $t^3/3 + C$
9. $F(x) = (3/2)x^2 - (4/x)$
11. $y = -x$
13. $V(t) = 64t - 16t^2$

Chapter 2, Review Section

1. $1/(2\sqrt{x+1})$
3. $-2x/(x^2+1)^2$
5. $-3/x^4$
7. 2
9. 11
11. $-1/(2a\sqrt{a})$
13. $-(2x+1)/(x^2+x)^2$
15. $f'(x) = 4x(x^2+2)$
17. $g'(z) = (1/z^2) - (2/z^3)$
19. 5.375 dollars

Chapter 3, Section 1

1. $f'(x) = 6x + 2$
3. $h'(z) = 3/(2\sqrt{z}) + 58z$
5. $f'(x) = -1/x^2$
7. $y' = a$
9. $f'(x) = 12x^2(x+1)$
11. $f'(x) = 6x(x^2+1)^2$
13. $h'(s) = 1 - 1/s^2$
15. $g'(t) = 2t + 3$
17. $G'(t) = 1/(2\sqrt{t}) - 1/t^2$
19. $w \to -1/w^2 - 5w^4$
21. $g'(t) = (4 + 12t)(2t + 3t^2)$
23. $K'(s) = -12s^2 - 12s - 4$

Chapter 3, Section 2

1. 7.5
3. 0
5. $a/3 + b/2 + c$
7. $2\sqrt{x} - 1/x + C$
9. -28
11. 2
13. $2a/5 + 2b/3 + 2c$
15. $(b^4 - a^4)/4 + (ab^2 - a^3)/2$

Chapter 3, Section 3

1. $f'(x) = 6x^2(x^3+2)$
3. $f'(x) = (5/2)x^{3/2}$
5. $k'(x) = (3x+2)/(2\sqrt{x})$
7. $g'(x) = (n+0.5)x^{n-0.5}$
9. $k'(x) = 3x^2 - (1/2)x^{-3/2}$
11. $g'(x) = 2a(ax+b)$
13. $f'(x) = 25x^4 + 10x$
15. $f'(z) = 6z(z^2+1)^2$

Chapter 3, Section 4

1. $f'(x) = -3/x^4$
3. $f'(t) = -4t/(t^2-1)^2$
5. $f'(t) = -6t^2(t^3+1)^{-3}$
7. $f'(x) = (\sqrt{2}-1)x^{\sqrt{2}-2}$
9. $h'(x) = -(5/2)x^{-7/2}$
11. $g'(z) = 2z - 4/z^3$
13. $f'(x) = -(15x^4+6)/(x^5+2x)^2$
15. $h'(z) = -(1/2)z^{-3/2}$
17. $f'(x) = 1/(1-x)^2$
19. $f'(x) = -(3x^4+4)/(x^3+2x)^2$
21. $f'(x) = (3x^4+4)/(3x^2+2)^2$
23. $h'(x) = \sqrt{x}/(\sqrt{x}-x)^2$

Chapter 3, Section 5

1. $f'(x) = 18x(3x^2 + 4)^2$
3. $g'(z) = 99(\sqrt{z^2 + 1} + z)^{99}/\sqrt{z^2 + 1}$
5. $f'(t) = t^2(t^3 - 1)^{-2/3}$
7. $F'(w) = 15(w^9 + 2w + 1)^{14}(9w^8 + 2)$
9. $f'(x) = (2x\sqrt{x^2 + 1} + x)/(2\sqrt{x^2 + 1}\sqrt{x^2 + \sqrt{x^2 + 1}})$
11. $h'(t) = (4\sqrt{t}\sqrt{t + \sqrt{t}} + 2\sqrt{t} + 1)/(8\sqrt{t}\sqrt{t + \sqrt{t}}\sqrt{t + \sqrt{t + \sqrt{t}}})$
13. $g'(z) = -(z^2 + 2z)/(z^2 + z + 1)^2$
15. $L'(s) = (s^3 + 4)^4(s^2 + 5)^6(29s^4 + 75s^2 + 56s)$
17. $T'(x) = (x + 2)^8(x - 1)^7(x - 3)^6(24x^2 - 37x - 35)$
19. $f'(t) = 14t[(t^2 + 1)^3 + (t^2 + 2)^4]^6[3(t^2 + 1)^2 + 4(t^2 + 2)^3]$

Chapter 3, Section 6

1. 5.1
3. 0.0101
5. 9.95
7. 1177.6
9. 353.5
11. $99/14 \approx 7.071$
13. $95/6 \approx 15.83$
15. $(1/3)(x^2 + 1)^{3/2} + C$
17. $(2\sqrt{2} + 2)a^5/15$
19. $(4/9)(x + 1)^{9/4} - (4/5)(x + 1)^{5/4} + C$
21. $\sqrt{2} - 1$
23. $(2/3)[(1/5)(x^3 + 1)^{5/2} - (1/3)(x^3 + 1)^{3/2}] + C$
25. $-2/(\sqrt{x} + 1) + C$
27. 2.25

Chapter 3, Section 7

1. $c = 2/\sqrt{3}$
3. Any c between 1 and 5.
5. $1/99 \approx 0.0101$, $|E| < 0.000003$
7. $(2.03)^{10} \approx 1177.6$, $|E| < 25$
9. $50^{3/2} \approx 353.500$, $|E| < 0.108$
11. $\sqrt{50} \approx 7.07143$, $|E| < 0.0008$
13. Pick numbers a and x. Then by the Mean Value Theorem, $f(x) = f(a) + f'(b)(x - a)$ for some b between x and a. But $f'(b) = 0$ by hypothesis, and hence $f(x) = f(a)$. If we let $c = f(a)$, then this shows that $f(x) = c$ for all x.

Chapter 3, Section 8

1. $f'(x) = -(6xy + 4y^3)/(3x^2 + 12xy^2)$
3. $y' = -y/x$
5. $y' = -(2x + y)/(x + 2y)$
7. $y' = -(1 + 2xy)/x^2$
9. $f'(x) = (x - y\sqrt{x^2 + y^2})/(x\sqrt{x^2 + y^2} - y)$
11. $y' = -\sqrt{y/x}$
13. $y = -4x + 7$
15. $y - 2 = (\sqrt{5}/2)(x - \sqrt{5})$

17.

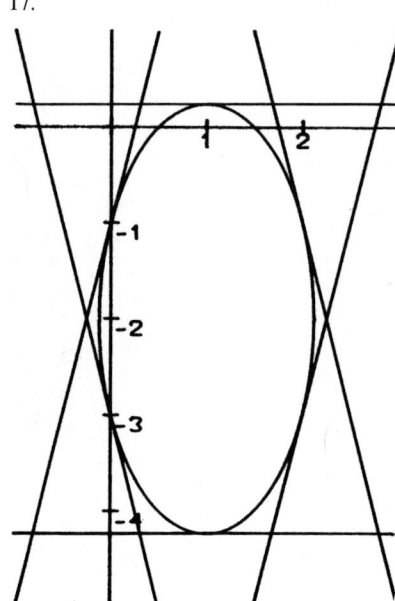

19. $\sqrt{26}$ miles

21. Inside

Chapter 3, Review Section

1. $f'(x) = 6x$
3. $y' = -(1/x^2) - 1$
5. $h'(s) = (3s + 3)/(2\sqrt{s})$
7. $x \to -(2x + 1)/(x^2 + x + 1)^2$
9. $g'(x) = -(2x + 1)/[2(x^2 + x + 1)^{3/2}]$
11. $C'(A) = -(A^2 + 2A)/(A^2 + A + 1)^2$
13. $P'(t) = -2/(t - 1)^2$
15. $R'(t) = -(1/t^2) - (2/t^3)$
17. $a'(z) = 56(z + 2)^6[1 + (z + 2)^7]^7$
19. $C'(k) = 2k + (1 + 6k - k^2)/(k^2 + 1)^2$
21. 3.5
23. 4
25. $(6^5 - 5^5)/5 = 930.2$
27. 28.8
29. $(2/11)(3^{11} - 2^{11}) \approx 31836.18$
31. 19 inches
33. 20,000 people
35. $4.00, 1000 liters
37. $C = 2x^2 + 3x + 20$ dollars

Chapter 4, Section 1

1. Increasing for x positive, decreasing for x negative
3. Always decreasing, $x \neq 0$
5. Increasing for $|x| \geq 1$, decreasing on $[-1, 1]$
7. Always increasing
9. Always decreasing, $x \neq 1$
11. Increasing for $x \leq -1 - \sqrt{2}$ or $x \geq -1 + \sqrt{2}$, decreasing elsewhere, $x \neq -1$

13. Decreasing for $x \geq 0$
15. Increasing for $x \geq 0$, decreasing for $x \leq 0$
17. Increasing for $x \geq 300$, decreasing on $[0, 300]$

Chapter 4, Section 2

1. Relative maximum at $x = -1$, relative minimum at $x = 1$
3. Maximum at $x = 0$
5. Relative minimum at $t = -1$, relative maximum at $t = 3$
7. Minimum at 0
9. No extreme values
11. Relative maximum at $t = -2 - \sqrt{3}$, relative minimum at $t = -2 + \sqrt{3}$
13. No extreme values

Chapter 4, Section 3

1. No extreme values, concave upward for $x > 0$, no inflection points

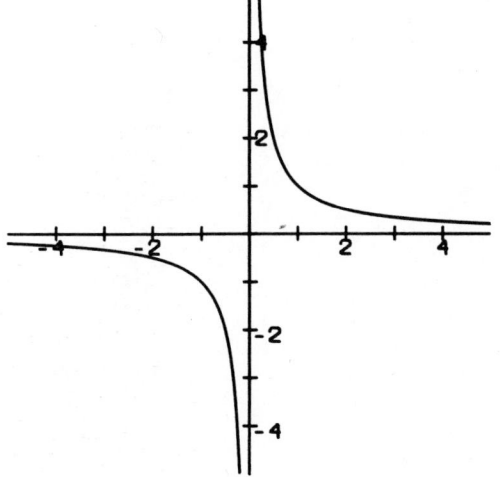

408 ANSWERS TO SELECTED PROBLEMS

3. No extreme values, concave downward for $x \geq 0$, no inflection points

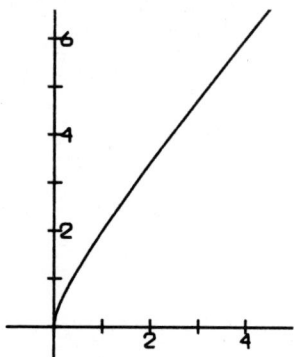

5. Relative maximum at $x = 2$, concave upward for $x \geq 3$, inflection point at $x = 3$

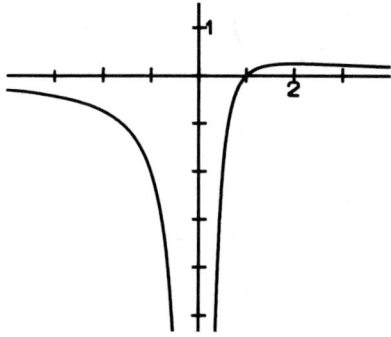

7. Relative minimum at $x = -1$, concave upward for $x \leq -2/3$ or $x \geq 0$, inflection points at $x = -2/3$ and $x = 0$

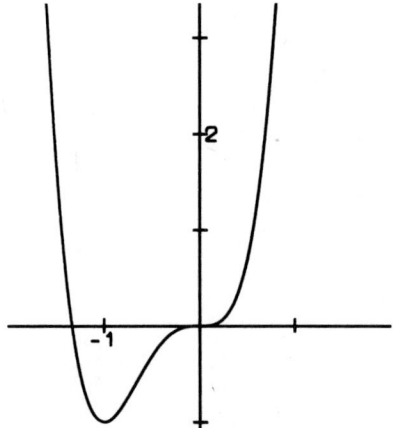

ANSWERS TO SELECTED PROBLEMS   409

9. No extreme values, concave upward for $x > 0$, no inflection points

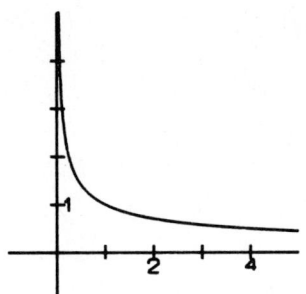

11. Relative maximum at $x = 0$, relative minimums at $x = (3 \pm 3\sqrt{33})/4$, concave upward for $x \geq 3$ or $x \leq -2$, inflection points at $x = 3$ and $x = -2$

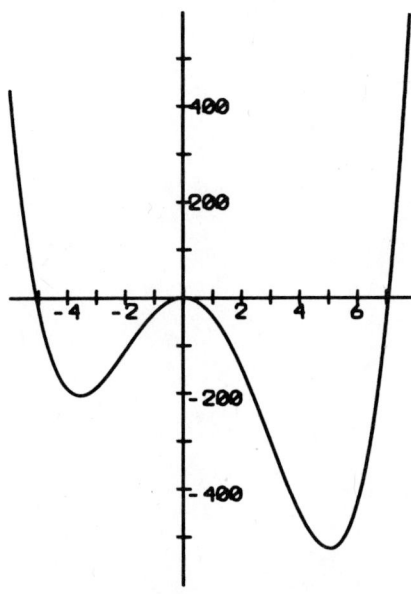

13. Relative minimum at $x = 0$, concave upward on $(-\sqrt{2/3}, \sqrt{2/3})$, inflection points at $x = \pm\sqrt{2/3}$

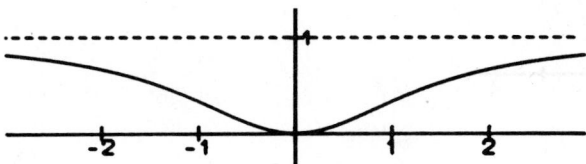

15. Relative maximum at $x = 0$, relative minimums at $x = \pm 1$, concave upward for $x \geq \sqrt{3}/3$ or $x \leq -\sqrt{3}/3$, inflection points at $x = \pm\sqrt{3}/3$

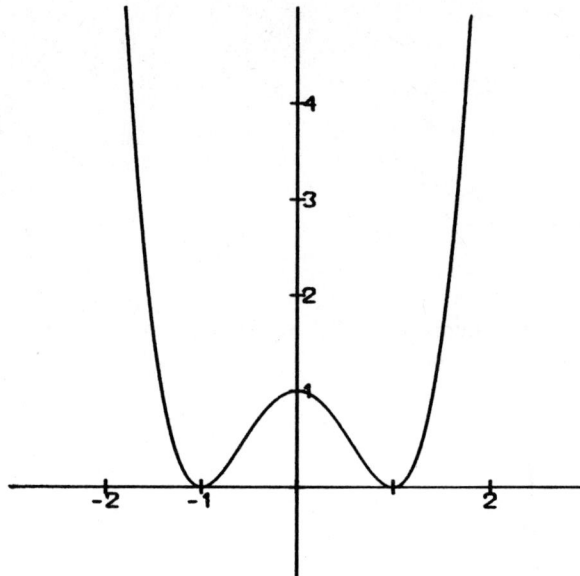

17. $f'''(x) = -6x^{-4}$

Chapter 4, Section 4

1.

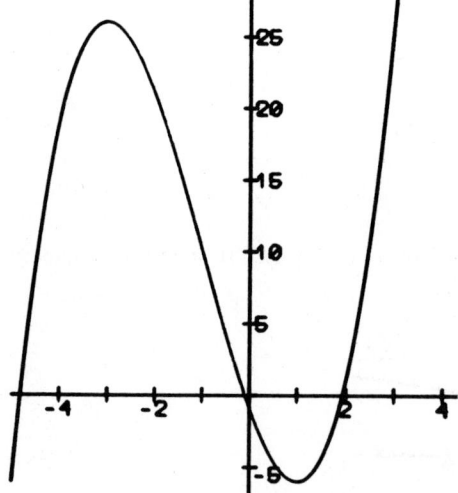

3.

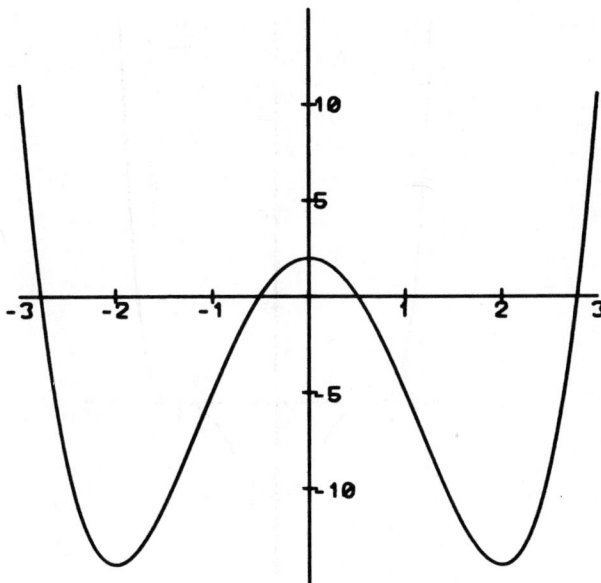

5.

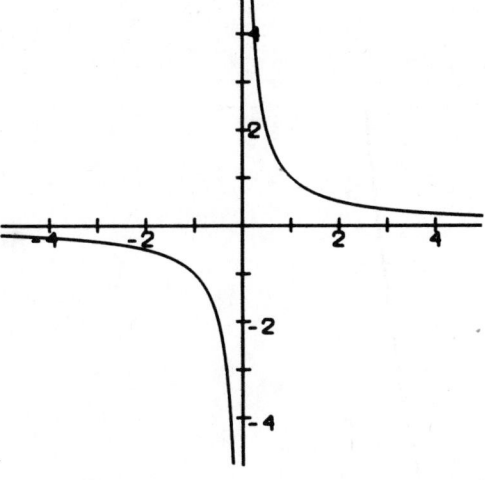

7.

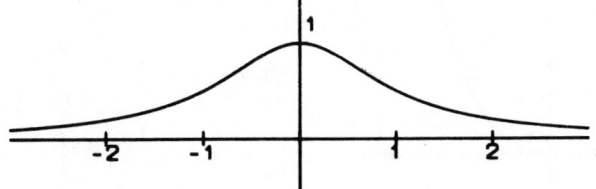

9.

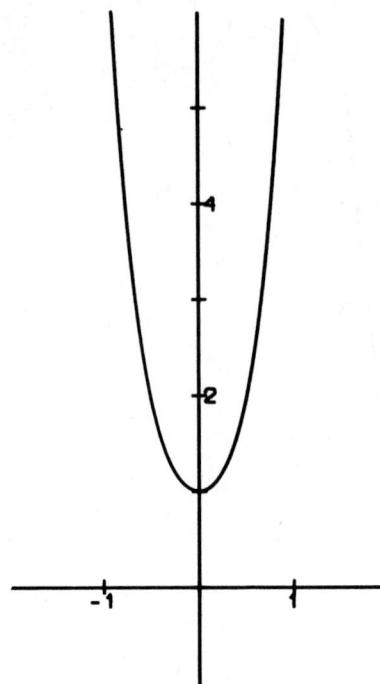

11.

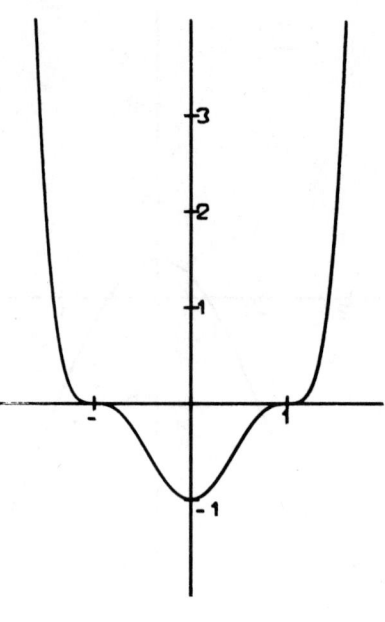

13.

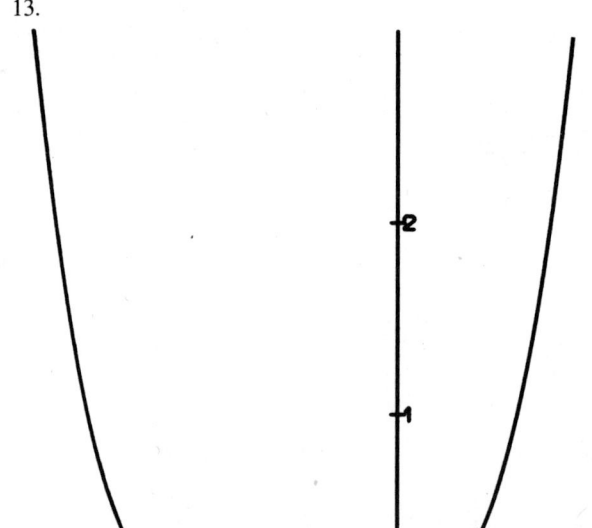

ANSWERS TO SELECTED PROBLEMS 413

15.

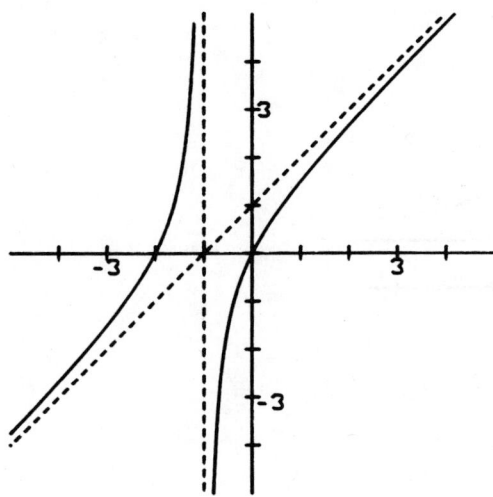

Chapter 4, Review Section

1.

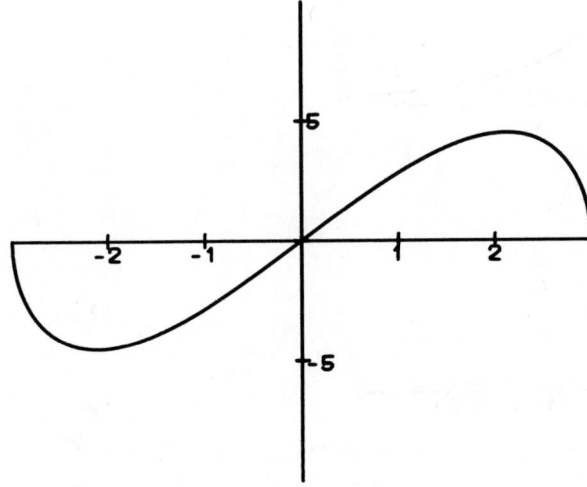

3.

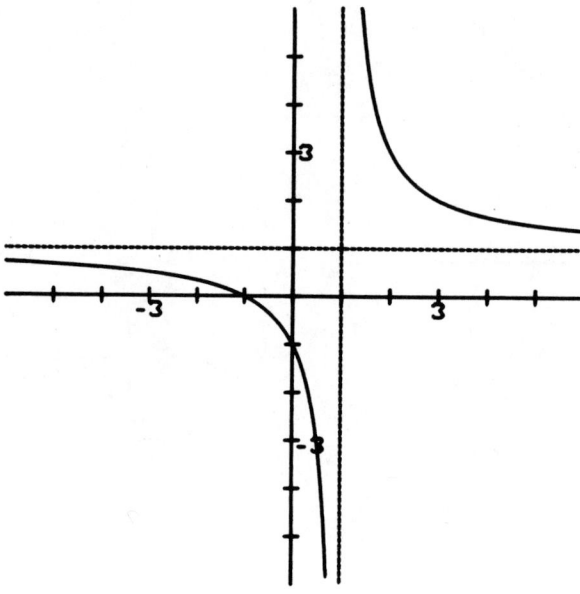

5.

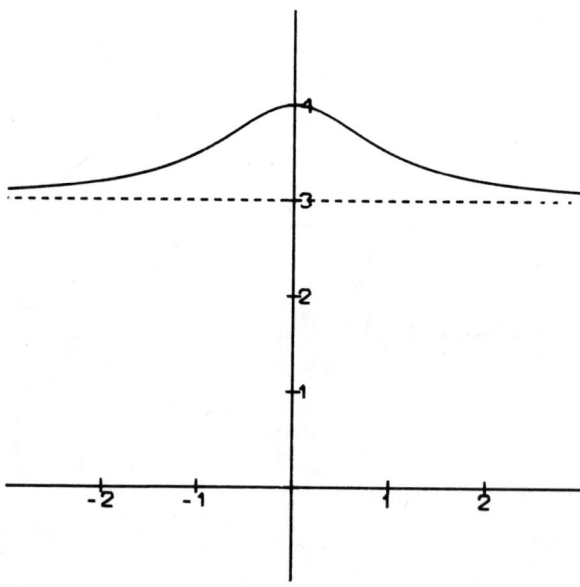

ANSWERS TO SELECTED PROBLEMS 415

7.

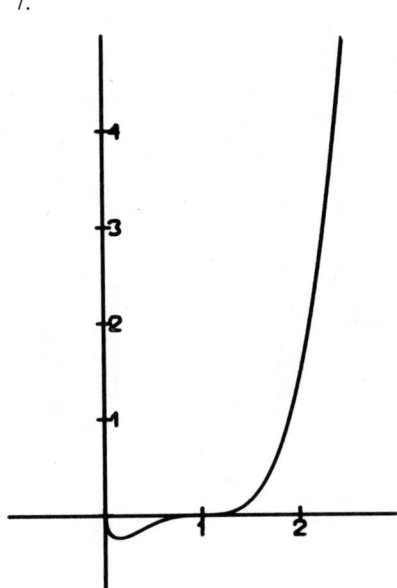

9.

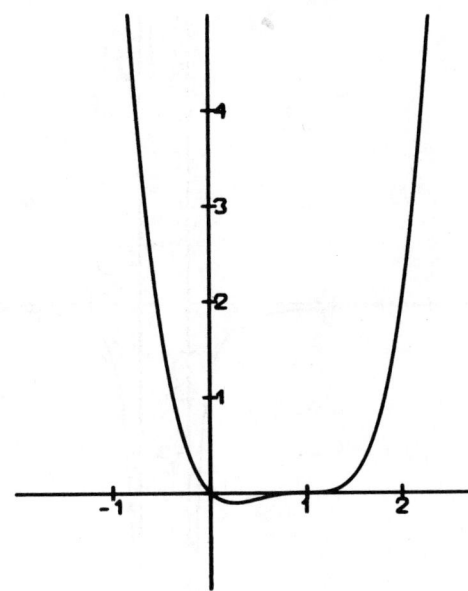

11.

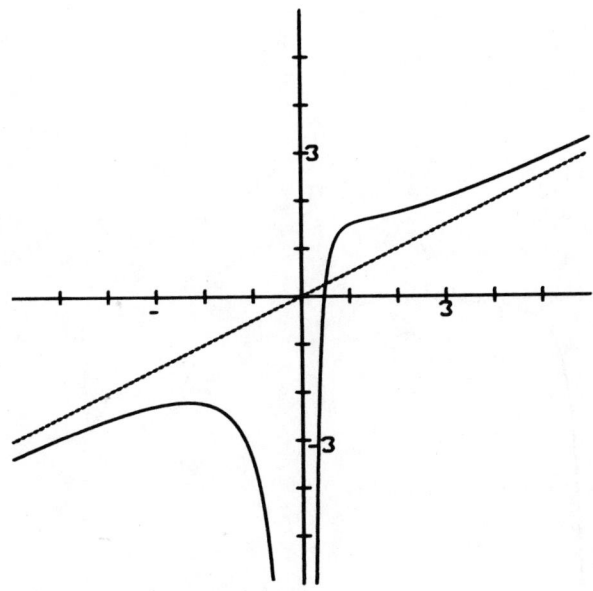

13.

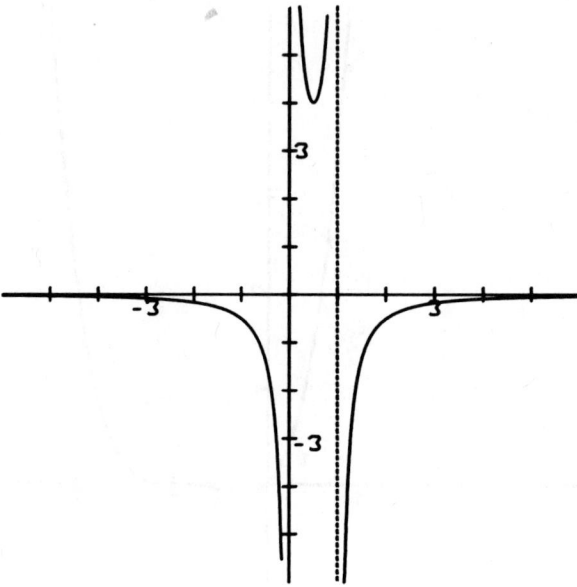

15.

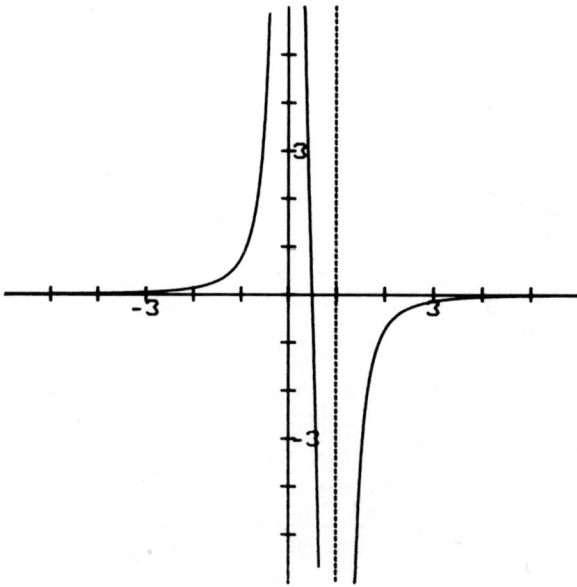

ANSWERS TO SELECTED PROBLEMS 417

17.

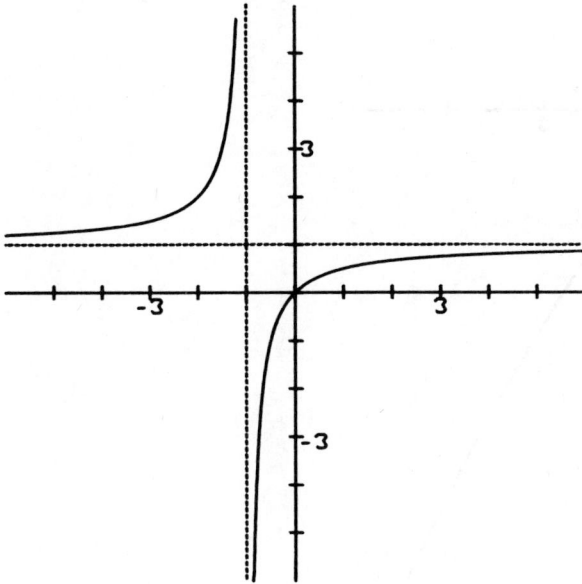

19.

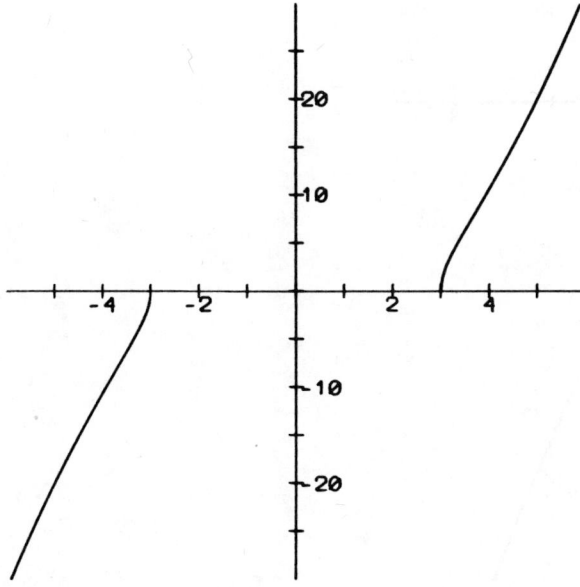

21.

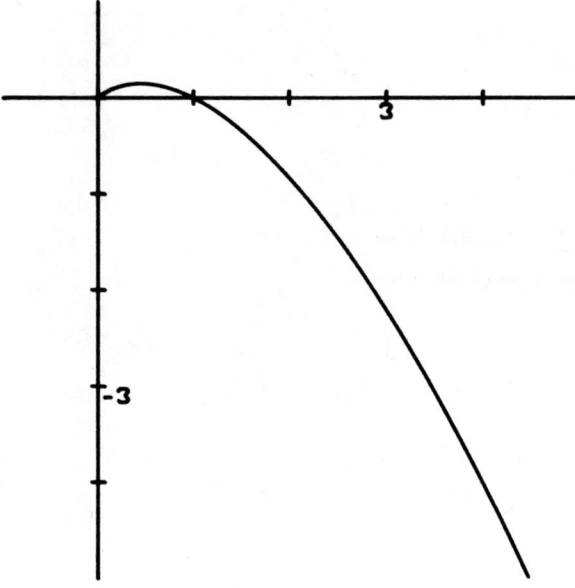

23.

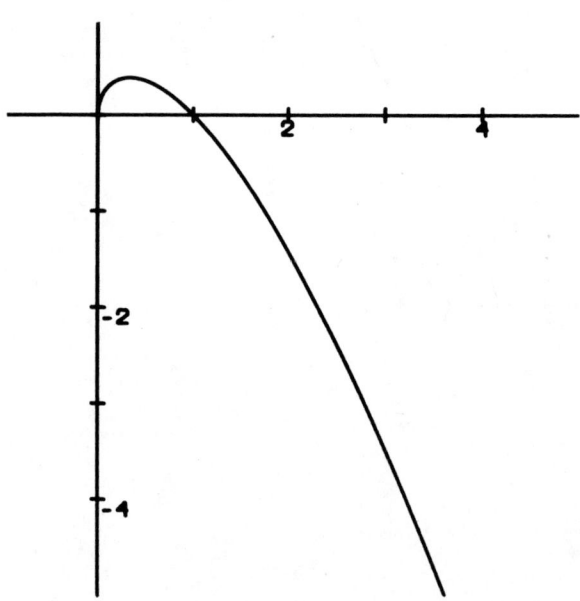

25.

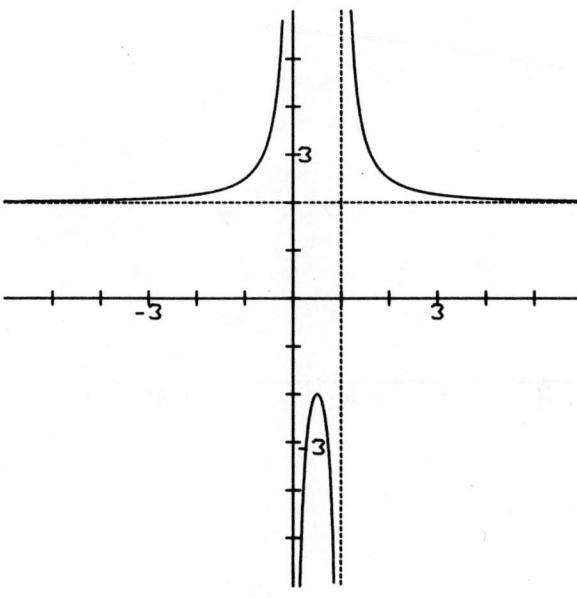

27.

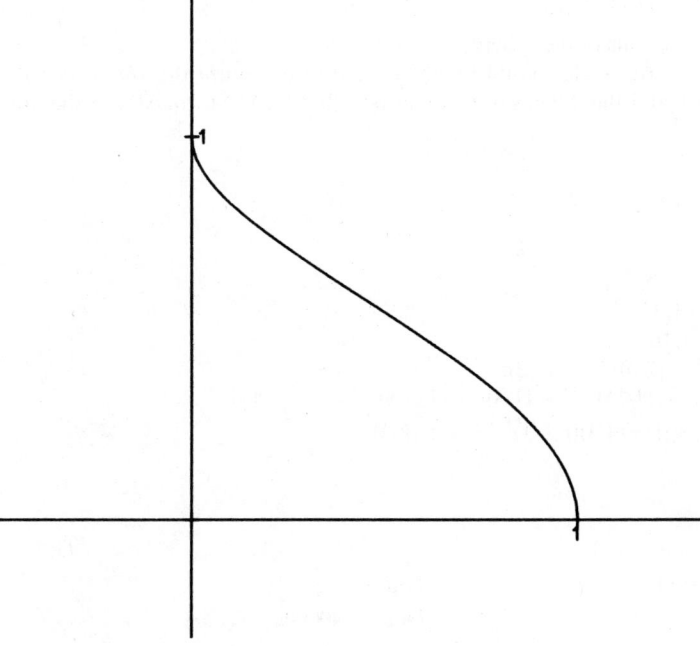

29.

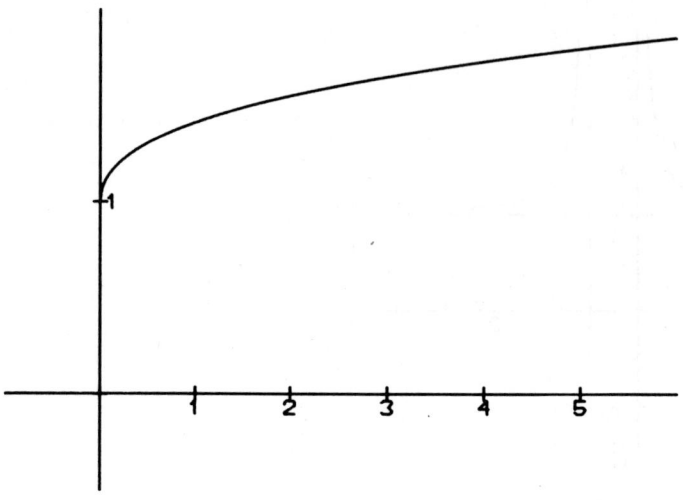

Chapter 5, Section 1

1. $(1, 1), (-1, 1)$ 3. 17   5. 1 to 1
7. Cut a square of dimensions x by x out of each corner, where
$$x = (a + b - \sqrt{a^2 + b^2 - ab})/6.$$
9. $1/2$ 11. 1
13. At 1:14 P.M., $2/\sqrt{13} \approx 0.5547$ miles
15. $V = 4\pi a^3/(3\sqrt{3})$, $a =$ radius of the sphere
17. The dimensions of the rectangle should be $440/\pi$ by 0 to maximize the area inside the track (circular track), and the dimensions should be $220/\pi$ by 110 to maximize the area inside the rectangle.
19. 3600 units, $21,584

Chapter 5, Section 2

1. $v(5) = 13$, $a(5) = 2$
3. $v(4) = 0.25$, $a(4) = -1/32$
5. $v(3) = 9/8$, $a(3) = -1/16$
7. $v(t) = (2/3)(t + 1)^{3/2} - (2/3)t^{3/2} + (1/3)$,
$s(t) = (4/15)(t + 1)^{5/2} - (4/15)t^{5/2} + (1/3)t + (26/15)$
9. $v(t) = 2\sqrt{t+1} - 1$, $s(t) = (4/3)(t + 1)^{3/2} - t + (2/3)$
11. 62.5 seconds

Chapter 5, Section 3

1. $5/18\pi$ inches per second   3. $dx/dt = -3$
5. $dy/dt = 3.8$ 7. $dm/dt = 4000m_0/(3\sqrt{3}c)$

ANSWERS TO SELECTED PROBLEMS

Chapter 5, Section 4

1. 1.414
3. 3.873
5. 4.405
7. 0.7245
11. $x = 3/2$

Chapter 5, Some Special Algorithms

1. 1.224745
3. 3.316625
5. 2.924018
7. 1.316074
9. 1.174619
11. 2.888279

Chapter 5, Section 5

1. $1030\pi/3$
3. 3π
5. 625π
7. $8\pi/17$
9. $3\pi/2$
11. $3\pi/8$
13. 312.5π
15. $2125\pi/3$
17. $38/3$

Chapter 5, Section 6

1. $xy^2 + 2x^2 - 2 + Cx = 0$
3. $xy^2 - x^3 - 2x^2 + 2 = Cx$
5. $s(t) = (4/3)(t + 1)^{3/2} - (1/3)$
7. $\sqrt{x^2 + 1} + (2/3)\sqrt{y^3 + 1} = C$

Chapter 5, Review Section

1. 77 bicycles, $269.23
3. 120 people, $14,400, no, capacity ≤ 120
5. a.

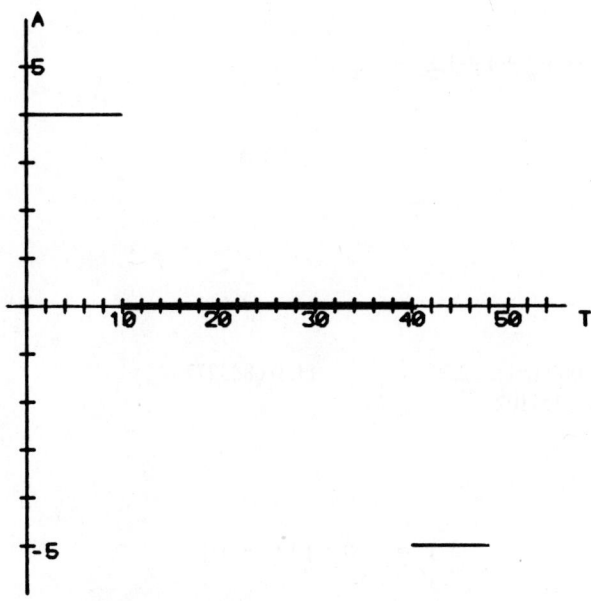

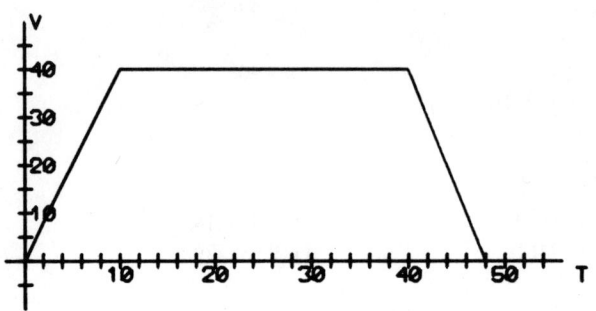

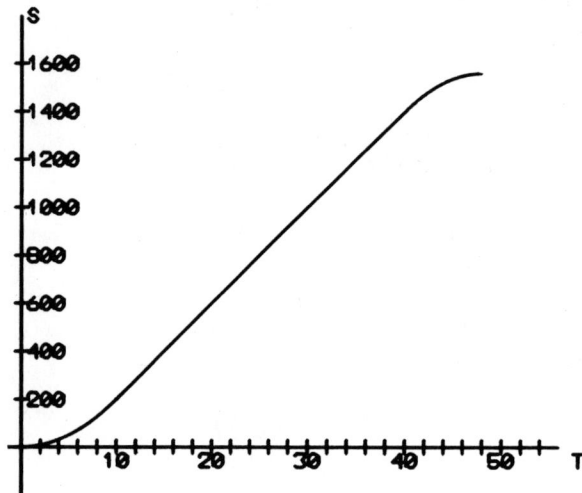

   b. 48 seconds
   c. 1560 feet
   d. 40 ft/sec
7. $1.5v_0$ 9. 1000 trees, $2.50   11. 0.6823277
13. 3.1038035   15. 0.3357102

Chapter 6, Section 1

1. $y' = 3/x$ 3. $y' = (2/x) + [3/(x + 1)]$
5. $y' = 1/(x \ln x)$ 7. $y' = 2/x$

9.

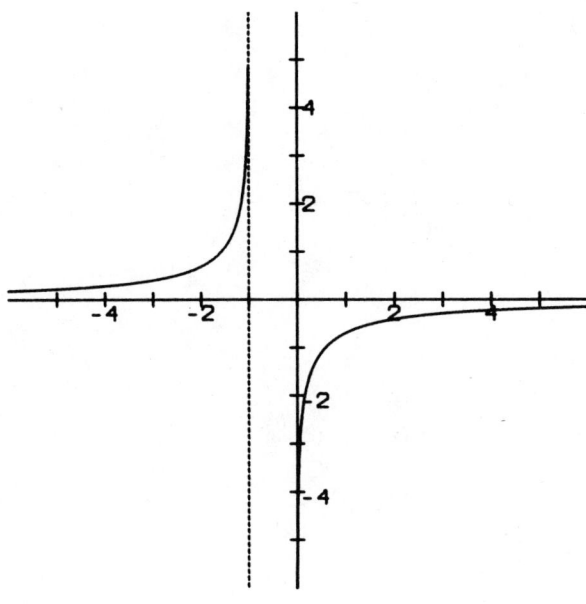

11.

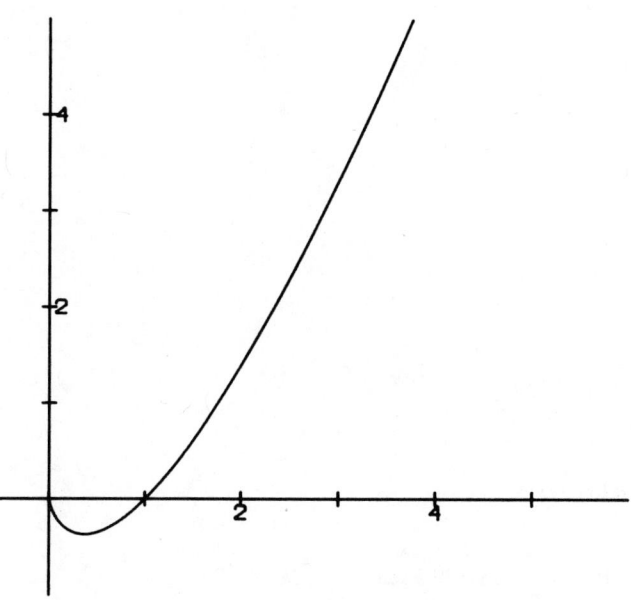

13. $f'(x) = \dfrac{x^{10}\sqrt{x^3}}{(x^2+1)^5}\left(\dfrac{10}{x} + \dfrac{3}{2x} - \dfrac{10x}{x^2+1}\right)$

15. $f'(x) = f(x)\left(\dfrac{18x}{x^2+1} + \dfrac{15x^2+10}{x^3+2x+3} + \dfrac{12x^3}{x^4+1}\right)$

17. $(1/2)\ln 2$
19. $\ln |\ln x| + C$
21. $\ln |x^3 + 2x + 4| + C$
23. $\ln 2 - (1/3)\ln 5$

Chapter 6, Section 2

1. $f'(x) = 2e^{2x}$
3. $f'(t) = 2te^{t^2}$
5. $f'(x) = 1$
7. $f'(x) = e^x/(2\sqrt{e^x+1})$
9. $e - 1$
11. $(1/4)e^{x^4} + C$
13. $(2/3)(e^x + 1)^{3/2} + C$
15. $\ln(e^x + 1) + C$

17. $y' = (x^2 + 1)^{2x}\left[\dfrac{4x^2}{x^2+1} + 2\ln(x^2+1)\right]$

19. $f'(x) = \left(x + \dfrac{1}{x}\right)^x\left[\dfrac{2x^2}{x^2+1} - 1 + \ln(x^2+1) - \ln x\right]$

21. $y' = (2^{\ln x}\ln 2)/x$

23. $y' = x^{x^x}x^x\left[\dfrac{1}{x} + \ln x + (\ln x)^2\right]$

25. $dI/dt = A^2 - I^2$
27. a. $t = (\ln b - \ln a)/(b - a)$

 b. $y = \dfrac{c}{b-a}\left[\left(\dfrac{b}{a}\right)^{-a/(b-a)} - \left(\dfrac{b}{a}\right)^{-b/(b-a)}\right]$

Chapter 6, Section 3

1. $y' = (2\ln x)/(x \ln a \ln b)$
3. $f'(x) = 1$
5. $y' = (2x \log_{10} e)/(x^2 + 1)$

7. $\log_b(c^d) = \dfrac{\ln c^d}{\ln b} = \dfrac{d \ln c}{\ln b} = d \log_b(c)$

9. 4
11. 2
13. $2x$
15. -3
17. 42, 107 or 108
19. 69.7 years, 139.3 years
21. 393,700

23.

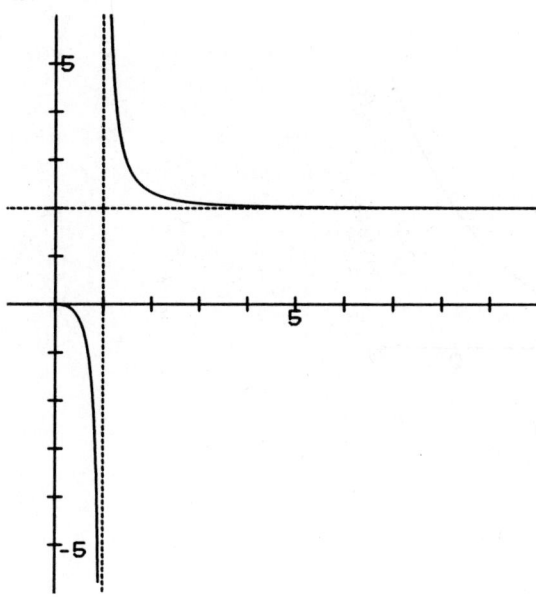

Chapter 6, For Hand-Held Calculators Only

1. a. 1.0897544   b. 1.0986122
3. 0.4342946
5. 12.18232 using $x = 0.5$, 12.182085 using $x = -0.5$

7. x	$\ln x$
0.5	-0.6930038
0.6	-0.5108072
0.7	-0.3566730
0.8	-0.2231434
0.9	-0.1053602
1.0	0.0000000
1.1	0.0953100
1.2	0.1823214
1.3	0.2623638
1.4	0.3364710
1.5	0.4054612

426 ANSWERS TO SELECTED PROBLEMS

Chapter 6, Review Section

1.

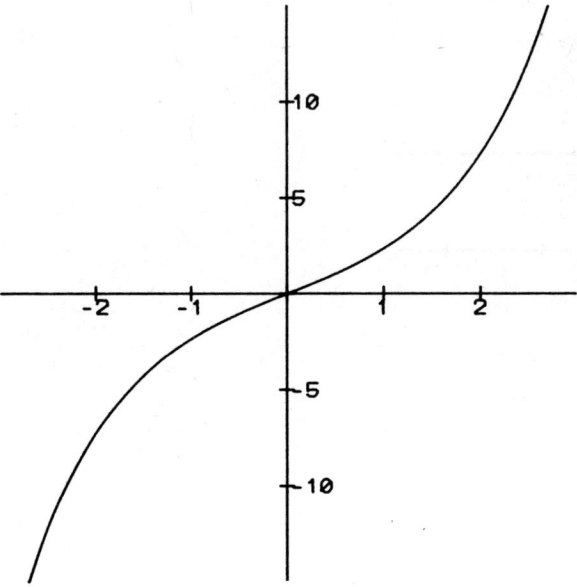

3.

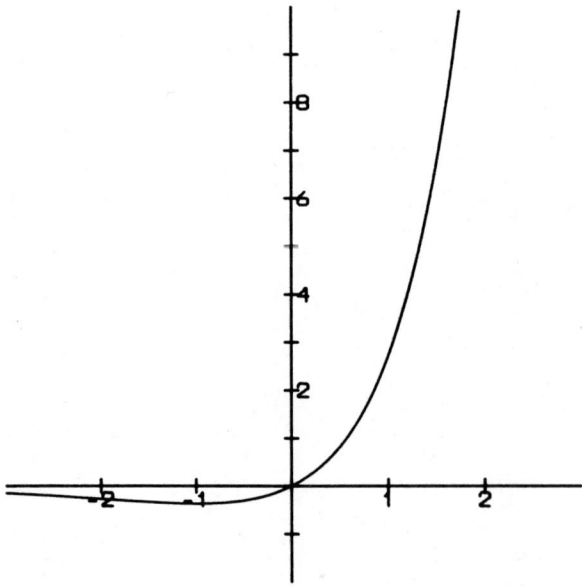

ANSWERS TO SELECTED PROBLEMS 427

5.

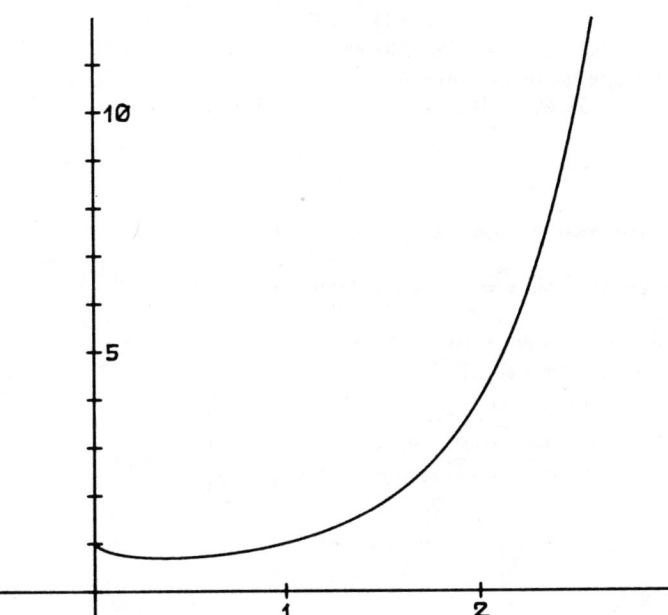

7.

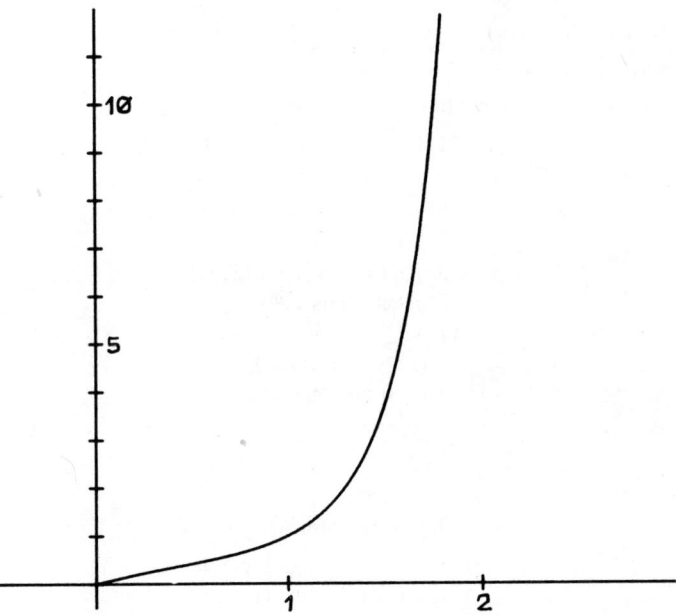

9. $(\pi/2)(e^2 - 1)$ 11. $(\pi/2)(\ln 2)^2$ 13. 8192
15. $x = (\ln 6)/(\ln 3)$ 17. $x = [\ln(8/3)]/[\ln(4/3)]$
19. 3.873 21. 1.845×10^{19}
23. 7562.4 25. $240,000
27. 99^{100} (Compute logarithms of the two numbers.)
29. $\ln 2 \approx 0.69315$ 31. 605 millibars 33. 1793 years

Chapter 7, Section 1

1. $\tan x = \sin x/\cos x = 1/(\cos x/\sin x) = 1/\cot x$

3. $\sin\left(\dfrac{\pi}{2} - x\right) = \sin\dfrac{\pi}{2}\cos x - \cos\dfrac{\pi}{2}\sin x = 1\cdot\cos x - 0\cdot\sin x = \cos x$

5. $\sin 2x = \sin(x + x) = \sin x \cos x + \cos x \sin x = 2\sin x \cos x$
7. $1 + \cos 2x = 1 + (\cos x \cos x - \sin x \sin x)$
$= 1 + \cos^2 x - \sin^2 x = 1 + \cos^2 x - (1 - \cos^2 x) = 2\cos^2 x$

9. $\tan(x + \pi) = \dfrac{\sin(x+\pi)}{\cos(x+\pi)} = \dfrac{\sin x \cos \pi + \cos x \sin \pi}{\cos x \cos \pi - \sin x \sin \pi} = \dfrac{-\sin x}{-\cos x} = \dfrac{\sin x}{\cos x} = \tan x$

11. $(1/2)(1 - \cos 2x) = (1/2)[1 - (\cos^2 x - \sin^2 x)]$
$= (1/2)[1 - (1 - \sin^2 x - \sin^2 x)] = (1/2)(2\sin^2 x) = \sin^2 x$

13. $\sin x \csc x = \sin x \dfrac{1}{\sin x} = 1$

15. $\tan(x + y) = \dfrac{\sin(x+y)}{\cos(x+y)} = \dfrac{\sin x \cos y + \cos x \sin y}{\cos x \cos y - \sin x \sin y}$

$= \dfrac{(\sin x \cos y + \cos x \sin y)/(\cos x \cos y)}{(\cos x \cos y - \sin x \sin y)/(\cos x \cos y)}$

$= \dfrac{(\sin x/\cos y) + (\sin y/\cos y)}{1 - (\sin x/\cos x)(\sin y/\cos y)} = \dfrac{\tan x + \tan y}{1 - \tan x \tan y}$

17. 1/2 19. 1 21. 1 23. $\sqrt{6}/2$

Chapter 7, Section 2

1. $y' = 2\cos 2x$
5. $y' = 0$
9. $y - (\sqrt{3}/2) = x - (\pi/6)$
13. $y = x + 1$
17. $\ln|\sin x| + C$

3. $f'(t) = (\sec^2 \sqrt{t})/(2\sqrt{t})$
7. $f'(x) = \cos x\, e^{\sin x}$
11. 3
15. $(x/2) - (1/4)\sin 2x + C$
19. $(1/3)\sin e^{3x} + C$

Chapter 7, Section 3

1. $y' = 2x/\sqrt{1 - x^4}$
5. $f'(x) = e^x/\sqrt{1 - e^{2x}}$
9. $y' = 1/(2\sqrt{x}\sqrt{1 - x})$

3. $y' = (2\arcsin x)/\sqrt{1 - x^2}$
7. $y' = -1/(x^2\sqrt{1 - x^2})$
11. $f'(x) = e^{\arctan x}/(1 + x^2)$

13. $y' = x^{\arcsin x}\left(\dfrac{1}{x}\arcsin x + \dfrac{\ln x}{\sqrt{1-x^2}}\right)$ 15. $f'(x) = \dfrac{\arctan x}{x\sqrt{x^2-1}} + \dfrac{\arcsec x}{x^2+1}$

17. $(1/2)\arcsec(x^2) + C$
19. $(1/3)\arctan(x^3) + C$
21. $-\sqrt{1-x^2} + C$
23. $(1/2)\ln(4+x^2) + C$
25. $\arctan\sqrt{u-1} + C$
31. $(5/6)\arctan(3/2)$

Chapter 7, Section 4

1. 6960 feet
3. $\sqrt{500 \cdot 550} \approx 524.4$ feet from the point on the street nearest the screen
5. $y = 3, y = -3, t = 8 + \pi + 4n\pi$
7. $\sqrt{6\sqrt[3]{2} + 3\sqrt[3]{4} + 5} = (2^{2/3} + 1)^{3/2} \approx 4.162$ meters
9. $-6y_0\pi$

Chapter 7, The Hyperbolic Functions

1. $\dfrac{d}{dx}\coth x = \dfrac{d}{dx}\dfrac{\cosh x}{\sinh x} = \dfrac{\sinh^2 x - \cosh^2 x}{\sinh^2 x} = \dfrac{-1}{\sinh^2 x} = -\csch^2 x$

3. $\sech^2 x + \tanh^2 x = \dfrac{1}{\cosh^2 x} + \dfrac{\sinh^2 x}{\cosh^2 x} = \dfrac{1+\sinh^2 x}{\cosh^2 x} = \dfrac{\cosh^2 x}{\cosh^2 x} = 1$

5. $\cosh a \cosh b + \sinh a \sinh b$
$= (1/2)(e^a + e^{-a})(1/2)(e^b + e^{-b}) + (1/2)(e^a - e^{-a})(1/2)(e^b - e^{-b})$
$= (1/4)(e^a e^b + e^a e^{-b} + e^{-a} e^b + e^{-a} e^{-b}) + (1/4)(e^a e^b - e^a e^{-b} - e^{-a} e^b + e^{-a} e^{-b})$
$= (1/2)(e^a e^b + e^{-a} e^{-b}) = (1/2)(e^{a+b} + e^{-(a+b)}) = \cosh(a+b)$

7. $\sech x = 4/5, \cosh x = 5/4, \sinh x = 3/4, \coth x = 5/3, \csch x = 4/3$ 9. 0

11. $\cosh^2 x + \sinh^2 x$
13. $\dfrac{-2x}{(1-x^2)^2}\sech^2\dfrac{1}{1-x^2}$

15. $-\sech x + C$
17. $(1/3)\tanh^3 x + C$
19. $-(1/2)\sech^2 x + C$
25. $(\sinh x)/\sqrt{1+\cosh^2 x}$
27. $\dfrac{\sinh(\arccoth x)}{1-x^2}$
29. $\arcsinh(\cosh x) + C$

31. $-(1/2)\arcsech x^2 + C$
33. $-(1/3)\arcsech\left(\dfrac{2x}{3}\right) + C$

Chapter 7, For Hand-Held Calculators Only

1. 0.0998334 3. 0.0996686 5. 0.9961948 7. 0.1763266 9. 1.1207963
11. $\cos(0.1) = 0.9950042, \sin(0.1) = 0.0998334, \cos^2(0.1) + \sin^2(0.1) = 1.0000000$

430 ANSWERS TO SELECTED PROBLEMS

Chapter 7, Review Section

1.

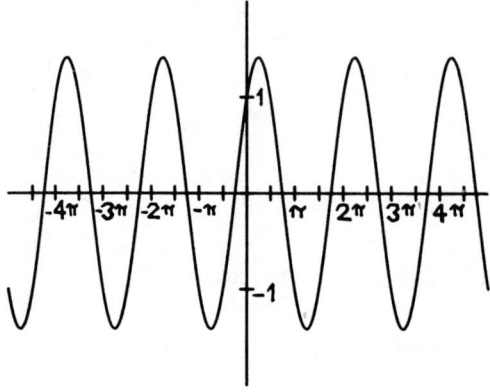

3.

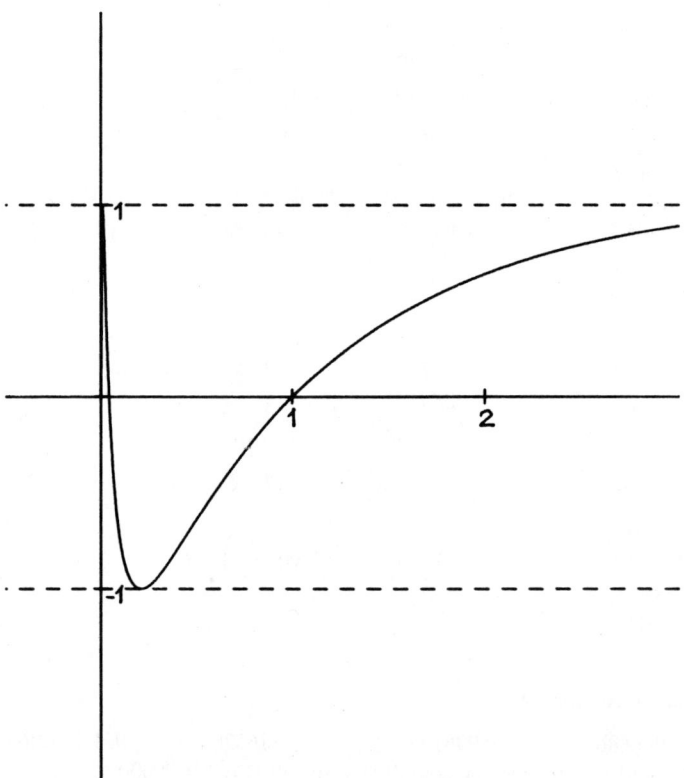

5.

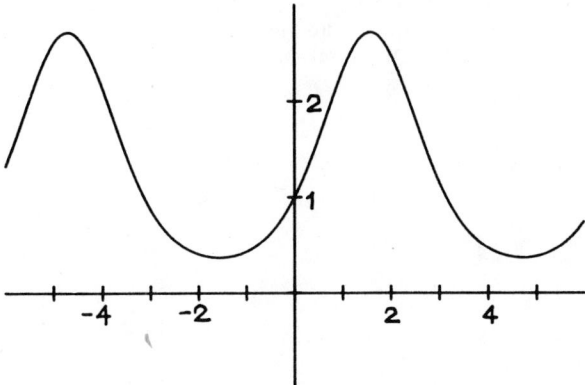

7.

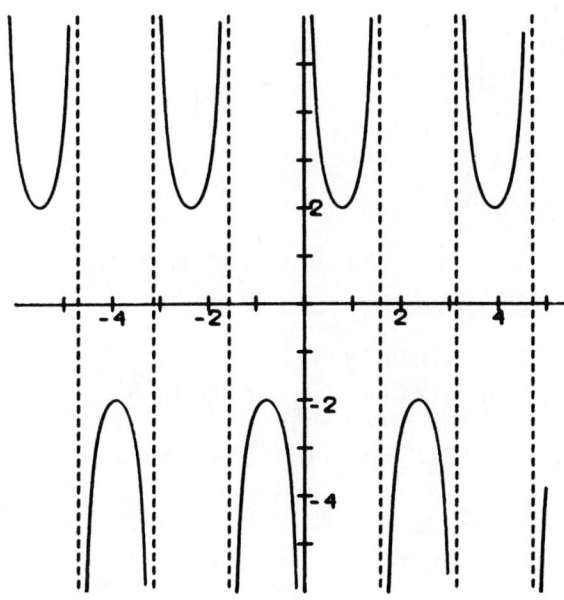

9. $-(1/2)\cos(x^2) + C$
13. $-2\ln|\cos\sqrt{t}| + C$
17. $-(1/2)\ln|\cos(x^2)| + C$
21. $(1/12)\arctan(3x/4) + C$

11. $(1/4)\sin^2(x^2) + C$
15. $(1/2)\sec(x^2 + 1) + C$
19. $-(\pi/4) + \arctan e$
23. $(x/2) + (1/4)\sin 2x + C$

Chapter 8, Section 1

1. a. $\int (u^2 - 1)2u^2\, du,\ u^2 = x + 3$

   c. $\int_1^{\sqrt{2}} u^2\, du,\ u^2 = x^2 + 1$

   e. $\int (u^2 + 4)u^2\, du,\ u^2 = x^2 - 4$

2. a. $\int_0^{\pi/4} \tan\theta\sqrt{1 + \tan^2\theta}\sec^2\theta\, d\theta$

   c. $\int_{\arcsin(\sqrt{5}/3)}^{\pi/2} \dfrac{3\sqrt{1 - \sin^2\theta}\, 3\cos\theta\, d\theta}{3\sin\theta}$

   e. $\dfrac{2}{9}\int \dfrac{\sec^2\theta\, d\theta}{\tan^2\theta\sqrt{1 + \tan^2\theta}}$

3. $(1/5)(x^2 - 4)^{5/2} + (4/3)(x^2 - 4)^{3/2} + C$

5. $-\sqrt{4x^2 + 9}/(9x) + C$

9. $(1/4)\arctan(x/2) - \dfrac{x}{2(x^2 + 4)} + C$

11. $(1/2)\arcsin(2x/3) + C$
15. $7/12$
21. $\sqrt{2}$
27. $\pi/12$

13. $\pi/24$
17. $61/27 \approx 2.259$
23. $\sqrt{3}/3$
29. $4 - 2\sqrt{2}\arctan\sqrt{2}$

19. $8/3$
25. 4
31. 379.47

33. $\int \csc x\, dx = \int \dfrac{dx}{\sin x} = \int \dfrac{dx}{2\sin(x/2)\cos(x/2)} = \int \dfrac{dx}{2\tan(x/2)\cos^2(x/2)}$

$= \int \dfrac{(1/2)\sec^2(x/2)\, dx}{\tan(x/2)} = \ln|\tan(x/2)| + C$

Chapter 8, Section 2

1. $(x^3/9)(3\ln x - 1) + C$
5. $(2/9)x\sqrt{x}(-2 + 3\ln x) + C$
9. $-(1/8)e^{-2x}(4x^3 + 6x^2 + 6x + 3)$
13. $(1/3)\sqrt{x^2 + 1}(x^2 - 2) + C$
17. $(2/9)x^3(-1 + 3\ln x) + C$
19. $x(\arcsin x)^2 + 2\sqrt{1 - x^2}\arcsin x - 2x + C$
21. $(1/2)(x^2 + 1)\arctan x - (1/2)x + C$
23. $x\ln(x^2 + 1) - 2x + 2\arctan x + C$

3. $(1/4)(e^2 + 1)$
7. $(x/2)[\sin(\ln x) - \cos(\ln x)] + C$
11. $(1/4)\ln(9 + x^4) + C$
15. $(2/75)(x^5 + 1)^{3/2}(3x^5 - 2) + C$

ANSWERS TO SELECTED PROBLEMS   433

Chapter 8, Section 3

1. $\pi/4$
3. $2\arcsin\sqrt{x} + C$
5. $(5\sqrt{5} - 1)/3$
7. The function $x \to \sqrt{x^2 + 4x}$ is not defined on $[-4, 0]$
9. 2π
11. $(243/8)\ln 3 + (25/2) 45.87$

Chapter 8, Section 4

1. $x + \dfrac{1}{x+1}$

3. $\dfrac{2}{x-2} + \dfrac{1}{x+1} - \dfrac{1}{(x+1)^2}$

5. $\dfrac{1}{x+1} + \dfrac{1}{(x+1)^3}$

7. $\dfrac{1}{x+1} + \dfrac{1}{x^3} - \dfrac{1}{x}$

9. $\dfrac{x+1}{(x^2+1)^2} + \dfrac{2x+1}{x^2+1}$

11. $x - \dfrac{1}{x^2} + \dfrac{x}{(x^2+1)^2}$

13. $\ln 2 + \arctan 2 - \pi/4$

15. $\ln|x| - \dfrac{2}{x} + \dfrac{3}{2}\ln(x^2+1) + \arctan x + C$

17. $\ln(x^2 + 2x + 5) - \ln|x - 1| - \dfrac{1}{2}\arctan\dfrac{x+1}{2} + C$

19. $\ln|x - 1| + \dfrac{1}{8}\ln|4x^2 - 8x + 13| + \dfrac{2}{3}\arctan\dfrac{2(x-1)}{3} + C$

21. $(1/2) + (3\pi/4)$

Chapter 8, Section 5

1. Does not exist
3. 2
5. Does not exist
7. 2
9. $\pi/4$
11. Does not exist
13. 0
15. Does not exist
17. $-1/4$
19. π
21. π

Chapter 8, Section 6

1. $6.5, |E| \le 0.5$
3. $1.117, |E| \le 0.084$
5. $1.5 + \sqrt{2} + \sqrt{3} \approx 4.646, |E| \le 0.063$
7. $77/72 \approx 1.069, |E| \le 0.25$
9. 0.628
11. 0.5246

Chapter 8, Section 7

1. 0.310
3. 3.097
5. $1.009, |E| \le 0.084, 1.000$
7. $6.785, |E| \le 1/48 \approx 0.021, (2/3)(5\sqrt{5} - 1) \approx 6.787$
9. 0.5236159, the Fundamental Theorem yields $\pi/6 \approx 0.5235987$
11. a. 98
b. 226
c. 205.33
d. 204.8

Chapter 8, Review Problems

1. $(1/6)\sin^2(x^3) + C$
3. $\arcsin(x/3) + C$
5. 2
7. $(1/4)\sqrt{4x^2 + 9} + C$
9. $(x/2)\sin^2 x - (x/4) + (1/8)\sin(2x) + C$
11. $2x(-1 + \ln x) + C$
13. $(e^x + 1)\ln(e^x + 1) - e^x + C$
15. Improper integral, does not exist
17. Improper integral, does not exist
19. $(1/2)(x-2)\sqrt{x^2 - 4x} - 2\ln|x - 2 + \sqrt{x^2 - 4x}| + C$
21. Improper integral, does not exist
23. $\ln|x| - \ln|x+1| + \dfrac{1}{x+1} + C$
25. $(x^2/2) - 2x + (12/5)\ln|x+2| + (3/10)\ln(x^2+1) + (4/5)\arctan x + C$
27. $-\dfrac{2}{125}\ln|x-1| - \dfrac{1}{25(x-1)} + \dfrac{2}{125}\ln|x+4| - \dfrac{1}{25(x+4)} + C$
29. $-8\sqrt{2/27}$
31. $5\ln 5 - 2\ln 2 - 3$
33. $\pi[2(\ln 2)^2 - 4\ln 2 + 2]$
35. $\dfrac{18}{13\pi}(1 + e^{2\pi/3})$
37. $e^{-x} + e^{-y} = C$
39. $\ln|\sec x + \tan x| - \ln|\cos y| = C$

Chapter 9, Section 1

1. $\ln(\sqrt{2} + 1)$
3. $\pi/2$
5. $14/3$
7. $3\sqrt{37} + (1/2)\ln(6 + \sqrt{37})$
9. $\sqrt{17} - \sqrt{2} + \ln(\sqrt{17} - 1) - \ln 4 - \ln(\sqrt{2} - 1)$
11. $53/6$

Chapter 9, Section 2

1. $\bar{x} = (1/4)(e^2 + 1)$, $\bar{y} = (e/2) - 1$
3. $\bar{x} = 1$, $\bar{y} = \pi/8$
5. $\bar{x} = 8/5$, $\bar{y} = 16/7$
7. $\bar{x} = 1.6$, $\bar{y} = 0.2131$
9. $\bar{x} = \bar{y} = 0.2$
11. $\bar{x} = (1/4)(e^2 + 1)$, $\bar{y} = e - 2$

Chapter 9, Section 3

1. 9
3. ln 2
5. 1
7. 0.5
9. $(4/3) + \ln 4$
11. $(1/2) + \ln 3$
13. $27500\delta\pi/3 \approx 28{,}797{,}932$ kilogram-meters
15. 2500 gram-centimeters

Chapter 9, Section 4

1. $\pi[e^2\sqrt{1+e^2} + \ln(e^2 + \sqrt{e^2+1}) - \sqrt{2} - \ln(1+\sqrt{2})]$
3. 2.866
5. $(\pi/4)(e^2 + 4 - e^{-2}) \approx 8.839$
7. $6\pi/5$
9. 8π
11. $1216\pi/3 \approx 1273.39$
13. $\sqrt{2} - (\sqrt{e^2+1}/e) + \ln(e + \sqrt{1+e^2}) - \ln(1+\sqrt{2})$
15. 1.5

Chapter 9, Review Section

1. a. $|x_2 - x_1|\sqrt{m^2+1}$ c. $\ln(1+\sqrt{2})$
   e. $-1 - \sqrt{2} + \sqrt{e^2+1} + \ln(-1+\sqrt{e^2+1}) + \ln(1+\sqrt{2}) \approx 2.0035$
2. a. $2\pi rh$
   c. $(\pi/6)[(4b+1)^{3/2} - (4a+1)^{3/2}]$
   e. $\pi[(\sqrt{3}/2) + \ln(1+\sqrt{3}) - (1/2)\ln 2] \approx 4.7814$
3. a. $\bar{x} = \bar{y} = 9/20$ c. $\bar{x} = 1, \bar{y} = \pi/8$
   e. $\bar{x} = 0, \bar{y} = 1.5$ g. $\bar{x} = 3/5, \bar{y} = 12/35$
5. $\pi[4\sqrt{17} - \sqrt{2} - \ln(\sqrt{17}+4) - \ln(1+\sqrt{2})]$
7. 2.5 centimeters
9. $(2^{2/3} + 3^{2/3})^{3/2} \approx 7.0235$
13. 18,000,000 foot-pounds, 19,500,000 foot-pounds
15. $\dfrac{4\sqrt{3}}{3}$ 17. 12.83 hours   19. $16/3$ ft^3
23. b. $c = mg/\bar{v}$ 25. $2\pi^2 r^2 c$

436 ANSWERS TO SELECTED PROBLEMS

Chapter 10, Section 1

1. $-b/a$

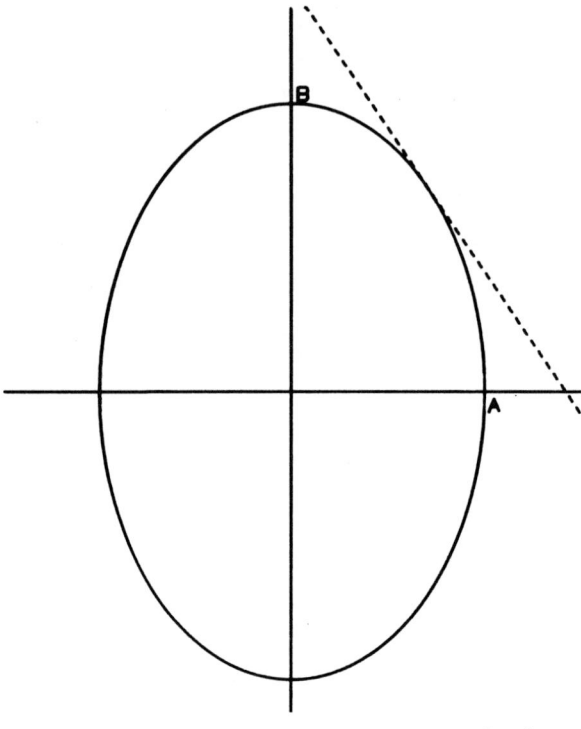

3. e

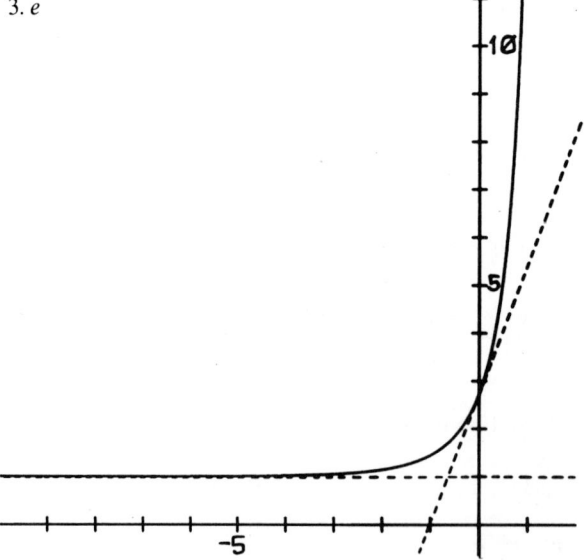

ANSWERS TO SELECTED PROBLEMS 437

5. 1

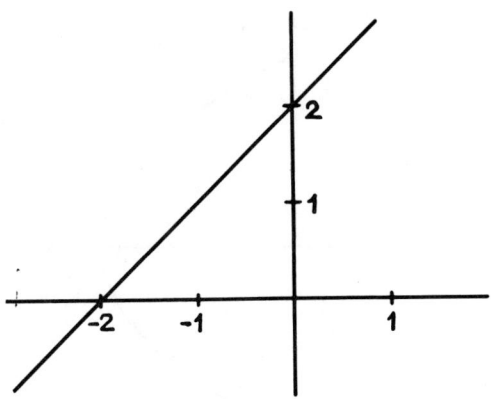

7. 27/40

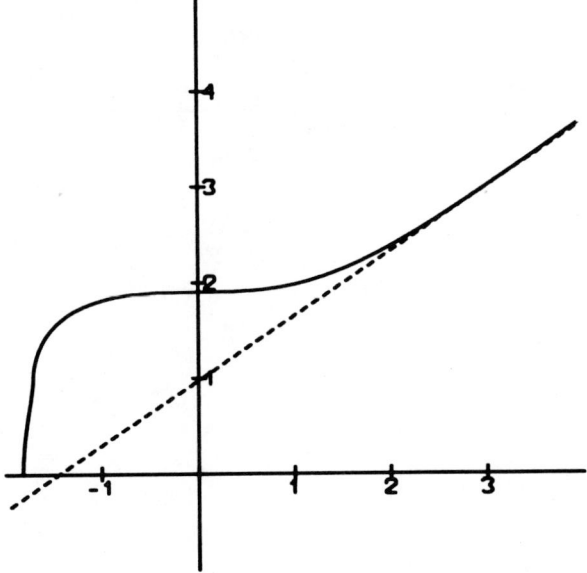

438　ANSWERS TO SELECTED PROBLEMS

9.

11.

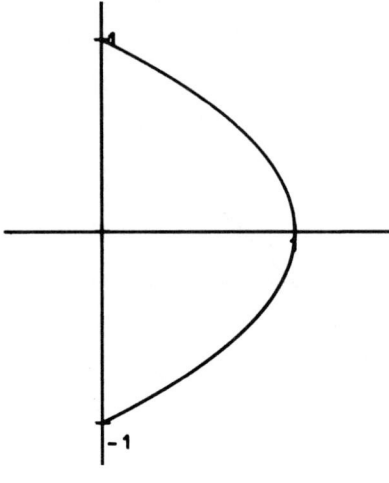

13.

15.

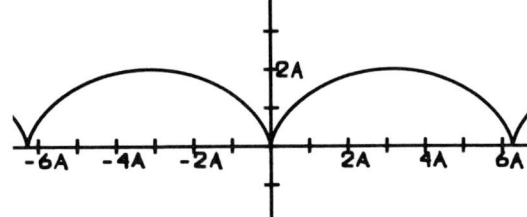

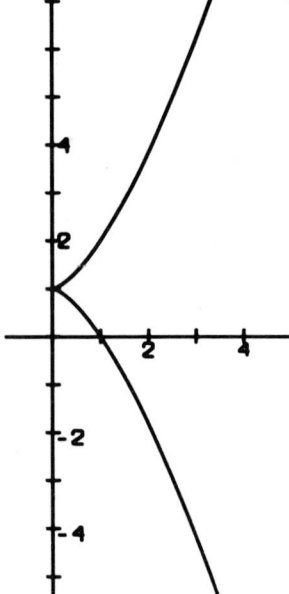

17. 45°   19. $\sqrt{2}$

Chapter 10, Section 2

1.
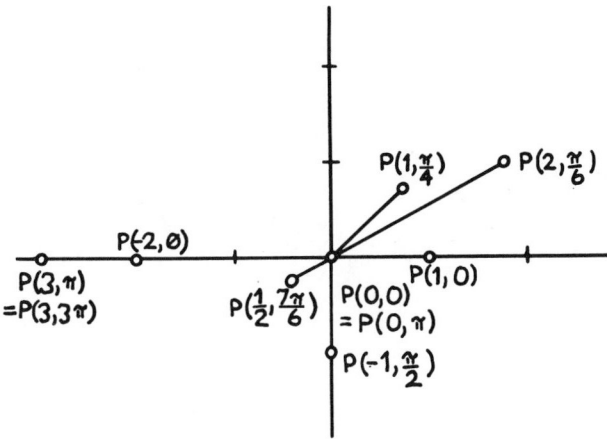

3. $P(\sqrt{2}, \pi/4)$, $P(\sqrt{2}, -\pi/6)$, $P(5, \arctan 4/3)$, $P(2, \pi/2)$, $P(1, \pi)$, $P(2, -\pi/4)$, $P(\sqrt{2}, 5\pi/4)$, $P(1, 3\pi/2)$, $P(0, 0)$, $P(\sqrt{2}, 3\pi/4)$

5.
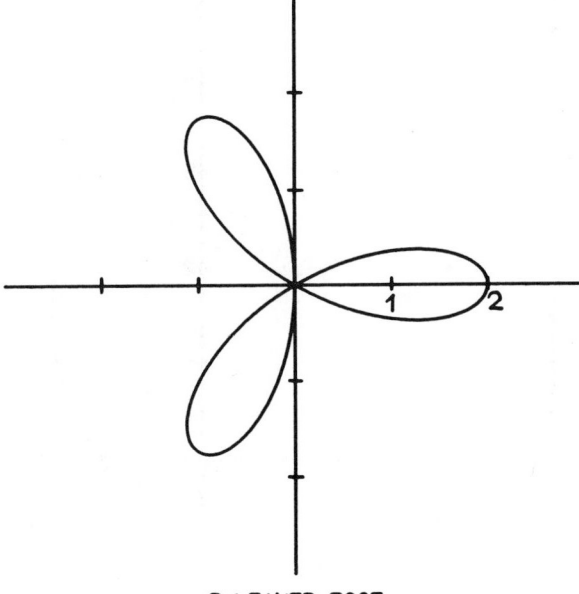

3 LEAVED ROSE

7.

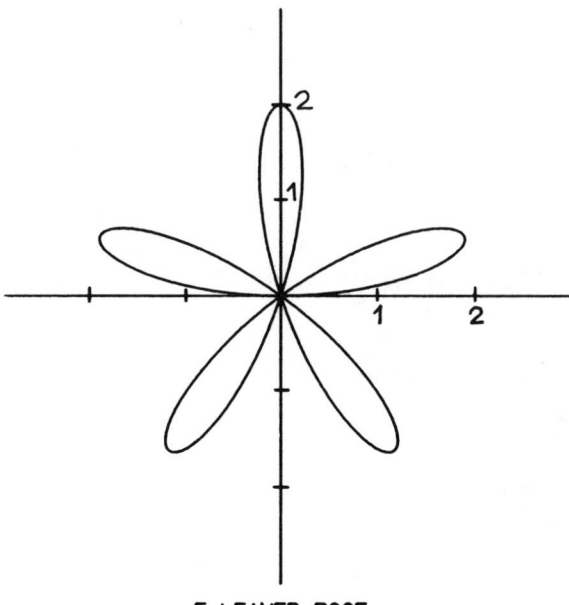

5 LEAVED ROSE

9.

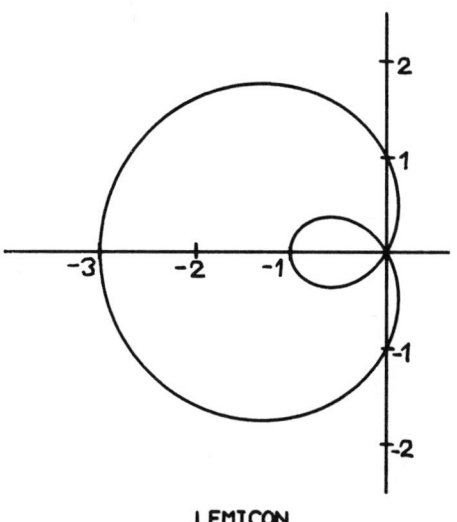

LEMICON

11.

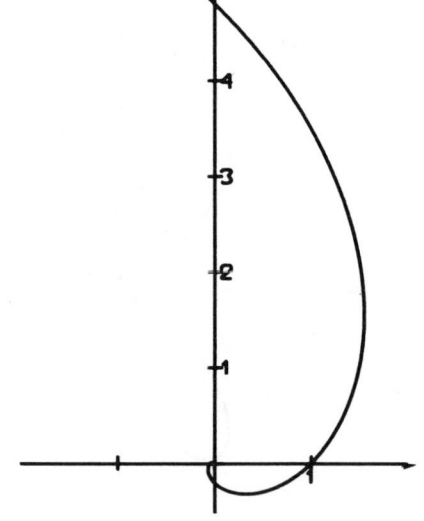

13. 8π 15. $\pi - (3\sqrt{3}/2)$ 17. $8a$ 19. 2

ANSWERS TO SELECTED PROBLEMS   441

Chapter 10, Section 3

1. Focus at (0, 5/2), directrix $y = -5/2$
3. $y^2 = x$
5. $y^2 = -4(1/16)x$

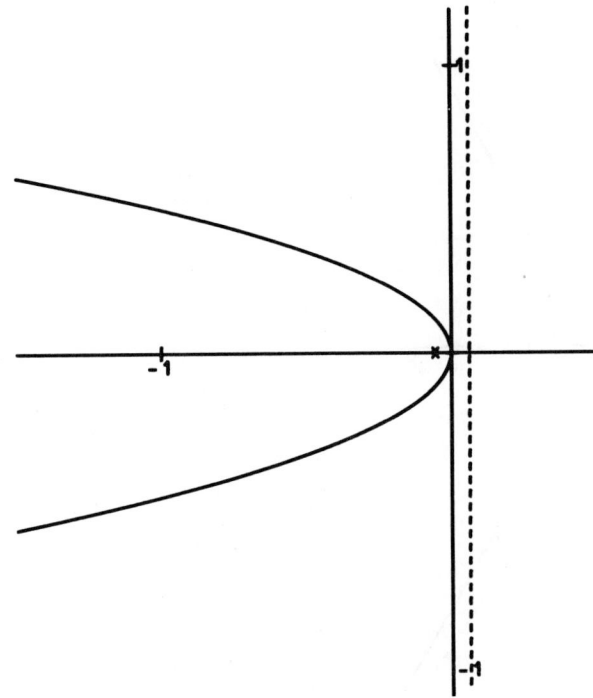

7. $(y - 2)^2 = 8(x - 3)$
9. $(0, 4)$ and $(0, -4)$
11. $\dfrac{x^2}{1} + \dfrac{y^2}{4} = 1$
13. $\dfrac{x^2}{13^2} + \dfrac{y^2}{5^2} = 1$

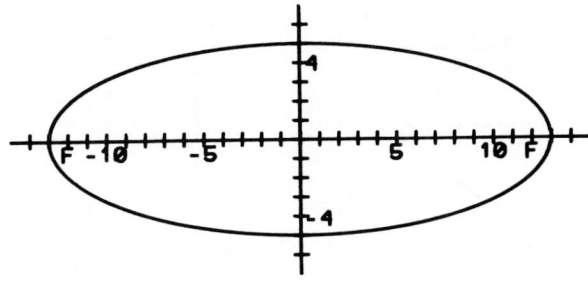

442 ANSWERS TO SELECTED PROBLEMS

15. $\dfrac{(x-2)^2}{5^2} + \dfrac{(y-1)^2}{4^2} = 1$

17. Foci at $(0, \sqrt{2})$ and $(0, -\sqrt{2})$, asymptotes $y = \pm x$

19.

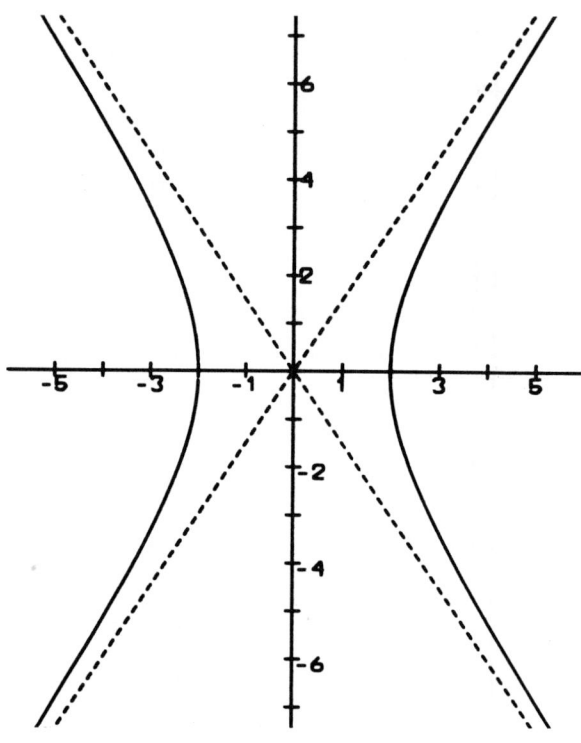

Chapter 10, Section 4

1. $\dfrac{(x-1)^2}{6^2} + \dfrac{(y+2)^2}{3^2} = 1$

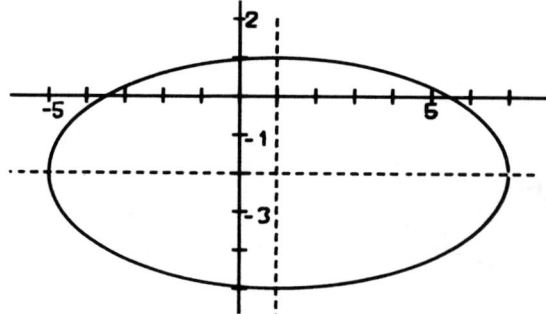

ANSWERS TO SELECTED PROBLEMS 443

3. $(y - 3)^2 = 8(x - 1)$

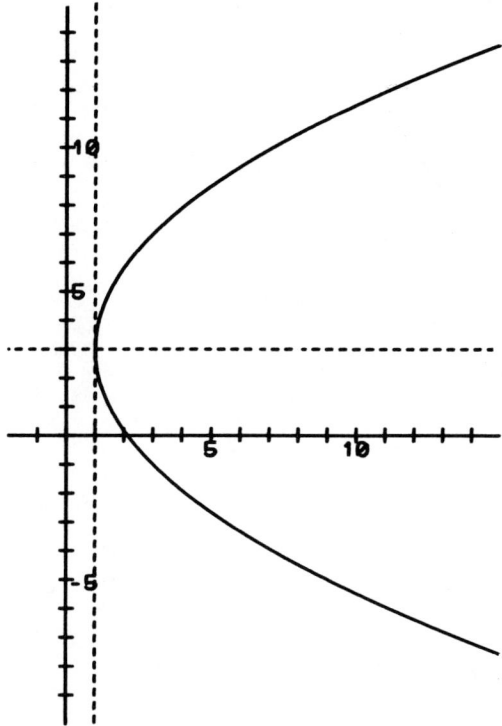

5. $(x - 1)^2 - y^2 = 1$

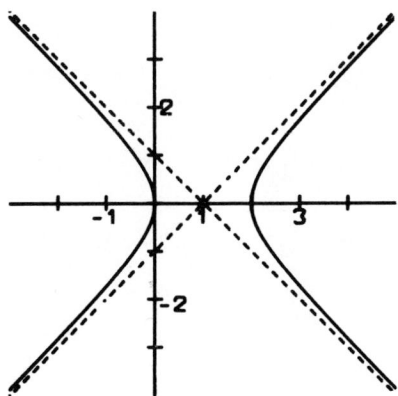

444 ANSWERS TO SELECTED PROBLEMS

7. $(x - 1)^2 = 12(y - 2)$

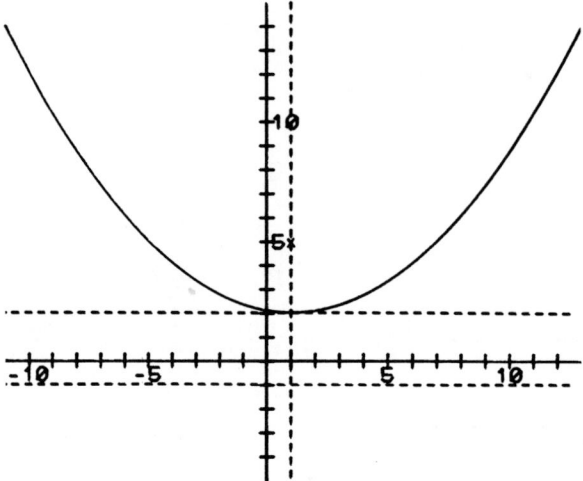

9. $\dfrac{(x + 1)^2}{2^2} + \dfrac{(y - 3)^2}{3^2} = 1$

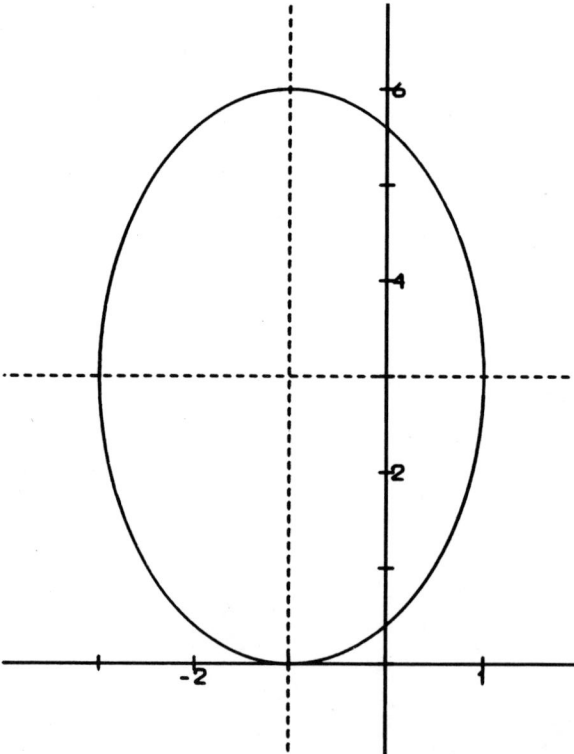

11. $\dfrac{(y-2)^2}{2} - \dfrac{(x+1)^2}{3} = 1$

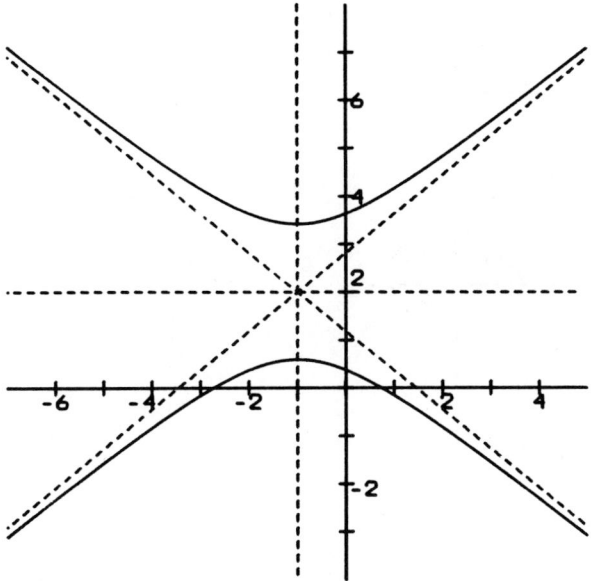

13. $\dfrac{(x-10)^2}{5^2} - \dfrac{(y+20)^2}{12^2} = 1$

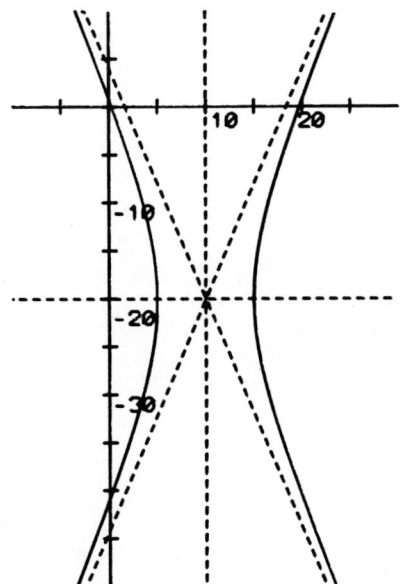

15. $\dfrac{(x-2)^2}{(1/3)^2} + \dfrac{(y+1)^2}{(1/5)^2} = 1$

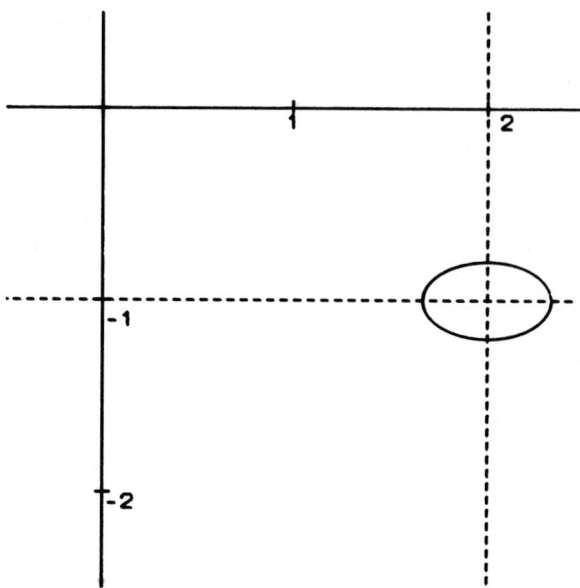

Chapter 10, Section 5

1. $(\sqrt{3}/2, -1/2), (1/2, \sqrt{3}/2), (\sqrt{3}, 1), (2, 0), ((\sqrt{3}+1)/2, (\sqrt{3}-1)/2)$
3. $Y = 4X + 2$
5. $Y = X^3$

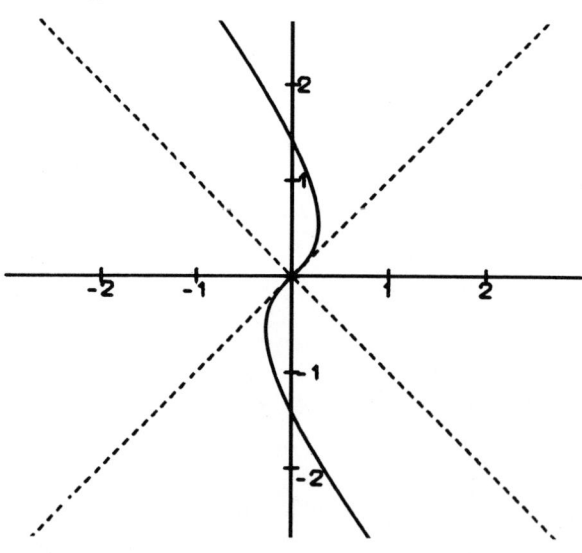

7.

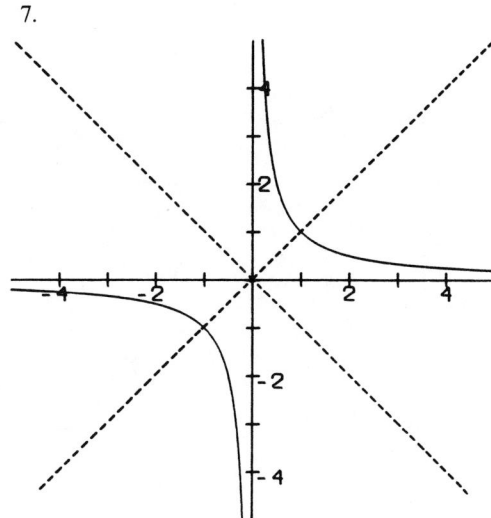

9. $(X + 1)^2 - (Y - 1)^2 = 1$

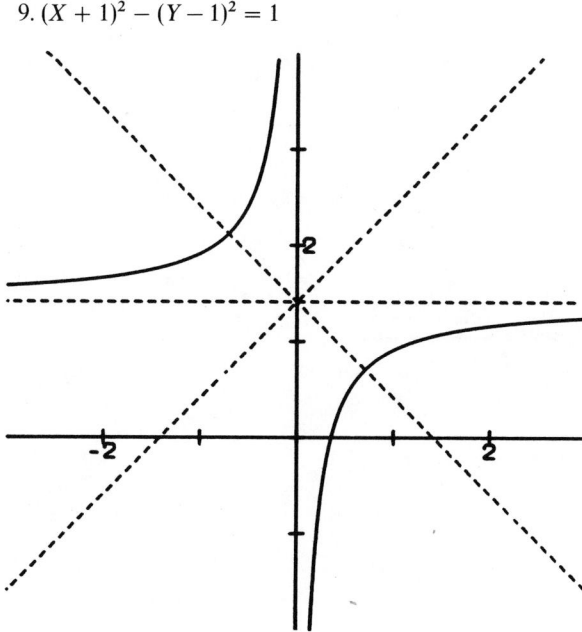

Chapter 10, Review Section

1. a. π
   c. $7\pi/2$
3. 1.5
5. $\pi a^2(2 - \sqrt{2})$
6. b. $(a/8)(4\pi - 3\sqrt{3})$
7. a. $r = 1 + 4\cos\theta$ c. $r = 2\sec(\theta - A)$ e. $r = a(1 - \sin\theta)$
8. b. $(x^2 + y^2)^2 = 2a^2xy$
9. a.

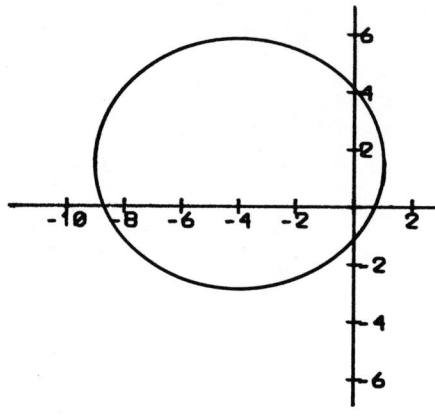

c.

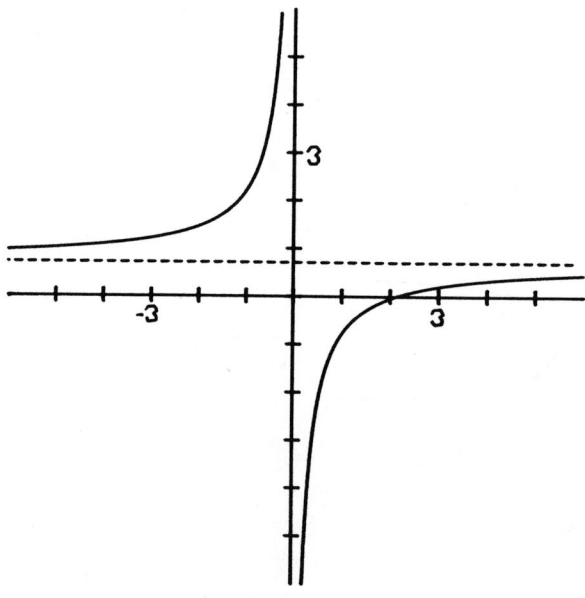

e.

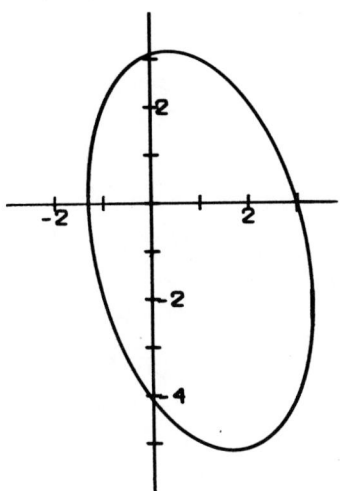

10. b.

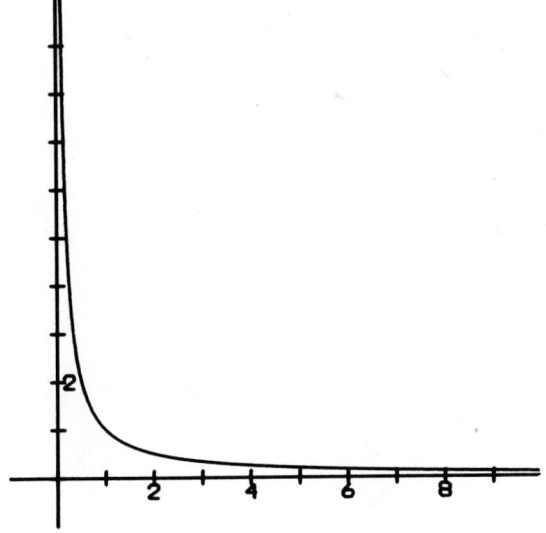

d.

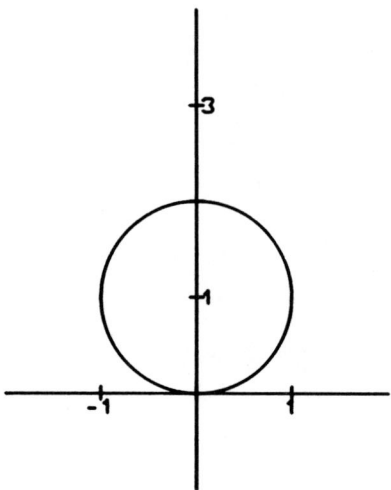

f.

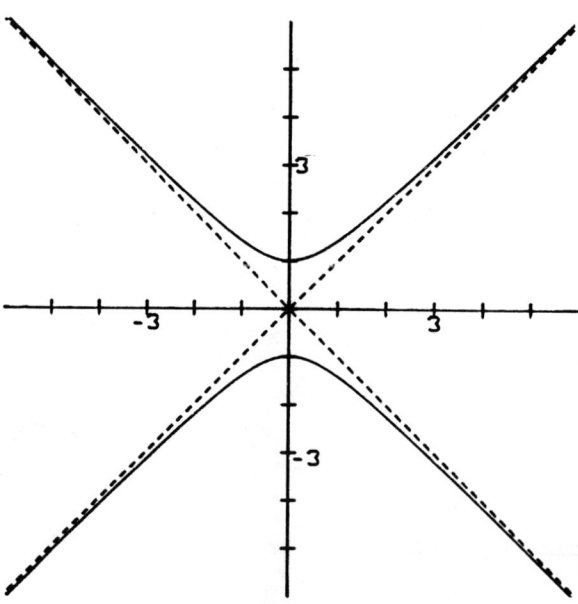

ANSWERS TO SELECTED PROBLEMS 451

h.

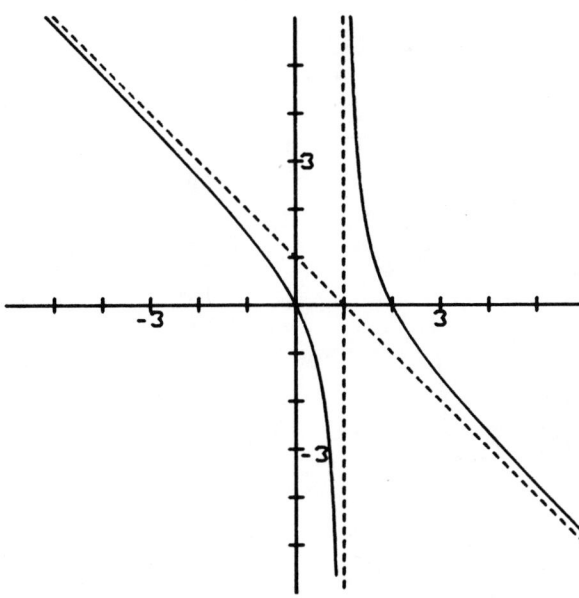

11. $y = -\dfrac{2\sqrt{3}}{9}x + \dfrac{4\sqrt{3}}{3}$

13. $\left(x - \dfrac{b}{2}\right)^2 + \left(y - \dfrac{a}{2}\right)^2 = (a^2 + b^2)/4$

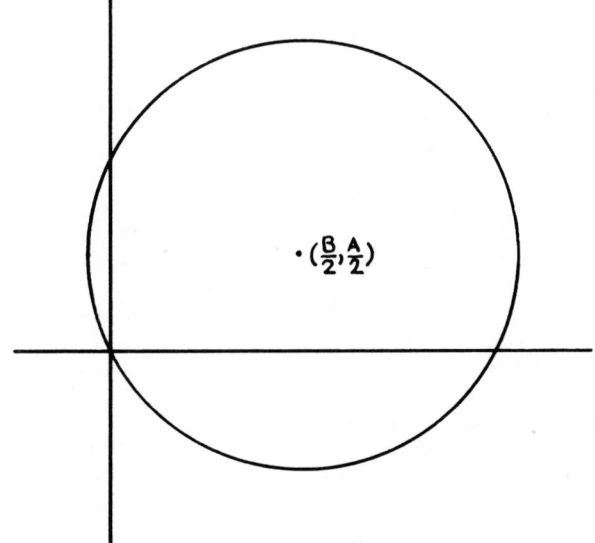

15. $3\pi/4$ 17. π 19. $4\pi ab^2/3$

23. $\dfrac{(x-4)^2}{16} - \dfrac{y^2}{9} = 1$ 25. $\arctan(3/4)$

27. The tangent line is vertical at $t = \pi/2$

Chapter 11, Section 1

1. $\sum_{i=0}^{k} (-1)^i [x^{2i}/(2i)!]$, $n = 2k$ or $n = 2k+1$

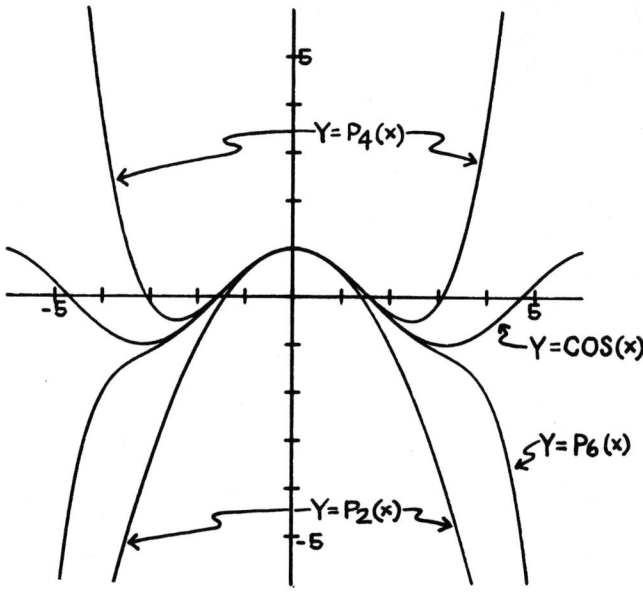

3. $\sum_{i=0}^{n} (-1)^i (x-1)^i$
5. $3x^3 + 5x^2 + 4x + 1$, $n \geq 3$
7. 7.0710678, error $\leq 5 \times 10^{-8}$
9. 0.0174524, error $< \pi^5/[5!(180)^5] \approx 1.35 \times 10^{-11}$
11. $a = \pi/6$, $\cos(28°) \approx 0.8829475$, error $< \pi^4/[24(90)^4] \approx 6.2 \times 10^{-8}$
13. a. $1 - x^2 + (1/3)x^4$
    b. $1 - (x^2/2) + (x^4/24)$, $1 - x^2 + (1/3)x^4 - (1/24)x^6 + (1/576)x^8$
    c. $1 - x^2 + (1/3)x^4 - (2/45)x^6 + (1/315)x^8$
17. $p(x) = -(1/20)x^2 + (11/20)$, $p(0) = 11/20$, $p(2) = 7/20$, $p(1/2) = 0.5375$

Chapter 11, Optional Section

1. $(((x^2/7/(-6) + 1)x^2/5/4 - 1)x^2/3/2 + 1)x$

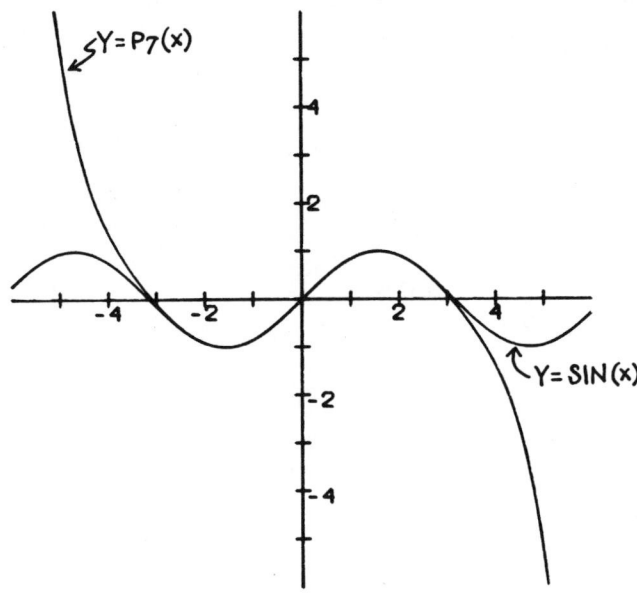

3. $((((((x/7 + 1)x/6 + 1)x/5 + 1)x/4 + 1)x/3 + 1)x/2 + 1)x + 1$
5. $-x - (x^2/2) - (x^3/3) - (x^4/4) - (x^5/5) - (x^6/6)$
7. $2[x + (x^3/3) + (x^5/5)]$ 9. 0.1823215

Chapter 11, Section 2

1. $1 + x - (x^2/2!) - (x^3/3!) + (x^4/4!) + (x^5/5!) - (x^6/6!) - (x^7/7!) + \cdots$
3. $\sum_{n=0}^{\infty} x^{2n}$ 5. $\sum_{n=1}^{\infty} x^n$
7. $\sum_{n=0}^{\infty} (-1)^n \dfrac{x^{2n+3}}{(2n+3)(2n+1)!} + C$
9. $\sum_{n=1}^{\infty} (-1)^{n-1} \dfrac{x^{2n}}{2n(2n-1)} + C$
11. $\sum_{n=0}^{\infty} \dfrac{x^{3n+1}}{(3n+1)n!} + C$ 17. $f(x) = \sin x$
19. $f(x) = \sum_{n=0}^{\infty} \dfrac{x^{2n+1}}{(2n+1)!} (= \sinh x)$

Chapter 11, Section 3

1. 1   3. 1   5. 0   7. ∞
9. Diverges   11. Converges   13. Converges   15. Converges

Chapter 11, Section 4

1. $\sum_{n=0}^{\infty} (-1)^n \frac{x^n}{n!}$ $(= e^{-x})$

3. $\sum_{n=0}^{\infty} \frac{x^{3n}}{(3n)!}$

5. $\sum_{n=0}^{\infty} \frac{x^{2n}}{n!}$ $(= e^{x^2})$

7. $y = 0$

9. $\sum_{n=0}^{\infty} \frac{x^{2n+1}}{2n+1}$ $(= \operatorname{arctanh} x)$

11. 0.5798

Chapter 11, Review Section

1. a. $\sum_{n=0}^{\infty} (n+1) 2^n x^n$

c. $\sum_{n=0}^{\infty} (-1)^n 10^{-n-1} (x - 10)^n$

e. $1 - \frac{x}{2} - \sum_{n=2}^{\infty} \frac{1 \cdot 3 \cdot 5 \cdots (2n-3) x^n}{2^n n!}$

g. $\sum_{n=0}^{\infty} (-1)^n (n+1) x^n$

i. $x + \frac{1}{2 \cdot 3} x^3 + \frac{1 \cdot 3}{2 \cdot 4 \cdot 5} x^5 + \frac{1 \cdot 3 \cdot 5}{2 \cdot 4 \cdot 6 \cdot 7} x^7 + \cdots$

k. $\sum_{n=0}^{\infty} (-1)^n 4^{-n-1} (x-3)^n$

2. b. Converges   d. Diverges   f. Converges
   h. Converges   j. Converges   l. Converges
3. a. 1.342   c. 0.9045   e. 0.764
4. b. $-e \leq x \leq e$ d. $0 < x < 6$ f. $-1 < x < 1$
5. a. 0.75

7. a. $\sum_{n=0}^{\infty} \frac{A(A-1)(A-2)\cdots(A-n+1)}{n!} x^n$

c. 1.1066
9. 1.0954462
10. b. Converges by comparison with $\sum a_n$ because $\ln(1 + a_n) < a_n$ for $a_n > 0$.

INDEX

Absolute value, 3
Acceleration, 161
Addition formulas:
  for cosine, 220
  for sine, 221
  for tangent, 223
Algebraic functions, 185
Algorithms:
  for nth roots, 174
  for square roots, 172
Analytic geometry:
  Cartesian coordinates, 4
  distance, 122
  lines, 6
  parametric equations, 326
  polar coordinates, 331
  rectangular coordinates, 4
  rotation of axes, 349
  translation of axes, 346
Angle, 219
Antiderivative, 94

Approximations:
  left-hand, 20
  midpoint, 21
  polynomial, 360
  right-hand, 21
  Simpson's rule, 293
  trapezoidal rule, 288
  using differentials, 114
  using hand-held calculators, 172, 212, 253, 369
Arc length, 300
  in parametric equations, 330
  in polar coordinates, 336
Arcsine, 232
Area, 13
  approximation using rectangles, 15, 20
  bounded by a graph of a function, 14
  bounded by the graphs of two functions, 46
  in polar coordinates, 334
  of a circle, 392

Area (*continued*)
  of a polygon, 14
  of a surface of revolution, 316
  of a trapezoid, 394
  of a triangle, 13, 393
  of the lateral surface of a cone, 395
  of the surface of a sphere, 395
Asymptote, 151
  horizontal, 152
  vertical, 151
Average cost, 184
Average rate of change, 60
Average value of a function, 85
Axes, 4

Binomial theorem, 74

Calculators:
  computations involving Taylor polynomials, 369
  computations of logs and exponentials, 212
  computations of values of the circular functions, 253
  nth roots and Newton's method, 172
Cardioid, 333
Cauchy, Augustin-Louis, 32
Center of mass, 304
Centroid, *see* Center of mass
Chain rule, 111
Change in f, 60
Change-of-basis formula, 203
Characteristic, 208
Circle, 122
  area, 392
  circumference, 393
  equation, 122
  sector, 393
Circular functions, 217
  cosecant, 221
  cosine, 217
  cotangent, 221
  derivatives, 223
  graphs, 228
  inverses, 231
  secant, 221
  sine, 217
  tables of values, 222
  tangent, 221
Closed interval, 4
Collapsing sum, 26

Common logarithms, 205
Complex number, 392
Composition of functions, 10
Composition rule, 111
Concave downward, 142
Concave upward, 142
Cone, 394
Conic sections, 338
  circle, 122
  degenerate, 349
  ellipse, 340
  hyperbola, 342
  parabola, 339
Constant sequence, 31
Continuous:
  at a point, 73
  on $[a, b]$, 37
  uniformly, 37
Control of error, 71
Convergence:
  of a sequence, 31
  of a series, 372, 376
  radius of, 379
  tests for, 372, 378, 380, 382
Cosine, 217
  derivative of, 226
  graph of, 228
  inverse of, 232
Critical point, 137
Cusp, 338
Cylinder, 394

Decreasing:
  at a point, 134
  function, 130
Definite integral, 41
Degree, 219
Derivative, 77
  first derivative test, 139
  of a composition, 110
  of a product, 104
  of a quotient, 107
  of a sum, 99
  second, 120
  second derivative test, 141
  slope of the tangent line, 65
Differentials, 113
Differential equations, 180
  separation of variables, 180
  series solutions, 384
Directrix, 340
Discontinuity, 73

Discriminant, 358
Distance formula, 122
Domain, 5

Elementary transcendental functions, 185
Ellipse, 340
Ellipsoid, 357
Error, 71, 120
    and limits, 71
    in Simpson's rule, 295
    in Taylor polynomials, 365
    in the trapezoidal rule, 290
Euler, Leonhard, 24
Even function, 147
Existence theorem, 119
Exponential:
    function, 193
    growth and decay, 209
Extreme values, 137
    endpoint, 137
    first derivative test for, 139
    second derivative test for, 143

Factorial, 75
First derivative test for extreme values, 136
Flow, 55
Focus:
    of an ellipse, 340
    of a hyperbola, 342
    of a parabola, 340
Function, 5
    algebraic, 37, 185
    composition of, 10
    continuous, 37, 73
    decreasing, 130
    elementary, 37, 185
    even, 147
    increasing, 130
    odd, 148
    product of, 8
    sum of, 8
Fundamental theorem of calculus, 90

Gas law, 168
Graph, 5
    concave upward, 142
    decreasing, 130
    increasing, 130
    inflection point, 144
    maximum, 136

Graph (*continued*)
    minimum, 136
    of a function, 5
    of a line, 6
Guessing, 95, 258

Half-life, 209
Hooke's law, 312
Hydrostatic force, 319
Hyperbola, 342
Hyperbolic functions, 245
Hypotenuse, 2, 393

Image, 5
Implicit differentiation, 122
Improper integrals, 283
Increasing:
    at a point, 134
    function, 130
Indefinite integral, 94
Induction, 28
Inequalities, 2
Inflection point, 144
Initial conditions, 182
Integer, 1
Integral:
    definite, 41
    improper, 283
    indefinite, 94
    of a sum, 46, 102
    of f from a to b, 41
    Riemann, 41
    test for convergence, 380
Integration methods:
    completing the square, 271
    partial fractions, 275
    parts, 267
    rationalizing substitutions, 116, 259
    Simpson's rule, 293
    substitution, 113
    trapezoidal rule, 288
    trigonometric substitution, 262
Intercept, 6
Interval, 4
Inverse functions, 196
    inverse circular functions, 231
    inverse hyperbolic functions, 249

Leibnitz, Gottfried Wilhelm, 41, 79
Less than, 2

Limit:
   of a function at a point, 68
   of a sequence, 32
   one-sided, 71
   theorem about, 32, 72
Line, 6
Logarithm:
   base a, 202
   change-of-basis formula, 203
   characteristic, 208
   common, 205
   mantissa, 208
   natural, 185
   table of values, 188, 206
Logarithmic differentiation, 191

Maclaurin series, 374
Major axis of an ellipse, 341
Mantissa, 208
Marginal cost, 98
Mathematical induction, 28
Mathematical modeling, 180
Maximum value, 136
Max-min problems, 139, 156
Mean value theorem:
   for derivatives, 119
   for integrals, 82
Minimum, 136
Minor axis of an ellipse, 341
Moment, 305

Natural logarithm, 185
   table of values, 188
Newton, Sir Isaac, 59, 79
Newton's method, 169
Numbers, 1

Odd function, 148
One-to-one function, 198
Ordered pair, 4

Parabola, 339 (*see also* Simpson's rule)
Parametric equations, 326
Partial fractions, 275
Periodic function, 8, 217
Polar coordinates, 331
Polynomial:
   approximations, 360
   linear, 277

Polynomial (*continued*)
   quadratic, 278
   Taylor, 364
Population growth, 209
Probability, 159
Product of functions, 8
Profit, 183
Pyramid, 178
Pythagorean theorem, 2, 393

Quadratic formula, 391
Quotient of functions, 8
Quotient rule for derivatives, 107

Radian, 219
Radioactive decay, 209
Radius of convergence, 379
Rates of change:
   average, 60
   instantaneous, 61, 163
Ratio test for convergence, 372, 378
Rational number, 1
Rational root theorem, 169
Rationalizing substitution, 116, 259
Real number, 1
Reciprocal function, 8
Reciprocal rule for derivatives, 108
Reflection, 197
Related rates, 165
Relative maximum, 137
   test for, 139, 143
Relative minimum, 137
   test for, 139, 143
Remainder term, 365
Riemann, Georg Bernard, 36
Riemann:
   integral, 41
   sum, 36
Root, 168, 391
Rotation of axes, 349

Scientific notation, 205
Secant, 221
Second derivative, 120
   test, 141
Sector of a circle, 393
Separation of variables, 180
Sequence, 30
   convergence of, 31
   divergence of, 32

INDEX 459

Sequence (*continued*)
   limit of, 32
   of partial sums, 372, 377
   term of, 31
Series, 376
   Maclaurin, 374
   solutions to differential equations, 384
   Taylor, 371
   tests for convergence, 372, 378, 380, 382
Sigma notation, 24
Simpson's rule, 292
Sine, 217
   derivative of, 225
   graph of, 228
   hyperbolic, 245
   inverse of, 232
Slope:
   of a line, 6
   of a tangent line, 62
Speed, 164
Sphere, 395
Spiral of Archimedes, 336
Substitution, 115, 258
   rationalizing, 116, 259
   trigonometric, 262
Sum of a series, 372
Sum of functions, 8
Summation:
   rules, 25
   sigma notation, 24
Surface area, 316
Symmetric:
   with respect to the origin, 148
   with respect to the y-axis, 147

Tables:
   of common logarithms, 206
   of natural logarithms, 188
   of values of the circular functions, 222
Tangent, 221
   derivative of, 227
   graph of, 229
Tangent line, 62

Tangent line (*continued*)
   as an aid in sketching, 67
   of a circle, 62
   slope of, 65
Taylor polynomials, 360
Taylor series, 371
Taylor's theorem, 364
Telescoping sum, 26
Tetrahedron, 179
Transcendental functions, 185
   circular functions, 217
   exponential functions, 193
   hyperbolic functions, 245
   inverse circular functions, 231
   natural logarithm, 185
Translation of axes, 346
Trapezoidal rule, 288
Trigonometric functions, 238 (*see also* Circular functions)
Trigonometric substitution, 262

Uniformly continuous, 37

Velocity, 55, 161
Vertical asymptote, 151
Volume:
   of a right circular cone, 394
   of a right circular cylinder, 394
   of a sphere, 395
   of revolution, 175, 318
Volumes of revolution:
   cylindrical shells, 319
   slices, 175

Weierstrass, Karl, 32
Work, 311

x-axis, 4

y-intercept, 6